INSTRUMENTATION AND MEASUREMENTS FOR ELECTRONIC TECHNICIANS

SECOND EDITION

INSTRUMENTATION AND MEASUREMENTS FOR ELECTRONIC TECHNICIANS

ROBERT B. GILLIES
Quinsigamond Community College

Merrill, an imprint of
Macmillan Publishing Company
New York

Maxwell Macmillan Canada
Toronto

Maxwell Macmillan International
New York Oxford Singapore Sydney

Cover photo: Courtesy of Tektronix, Inc.
Editor: Dave Garza
Production Editor: Jonathan Lawrence
Art Coordinator: Lorraine Woost
Cover Designer: Robert Vega
Production Buyer: Pamela D. Bennett

This book was set in Century Schoolbook by The Clarinda Company and was printed and bound by Semline, Inc., a Quebecor America Book Group Company. The cover was printed by Phoenix Color Corp.

Printed in the United States of America

Macmillan Publishing Company
866 Third Avenue
New York, New York 10022

Macmillan Publishing Company is part of the Maxwell Communication Group of Companies.

Maxwell Macmillan Canada, Inc.
1200 Eglinton Avenue East, Suite 200
Don Mills, Ontario M3C 3N1

Library of Congress Cataloging-in-Publication Data
Gillies, Robert B.
Instrumentation and measurements for electronic technicians / Robert B. Gillies. — 2nd ed.
p. cm.
Includes bibliographical references and index.
ISBN 0-02—343051-6
1. Electronic measurements. 2. Electronic instruments.
I. Title.
TK7878.G55 1993
621.381′548—dc20 92-7495
CIP

Printing: 1 2 3 4 5 6 7 8 9 Year: 3 4 5 6 7

Photo credits: pp. 56, 74, Simpson Electric Company; pp. 74, 78, 187, reproduced with permission of the John Fluke Manufacturing Co., Inc.; pp. 126, 134, Clarostat Manufacturing Company; pp. 153, 226, Gen Rad, Inc.; p. 154, Electro Scientific Industries; pp. 180, 181, Krohn-Hite Corporation; p. 182, Gigatronics, Inc.; pp. 182, 184, Systron Donner; pp. 185, 186, Marconi Instruments; p. 186, Sencore; p. 198, courtesy of Fenwall Electronics/APD; p. 200, courtesy of Motorola, Inc.; pp. 204, 205, F. W. Bell.

MERRILL'S INTERNATIONAL SERIES IN ENGINEERING TECHNOLOGY

INTRODUCTION TO ENGINEERING TECHNOLOGY

Pond, *Introduction to Engineering Technology, 2nd Edition*, 0-02-396031-0

ELECTRONICS TECHNOLOGY

Electronics Reference

Adamson, *The Electronics Dictionary for Technicians*, 0-02-300820-2
Berlin, *The Illustrated Electronics Dictionary*, 0-675-20451-8
Reis, *Becoming an Electronics Technician: Securing Your High-Tech Future*, 0-02-399231-X

DC/AC Circuits

Boylestad, *DC/AC: The Basics*, 0-675-20918-8
Boylestad, *Introductory Circuit Analysis, 6th Edition*, 0-675-21181-6
Ciccarelli, *Circuit Modeling: Exercises and Software, 2nd Edition*, 0-02-322455-X
Floyd, *Electric Circuits Fundamentals, 2nd Edition*, 0-675-21408-4
Floyd, *Electronics Fundamentals: Circuits, Devices, and Applications, 2nd Edition*, 0-675-21310-X
Floyd, *Principles of Electric Circuits, 4th Edition*, 0-02-338501-4
Floyd, *Principles of Electric Circuits: Electron Flow Version, 3rd Edition*, 0-02-338531-6
Keown, *PSpice and Circuit Analysis*, 0-675-22135-8
Monssen, *PSpice with Circuit Analysis* 0-675-21376-2
Tocci, *Introduction to Electric Circuit Analysis, 2nd Edition*, 0-675-20002-4

Devices and Linear Circuits

Berlin & Getz, *Fundamentals of Operational Amplifiers and Linear Integrated Circuits*, 0-675-21002-X
Berube, *Electronic Devices and Circuits Using MICRO-CAP II*, 0-02-309160-6
Berube, *Electronic Devices and Circuits Using MICRO-CAP III*, 0-02-309151-7
Bogart, *Electronic Devices and Circuits, 3rd Edition*, 0-02-311701-X
Tocci, *Electronic Devices: Conventional Flow Version, 3rd Edition*, 0-675-21150-6
Floyd, *Electronic Devices, 3rd Edition*, 0-675-22170-6
Floyd, *Electronic Devices: Electron Flow Version*, 0-02-338540-5
Floyd, *Fundamentals of Linear Circuits*, 0-02-338481-6
Schwartz, *Survey of Electronics, 3rd Edition*, 0-675-20162-4
Stanley, *Operational Amplifiers with Linear Integrated Circuits, 2nd Edition*, 0-675-20660-X
Tocci & Oliver, *Fundamentals of Electronic Devices, 4th Edition*, 0-675-21259-6

Digital Electronics

Floyd, *Digital Fundamentals, 4th Edition*, 0-675-21217-0
McCalla, *Digital Logic and Computer Design*, 0-675-21170-0
Reis, *Digital Electronics through Project Analysis* 0-675-21141-7
Tocci, *Fundamentals of Pulse and Digital Circuits, 3rd Edition*, 0-675-20033-4

Microprocessor Technology

Antonakos, *The 68000 Microprocessor: Hardware and Software Principles and Applications, 2nd Edition*, 0-02-303603-6
Antonakos, *The 8088 Microprocessor*, 0-675-22173-0
Brey, *The Advanced Intel Microprocessors*, 0-02-314245-6
Brey, *The Intel Microprocessors: 8086/8088, 80186, 80286, 80386, and 80486: Architecture, Programming, and Interfacing, 2nd Edition*, 0-675-21309-6
Brey, *Microprocessors and Peripherals: Hardware, Software, Interfacing, and Applications, 2nd Edition*, 0-675-20884-X
Gaonkar, *Microprocessor Architecture, Programming, and Applications with the 8085/8080A, 2nd Edition*, 0-675-20675-6
Gaonkar, *The Z80 Microprocessor: Architecture, Interfacing, Programming, and Design, 2nd Edition*, 0-02-340484-1
Goody, *Programming and Interfacing the 8086/8088 Microprocessor: A Product- Development Laboratory Process*, 0-675-21312-6
MacKenzie, *The 8051 Microcontroller*, 0-02-373650-X
Miller, *The 68000 Family of Microprocessors: Architecture, Programming, and Applications, 2nd Edition*, 0-02-381560-4
Quinn, *The 6800 Microprocessor*, 0-675-20515-8
Subbarao, *16/32 Bit Microprocessors: 68000/68010/68020 Software, Hardware, and Design Applications*, 0-675-21119-0

Electronic Communications

Monaco, *Introduction to Microwave Technology*, 0-675-21030-5
Monaco, *Preparing for the FCC Radio-Telephone Operator's License Examination*, 0-675-21313-4
Schoenbeck, *Electronic Communications: Modulation and Transmission, 2nd Edition*, 0-675-21311-8
Young, *Electronic Communication Techniques, 2nd Edition*, 0-675-21045-3
Zanger & Zanger, *Fiber Optics: Communication and Other Applications*, 0-675-20944-7

Microcomputer Servicing

Adamson, *Microcomputer Repair*, 0-02-300825-3
Asser, Stigliano, & Bahrenburg, *Microcomputer Servicing: Practical Systems and Troubleshooting, 2nd Edition*, 0-02-304241-9
Asser, Stigliano, & Bahrenburg, *Microcomputer Theory and Servicing, 2nd Edition*, 0-02-304231-1

Programming

Adamson, *Applied Pascal for Technology*, 0-675-20771-1
Adamson, *Structured BASIC Applied to Technology, 2nd Edition*, 0-02-300827-X
Adamson, *Structured C for Technology*, 0-675-20993-5
Adamson, *Structured C for Technology (with disk)*, 0-675-21289-8
Nashelsky & Boylestad, *BASIC Applied to Circuit Analysis*, 0-675-20161-6

Instrumentation and Measurement

Berlin & Getz, *Principles of Electronic Instrumentation and Measurement*, 0-675-20449-6
Buchla & McLachlan, *Applied Electronic Instrumentation and Measurement*, 0-675-21162-X
Gillies, *Instrumentation and Measurements for Electronic Technicians, 2nd Edition*, 0-02-343051-6

Transform Analysis

Kulathinal, *Transform Analysis and Electronic Networks with Applications*, 0-675-20765-7

Biomedical Equipment Technology

Aston, *Principles of Biomedical Instrumentation and Measurement*, 0-675-20943-9

Mathematics

Monaco, *Essential Mathematics for Electronics Technicians*, 0-675-21172-7
Davis, *Technical Mathematics*, 0-675-20338-4
Davis, *Technical Mathematics with Calculus*, 0-675-20965-X

INDUSTRIAL ELECTRONICS/INDUSTRIAL TECHNOLOGY

Bateson, *Introduction to Control System Technology, 4th Edition*, 0-02-306463-3
Fuller, *Robotics: Introduction, Programming, and Projects*, 0-675-21078-X
Goetsch, *Industrial Safety: In the Age of High Technology*, 0-02-344207-7
Goetsch, *Industrial Supervision: In the Age of High Technology*, 0-675-22137-4
Horath, *Computer Numerical Control Programming of Machines*, 0-02-357201-9
Hubert, *Electric Machines: Theory, Operation, Applications, Adjustment, and Control*, 0-675-20765-7
Humphries, *Motors and Controls*, 0-675-20235-3
Hutchins, *Introduction to Quality: Management, Assurance, and Control*, 0-675-20896-3
Laviana, *Basic Computer Numerical Control Programming* 0-675-21298-7
Reis, *Electronic Project Design and Fabrication, 2nd Edition*, 0-02-399230-1
Rosenblatt & Friedman, *Direct and Alternating Current Machinery, 2nd Edition*, 0-675-20160-8
Smith, *Statistical Process Control and Quality Improvement*, 0-675-21160-3
Webb, *Programmable Logic Controllers: Principles and Applications, 2nd Edition*, 0-02-424970-X
Webb & Greshock, *Industrial Control Electronics, 2nd Edition*, 0-02-424864-9

MECHANICAL/CIVIL TECHNOLOGY

Keyser, *Materials Science in Engineering, 4th Edition*, 0-675-20401-1
Kraut, *Fluid Mechanics for Technicians*, 0-675-21330-4
Mott, *Applied Fluid Mechanics, 3rd Edition*, 0-675-21026-7
Mott, *Machine Elements in Mechanical Design, 2nd Edition*, 0-675-22289-3
Rolle, *Thermodynamics and Heat Power, 3rd Edition*, 0-675-21016-X
Spiegel & Limbrunner, *Applied Statics and Strength of Materials*, 0-675-21123-9
Wolansky & Akers, *Modern Hydraulics: The Basics at Work*, 0-675-20987-0
Wolf, *Statics and Strength of Materials: A Parallel Approach to Understanding Structures*, 0-675-20622-7

DRAFTING TECHNOLOGY

Cooper, *Introduction to VersaCAD*, 0-675-21164-6
Goetsch & Rickman, *Computer-Aided Drafting with AutoCAD*, 0-675-20915-3
Kirkpatrick & Kirkpatrick, *AutoCAD for Interior Design and Space Planning*, 0-02-364455-9
Kirkpatrick, *The AutoCAD Book: Drawing, Modeling, and Applications, 2nd Edition*, 0-675-22288-5
Lamit and Lloyd, *Drafting for Electronics, 2nd Edition*, 0-02-367342-7
Lamit and Paige, *Computer-Aided Design and Drafting*, 0-675-20475-5
Maruggi, *Technical Graphics: Electronics Worktext, 2nd Edition*, 0-675-21378-9
Maruggi, *The Technology of Drafting*, 0-675-20762-2
Sell, *Basic Technical Drawing*, 0-675-21001-1

TECHNICAL WRITING

Croft, *Getting a Job: Resume Writing, Job Application Letters, and Interview Strategies*, 0-675-20917-X
Panares, *A Handbook of English for Technical Students*, 0-675-20650-2
Pfeiffer, *Proposal Writing: The Art of Friendly Persuasion*, 0-675-20988-9
Pfeiffer, *Technical Writing: A Practical Approach*, 0-675-21221-9
Roze, *Technical Communications: The Practical Craft*, 0-675-20641-3
Weisman, *Basic Technical Writing, 6th Edition*, 0-675-21256-1

To my family

Preface

The purpose of this book is to introduce the student to fundamentals of electronic measurement techniques and instrumentation. In today's highly technological industries, engineers design equipment and solve problems. Technicians build, test, install, maintain, and service equipment from prototype to finished product. This book is intended to help the future technician develop an understanding of test equipment while stressing its use, application, and maintenance.

This book was designed to serve as the text for a one-semester course. However, it can easily be used for a two-semester course with a break-out after Chapter 8. The presentation of topics tends to be from review and simple measuring devices to more complex signal processors and measuring systems. If desired, many chapters can be presented out of sequence without seriously altering the flow of the text.

The material presented presumes that the student has a working knowledge of basic electricity and electronics including dc, ac, Ohm's law, Kirchhoff's laws, Thevenin's theorem, open circuits, short circuits, grounds, and solid-state devices. A general knowledge of digital circuits is helpful but not required. The mathematics presented is algebra based with some trigonometry; previous knowledge of algebra is required.

Chapters 1 through 3 are basically review chapters, with emphasis on safety and correct lab techniques, handling of data and errors, and operational amplifiers. If the student has already studied op-amps, Chapter 3 may be omitted, although some of the applications it covers may be unfamiliar and would be helpful for review or for new experience. Included in Chapter 1 is a reference to logbooks, an example of which appears in Appendix F. Although some instructors require students to keep a logbook, this approach may be omitted.

Chapters 4 through 6 cover a variety of analog dc and ac meters as well as digital multimeters. Chapters 7 and 8 present the theory and operation of cathode ray oscilloscopes, stressing the importance of these devices as a technician's tool. Chapters 9 through 11 cover electronic devices that can be used in a variety of measurements including the potentiometer, bridges, and signal-processing circuits. Chapter 12 presents measurements that incorporate instruments such as signal sources and other devices and circuits presented in the previous chapters. Chapter 13 describes transducer characteristics and operation. Chapter 14 introduces the principles of fiber optics. Chapter 15 introduces automatic test equipment. To give students additional hands-on experience, most chapters conclude with one or two lab problems.

NEW TO THIS EDITION

This edition has incorporated feedback from numerous students and teachers who used the first edition. Where necessary, the text has been revised to improve accuracy and clarity. In addition, new information is presented on a number of topics. In Chapter 1, a detailed discussion of electrostatic discharge (ESD) has been added, including techniques that can be used to avoid damage to electronic components caused by ESD. Chapter 12 includes information on the Fluke Scopemeter™, a new integrated system that combines the features of an oscilloscope, a multimeter, a sine/square generator, and a component tester. And Chapter 13 now includes an in-depth discussion of Hall-effect transducers and the new hand-held devices that use them to measure current without damaging the conductor or its insulation. For the reader's convenience, Appendix G has been added as a compilation of the most important formulas used throughout the text.

ACKNOWLEDGMENTS

I would like to acknowledge the dedicated work of my colleagues Sharron Lamoureux, who made this book easy to read, and Ronald Josephson, Ronald Fournier, Adrian Comeaux, Dadbeh Bigonahy, Kenneth Dwyer, and Lois E. Brown, who assisted me with technical reviews and friendly support over a long period of time.

I would also like to thank the following reviewers, whose comments were helpful in preparing this edition: James R. Carstens, Michigan Technological University; Gary J. Carter, Nashville State Technical Institute; Kenneth P. DeLucca, Millersville University; Steven R. Duginski, Northeast Wisconsin Technical College; Earl T. Farley, Texas Tech University; Demetrios K. Kostopolous, State University of New York; Harvey Laabs, North Dakota State College of Science; Mohsin Mehtar, Greater New Haven State Technical College; Lee Rosenthal, Fairleigh Dickinson University; Chandra Sekhar, Purdue University—Calumet; and Ulrich E. Zeisler, Salt Lake Community College.

Contents

1 Safety, Lab Techniques, Troubleshooting, and Information Referencing 1

SAFETY 1
- ☐ Personal Safety 1
- ☐ Workbench Safety 3
- ☐ Handling and Servicing Equipment 5

LAB PROCEDURES 5

TROUBLESHOOTING 7
- ☐ Signal Tracing 8
- ☐ Signal Injection 9

REFERENCES 10

SUMMARY 11

EXERCISES 11

2 Measurement and Errors 13

MEASUREMENT 13
- ☐ The National Bureau of Standards 14
- ☐ Calibration Curves 14
- ☐ Accuracy versus Precision 14
- ☐ Standards 15
- ☐ Technique 15
- ☐ Interpreting Data 15

ERRORS 15
- ☐ Gross Errors 15
- ☐ Systematic Errors 15
- ☐ Random Errors 15

HANDLING GROSS ERRORS 15
- ☐ Parallax 16
- ☐ Interpolation 16
- ☐ Significant Figures 16
- ☐ Rounding Off 17

STATISTICAL ANALYSIS 17
- ☐ Deviation and Standard Deviation 17
- ☐ Probable Error 19
- ☐ Handling Probable Errors 19

SUMMARY 21

EXERCISES 21

PROGRAMMING PROBLEM 2–1—Statistical Analysis 22

3 Operational Amplifiers 23

HISTORY OF THE OPERATIONAL AMPLIFIER 24

USES OF OP-AMPS 24

CHARACTERISTICS OF A TYPICAL OP-AMP 24

OPEN-LOOP AMPLIFICATION 25

NONINVERTING OP-AMP WITH A FEEDBACK LOOP 27

INVERTING OP-AMP WITH A FEEDBACK LOOP 28

THE OP-AMP AS A BUFFER AMPLIFIER 29

OP-AMPS IN MATHEMATICAL OPERATIONS 30
□ Multiplication 30
□ Division 30
□ Addition 30
□ Subtraction 31
□ Solving Systems of Equations 31
□ Differentiation 33
□ Integration 34
SUMMARY 35
EXERCISES 35
LAB PROBLEM 3–1—Operational Amplifiers 37

4 dc Analog Meters 45

THE BASIC METER MOVEMENT 46
INTERNAL RESISTANCE OF THE METER MOVEMENT 46
FULL-SCALE DEFLECTION 47
SHUNTING 48
MULTIRANGE AMMETER 48
AYRTON SHUNT 49
VOLTMETERS 51
MULTIRANGE VOLTMETERS 51
SENSITIVITY 52
LOADING PROBLEMS 53
SERIES-TYPE OHMMETER 54
SHUNT-TYPE OHMMETER 56
SUMMARY 56
EXERCISES 57
LAB PROBLEM 4–1—Measuring Internal Resistance 59
LAB PROBLEM 4–2—Ayrton Shunt 59
PROGRAMMING PROBLEM 4–1—Two-range Ayrton Shunt 61
PROGRAMMING PROBLEM 4–2—Three-range Ayrton Shunt 61

5 ac Analog Meters 63

REVIEW OF ac TERMINOLOGY 63
□ Peak 63
□ Peak-Peak 64
□ rms 64
□ Average (Effective or dc Equivalent) 64
RECTIFIERS 65
LINEARITY 66
LEAKAGE 67
PEAK DETECTOR 67
PEAK-PEAK DETECTOR 68
TRUE rms DETECTOR 68
SENSITIVITY 69
FREQUENCY CONSIDERATIONS 69
BALANCED-BRIDGE ELECTRONIC VOLTMETER 70
SUMMARY 71
EXERCISES 71
LAB PROBLEM 5–1—ac Detectors and Bandwidth Measurement 72

6 Digital Multimeters 73

HISTORY OF DIGITAL MULTIMETERS 74
□ Tube-Driven Models 74
□ Transistorized Models 74
□ Integrated Circuit Models 74
□ Large-Scale Integration Models 74
CRITERIA FOR SELECTING A DMM 74
TYPES OF A/D CONVERSION 75
□ Ramp or Counter Type 75
□ Successive Approximation Type 75
□ Dual-Slope/Integrator Method 76
A TYPICAL DMM (THE FLUKE 8060A) 77
□ Functional Description 77
□ The Microcomputer 77
□ The Measurement Acquisition Chip 79
□ A/D Conversion Cycle 79
□ Voltage Measurement 79
□ Current Measurement 80
□ Resistance Measurement 81
□ Conductance Measurement 81
□ Continuity Indication 81
□ Frequency Measurement 81
MULTIMETER SAFETY 82
FREQUENCY CONSIDERATIONS 82
SUMMARY 82
EXERCISES 82
LAB PROBLEM 6–1—Digital Multimeters 83

7 Oscilloscopes—Theory of Operation 87

WHAT IS A CATHODE RAY OSCILLOSCOPE? 88
- The Electron Beam 88
- The Beam Finder 89
- The Phosphor 90
- Graticules 91
- The Display 91

THE VERTICAL SECTION 92
- The Probe 92
- Picking a Probe 93
- Circuit Loading 93
- The Vertical Attenuator (Range Selector) 93
- Vertical Amplifiers/Phase Splitters 94
- The Delay Line 94

THE HORIZONTAL SECTION 94
THE TRIGGERING SECTION 94
- Internal Triggering 96
- Line Triggering 96
- External Triggering 96

LINEARITY OF SWEEP 96
THE HORIZONTAL SWEEP 97
DUAL TRACE 97
DELAYED SWEEP 98
SAMPLING OSCILLOSCOPES 99
DIGITAL STORAGE OSCILLOSCOPES 99
ANALOG STORAGE OSCILLOSCOPES 99
SUMMARY 100
EXERCISES 100
LAB PROBLEM 7–1—Oscilloscope Characteristics and Measurements 101

8 Oscilloscopes—Measurement 105

TURNING ON, AND CHECKING THE CALIBRATION 106
dc MEASUREMENT WITH AN OSCILLOSCOPE 106
ac MEASUREMENT WITH AN OSCILLOSCOPE 107
TIME AND FREQUENCY MEASUREMENTS 109
DUAL-TRACE METHOD OF PHASE MEASUREMENT 110
LISSAJOUS FIGURES 111
- Phase Measurements Using Lissajous Figures 111
- Frequency Measurements Using Lissajous Figures 112

A + B AND A − B 114
USING THE DELAYED SWEEP MODE 115
z-AXIS MODULATION 116
NOTES ON FIELD SERVICE 116
SUMMARY 116
EXERCISES 117
LAB PROBLEM 8–1—Oscilloscope Familiarization 120

9 Potentiometers and Potentiometric Bridges 125

POTENTIOMETERS 126
- What is a Potentiometer? 126
- Potentiometer across a Single Source of emf 126
- Thevenin Equivalent of a Potentiometer across a Single emf Source 127
- Potentiometer across Two emf Sources 128
- Thevenin Equivalent of a Potentiometer across Two emf Sources 128
- Slightly Unbalanced Bridge 130

POTENTIOMETRIC (SLIDE-WIRE) BRIDGES 131
- Slide-Wire Potentiometer 131
- Potentiometric Bridge 132
- Normalizing the Bridge 132
- The Sensitivity Control 132
- Measuring Potentials 132
- Thevenin Equivalent of the Bridge 133
- Methods of Using the Bridge 134
- Advantages and Disadvantages 134

SUMMARY 134
EXERCISES 135
LAB PROBLEM 9–1—Slightly Unbalanced Potentiometer 137
LAB PROBLEM 9–2—Potentiometric Bridges 139

10 dc and ac Bridges 141

THE WHEATSTONE BRIDGE 141
THEVENIN EQUIVALENT OF A WHEATSTONE BRIDGE 142
- Unbalanced Bridge 142
- A Special Design 144

☐ Thevenin Equivalent of a Slightly Unbalanced Bridge 144

ac BRIDGES 146
SOME IMPEDANCE CONVERSION FORMULAS 148
OTHER BRIDGE CIRCUITS 152
IMPEDANCE BRIDGES 153
☐ Manual Impedance (*LRC*) Bridges 153
☐ Automatic Impedance (*LRC*) Bridges 154
☐ Theory of Operation of Automatic *LRC* Bridges 154
SUMMARY 155
EXERCISES 155
LAB PROBLEM 10–1—Impedance Bridge 156
LAB PROBLEM 10–2—Capacitance Comparison Bridge 158

11 Signal-Processing Circuits 161

CLIPPERS AND CLAMPERS 161
AMPLIFIERS 163
ATTENUATORS 163
FREQUENCY MULTIPLIERS 164
FREQUENCY DIVIDERS 165
ANALOG-TO-DIGITAL CONVERTERS 166
DIGITAL-TO-ANALOG CONVERTERS 166
MODULATION TECHNIQUES 168
☐ Amplitude Modulation 168
☐ Frequency and Phase Modulation 169
☐ Pulse Modulation 170
SEPARATION OF TWO SIGNALS 170
SUMMARY 173
EXERCISES 173
LAB PROBLEM 11–1—D/A Converters 173
LAB PROBLEM 11–2—Input Attenuator 176

12 Measuring Systems and Analyzers 179

IEEE AND GPIB BUSSES 180
SIGNAL SOURCES 180
FREQUENCY SYNTHESIZERS 182
LOGIC PROBES AND PULSERS 183
FREQUENCY COUNTERS 183
MICROPROCESSOR CONTROL 184
SWEEP GENERATORS 184
MICROWAVE 185
ANALYZERS 185
☐ Logic Analyzers 185
☐ Network Analyzers 185
☐ Network/Spectrum Analyzers 186
☐ Waveform Analyzers 186
☐ System Analyzers 186
☐ Intregrated System 187
SUMMARY 187
EXERCISES 187
LAB PROBLEM 12–1—Calibrating an Audio Oscillator 188
LAB PROBLEM 12–2—Test System Simulation 191

13 Transducers 193

MECHANICAL TRANSDUCERS 194
☐ Strain Gages 194
☐ Pressure Transducers 195
☐ Accelerometers 196
☐ Flow Meters 196
THERMAL TRANSDUCERS 196
☐ Thermocouples 196
☐ Resistance Thermometers 197
☐ Thermistors 198
☐ Infrared Detectors 199
OPTICAL TRANSDUCERS 199
☐ Photoconductive Cells 199
☐ Photovoltaic Cells 200
ACOUSTICAL TRANSDUCERS 201
☐ Dynamic Microphones 201
☐ Condenser (Capacitor) Microphones 202
☐ Ribbon Microphones 202
☐ Carbon Button Microphones 202
☐ Microphone Ratings 203
ULTRASONIC RANGING SYSTEMS 203
MAGNETIC TRANSDUCERS 203
☐ Differential-Pressure Transducer 203
☐ Linear-Variable Differential Transformer 204
☐ Hall-Effect Transducer 204
☐ Selsyn Generator 205
CHEMICAL TRANSDUCERS 205
BIOLOGICAL TRANSDUCERS 205

NUCLEAR TRANSDUCERS (RADIATION DETECTORS) 206
☐ Germanium Gamma Ray Detector 207
☐ Photomultiplier Tube 207
SUMMARY 208
EXERCISES 208

14 Fiber Optics 209
WHY LASERS? 210
LIGHT WAVE PARAMETERS 211
OPTICAL FIBER LINKS 211
A BASIC COMMUNICATIONS SYSTEM 211
TRANSMISSION LINE CHARACTERISTICS 212
ADVANTAGES OF OPTICAL FIBERS 212
DISADVANTAGES OF OPTICAL FIBERS 213
SPACE OPTICAL SYSTEMS 213
LIGHT SOURCES 214
LIGHT DETECTORS 214
USE OF OPTICAL FIBERS 214
AMPLITUDE MODULATION VERSUS PULSE AMPLITUDE MODULATION 215
SUMMARY 216
EXERCISES 216
LAB PROBLEM 14–1—Free-Space Transmission 217

15 Automatic Test Equipment (ATE) 219
AUTOMATIC TEST EQUIPMENT 219
WHY ATE? 220
TYPES OF FAULTS 220
MANUAL TESTING 221
ATE IN-CIRCUIT TESTERS 221
TEST EQUIPMENT REQUIRED 222
DIGITAL TESTING 223
COMPONENT ISOLATION 224
☐ Analog Testing 224
☐ Digital Testing 224
FLIP-FLOPS 225
BUSSED DEVICES 226
SUMMARY 226
EXERCISES 227
LAB PROBLEM 15–1—ATE Isolation 227

Appendix A Thevenin's Theorem—Three Applications 229

Appendix B SI Units and Dimensional Analysis 235

Appendix C Derivation of Amplitude Modulation and Frequency Modulation Formulas 241

Appendix D Impedance Bridge Derivation 245

Appendix E BASIC Syntax 249

Appendix F The Laboratory Notebook 251

Appendix G Important Formulas 255

Glossary 257

Solutions to Odd-Numbered Problems 263

Index 267

1 Safety, Lab Techniques, Troubleshooting, and Information Referencing

OUTLINE

OBJECTIVES

SAFETY

- ☐ Personal Safety
- ☐ Workbench Safety
- ☐ Handling and Servicing Equipment

LAB PROCEDURES

TROUBLESHOOTING

- ☐ Signal Tracing
- ☐ Signal Injection

REFERENCES

SUMMARY

EXERCISES

OBJECTIVES

After completing this chapter, you will be able to:

- ☐ List common safety procedures.
- ☐ Apply proper lab techniques in several basic areas.
- ☐ Apply some basic troubleshooting techniques.
- ☐ Locate specific kinds of technical data.
- ☐ Derive certain technical information from publications.

Safety procedures, proper lab techniques, information referencing, and some basic troubleshooting techniques are essential to successful work habits. This chapter presents various ways in which the technician may develop these work habits.

SAFETY

Safety precautions are intended to protect the technician, the test equipment, and the device being tested. Most procedures are simple and could be summed up as "common sense," but simple as they may seem, they should not be bypassed or taken for granted.

Most accidents in electronics are caused by impatience, carelessness, or improper techniques. A good understanding of how equipment operates is basic to good technique. With experience, most technicians develop individualized routines and shortcuts. However, safety should never be ignored when taking shortcuts.

The safety precautions covered in the following discussion fall into three categories: personal safety, workbench safety, and handling and servicing equipment.

Personal Safety

1. Don't wear rings, watches, or other metal jewelry. These items can be hazardous if they come in contact with electrical conductors.

2. Avoid loose-fitting clothing since it can catch on equipment and drag the equipment together or onto the floor.

3. Don't touch exposed connections and components when the power is on. If possible keep one hand in your pocket. If you are probing a circuit with your hand and a short circuit occurs, the current may flow into your fingers. By using only one hand, you will restrict the current path to that hand (Figure 1–1A). If you use both hands and a potential causes current to flow into your body, one possible current path is across your chest where your heart is located (Figure 1–1B). Sufficient current can cause fibrillation (interference with heartbeat rhythm) or worse.

4. If you think a component may be hot, approach it with the back of your hand or finger (Figure 1–2). When exposed to heat or electrical shock the muscles of the hand will contract. If the back of the hand is used, the muscle contraction will cause the hand to be pulled away from the component. If the front of the hand is used, the contraction will cause the hand to come in contact with the component. This reaction explains why a person seems unable to let go of an electrical conductor: The muscles of the hand are forced into contraction by the electricity.

Table 1–1 illustrates the harmful effect of serious levels of electric current as it passes through the body. A current of only 0.01 A would come as a surprise, which might cause a reaction that could lead to injury. A current as small as 0.1 A can cause death. All effects listed might result in an involuntary muscle spasm that could hurl the body or even an arm against a sharp projection or an even more dangerous voltage source.

5. Heed safety warnings on equipment and in operation manuals.

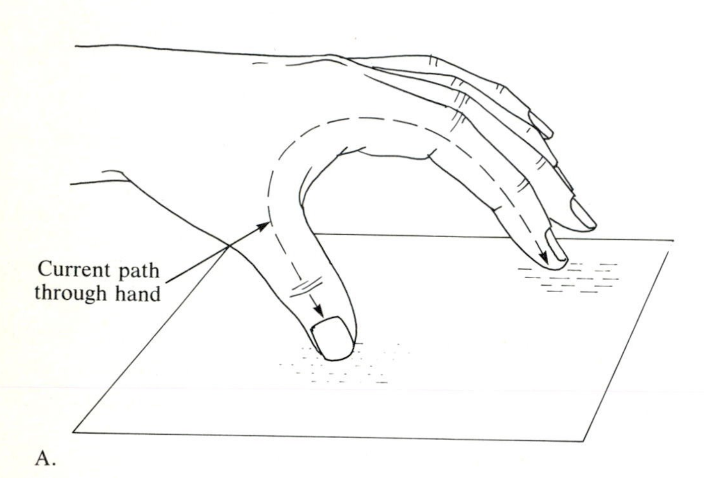

A.

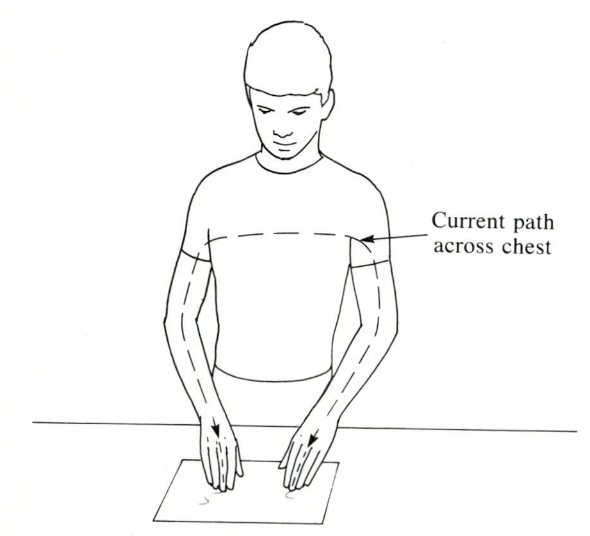

Keep one hand in your pocket to prevent this!

B.

FIGURE 1–1
Use one hand rather than two.

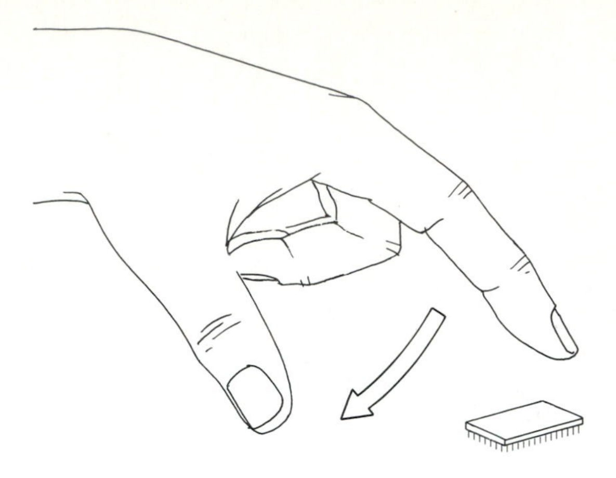

Muscle contraction pulls finger toward chip.

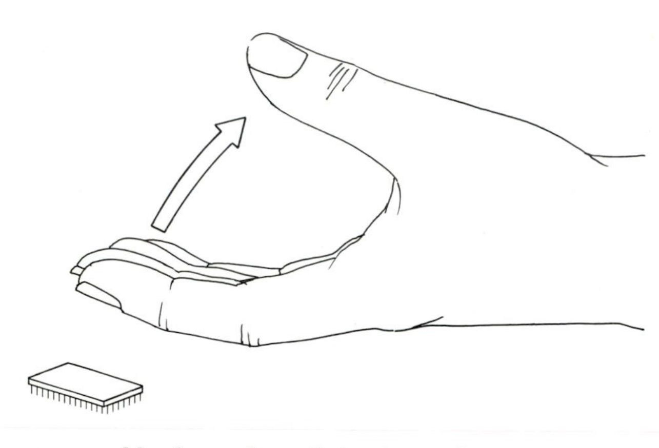

Muscle reaction pulls hand away from chip.

FIGURE 1–2
Use the back of your hand.

TABLE 1–1
Effect of electric current on the human body

Amperes	Effect
0.001	Threshold of sensation Mild sensation
↓	
0.01	Painful sensation Muscular paralysis Inability to let go of component Severe shock Interrupted and labored breathing Extreme breathing difficulties
↓	
0.1	Death possible
↓	
1.0	Severe burns Cessation of breathing Death imminent

6. Wear safety glasses whenever servicing or repairing any equipment. Some defective components (e.g., capacitors and chips) may explode when power is applied. If wire is stripped of its insulation or cut, tiny pieces of insulation or wire may fly into the eye, causing damage.

7. Hot solder irons and guns usually don't look hot but can remain hot for a long time after being disconnected or turned off. Be careful.

8. Flying bits of solder can burn skin or clothing.

9. Splaying of braided wire for ground connections on coaxial or shielded cable can expose sharp ends, which can puncture fingers (Figure 1–3).

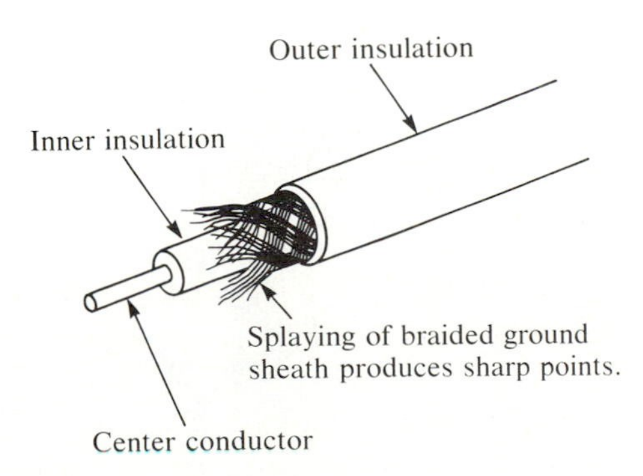

FIGURE 1–3

Workbench Safety

1. The workbench should not be a conducting one built of or covered with sheet metal.

2. Move all unused wires, instruments, and conducting tools away from the work area since these represent a potential short-circuit path and could harm you or damage equipment in the work area.

3. Make sure all tools are insulated or made of nonconducting materials such as plastic.

4. Clothing placed on a workbench can constitute a hazard since it takes up space.

5. Make sure service outlets are properly wired. Proper wiring will eliminate the possibility of a *floating ground.* A floating ground will occur when a two-pronged plug is inserted backward into a service outlet, causing a high potential instead of ground to be applied to the chassis and case (and probe ground). The resulting potential difference between the ground of the probe from one device and the ground of the probe from another will then equal the entire line voltage. When the two grounds are connected together, instantaneous disaster results. If the test apparatus is properly equipped with three-conductor power cords, or if polarized two-pronged plugs are used in proper outlets, the problem can be avoided. The potential problem can be detected by using a neon tester between points that should be at the same (ground) potential. The tester consists of a circuit with a neon bulb and two leads. If a potential difference exists between two points, the bulb will light up. An ac voltmeter can also be used to detect a floating ground.

Equipment chassis (except some special hospital and military equipment) and all television sets manufactured today use transformerless or "hot" chassis power supplies to meet demands for improved energy savings and lower prices. At some time in your career, you will probably run into a transformerless chassis. The three power supplies illustrated in Figures 1–4 through 1–6 are transformerless and therefore require isolation transformer protection for safe testing and servicing.

Connecting grounded test equipment to the half-wave rectifier chassis will not cause a problem if the ac cord is plugged in so that the chassis is at ground potential through the common side of the ac line. To check for this, measure the potential between the chassis and the grounded test equipment. A high ac voltage reading indicates that the "hot" side of the line is feeding the chassis, a situation that can be remedied by simply reversing the ac plug.

Reversing the polarity of the ac line voltage will not work with the bridge or switching power supplies. In these cases, the chassis is at half the ac

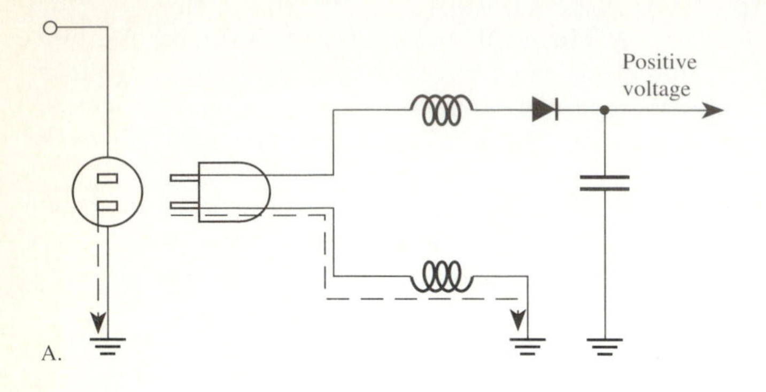

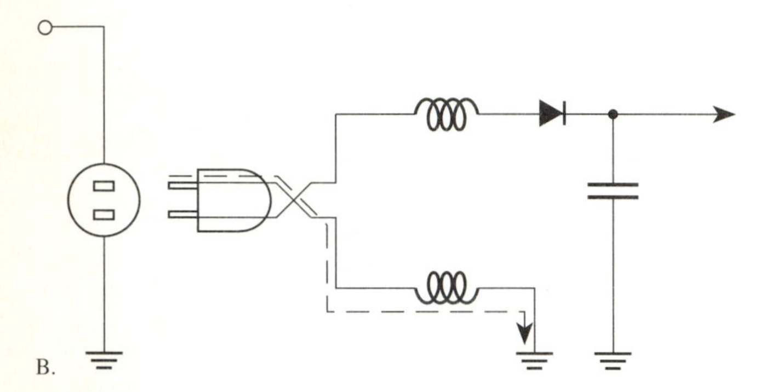

FIGURE 1–4
Half-wave rectified power supply. A. One side of the ac line is connected directly to the chassis in a half-wave hot grounded power supply. B. The reversed plug connects the ac voltage line directly to ground.

line regardless of the polarity of the ac plug. Connecting any grounded test instrument will short out half of the power supply as shown in Figure 1–7. If this happens, costly damage can result to test equipment or the equipment being tested. Although breaking the ground path causing the problem may appear to be a solution, removing the test equipment's third-wire safety ground can cause other serious problems.

The third-wire ground on many test instruments is needed to properly shield sensitive circuits, such as microprocessors and high-impedance amplifiers, in order to prevent erroneous readings. This ground wire also prevents a serious shock hazard from forming between the instrument's ground and any metal object that may be connected to earth ground. Placing yourself across this ac potential difference could cause serious injuries. These problems can be avoided if you ALWAYS plug EVERY chassis into an isolation transformer before attempting to test or service it.

6. One way to avoid accidents is to have a clean, efficient work area. A good step in this direction is to carefully select the tools to be used. Make sure they are appropriate for the job and are clean and in proper working condition. Tools should be inspected periodically to determine if they need sharpening or replacement. Metal tools that are damaged can have improper sharp metal edges or protrusions that can cut the user. And tools that are not clean could have conducting dirt that might prove electrically dangerous, or oily substances that could make them slippery.

7. Avoid water. Water can cause corrosion and other deterioration of components. In addition, certain liquids contain materials that can impede mechanical action. A spilled can of soda can destroy a computer keyboard and produce short circuits.

8. Make sure benches, tables, and stands can safely and stably support equipment. Equipment that falls to the floor can be seriously damaged and can cause injury.

9. Don't overload the power service at your test site.

10. Probes should be insulated. Don't use alligator clips that have no protective shrouds.

11. Briefly touch alligator clips to the contact point before clipping them in place. If anything goes wrong, you can remove the clip more easily if it is not attached to the test point.

12. If you need to place a test probe in, or inject a signal into, a place that is relatively inaccessible (because the probe is too fat or too short), use a clip lead from the probe tip to the barrel of a narrow screwdriver. You can then carefully insert the metallic barrel of the screwdriver to locate the desired test point.

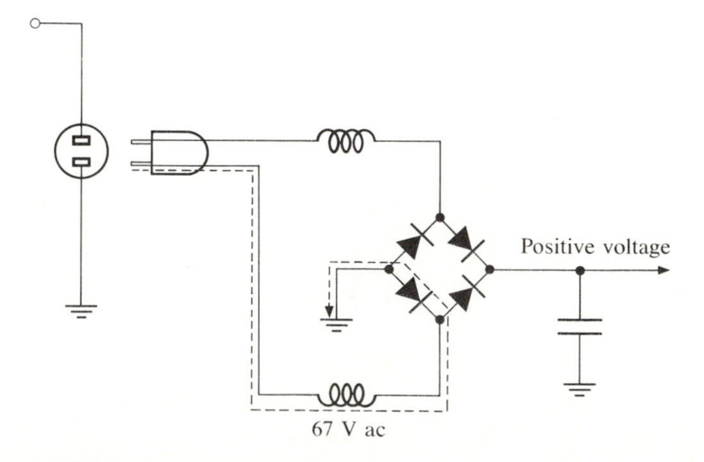

FIGURE 1–5
Full-wave rectified power supply. A 67 V potential always occurs between the chassis and earth grounds in a bridge rectifier power supply.

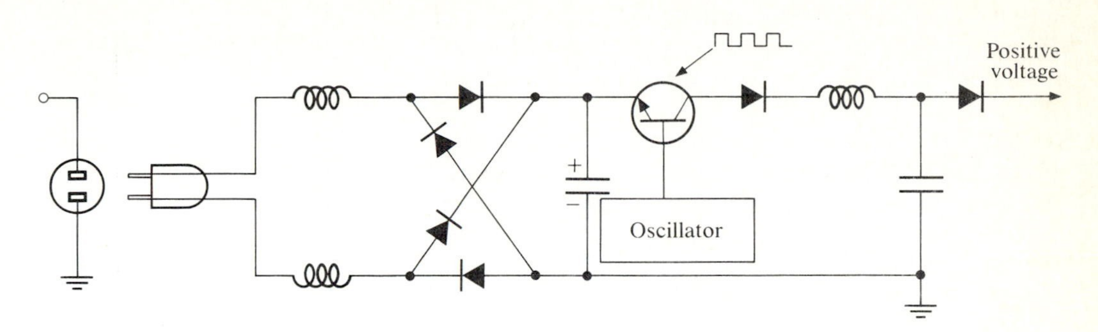

FIGURE 1–6
Switching power supply. The switching transistor converts the dc output from the bridge into a square wave, which is then filtered for a dc output. Switching power supplies are found in many new equipment chassis.

Handling and Servicing Equipment

1. Older pieces of equipment may have metal cases, which are usually connected to a metal chassis that is assumed to be at ground potential. Make sure this equipment has a properly wired three-pronged plug that will provide the expected ground.

2. Don't service electrical devices by yourself. If an electrical accident occurs, it is important to have someone present who can seek assistance.

3. Know the symbols for dangerous circuits and observe the safety instructions published for the equipment you are using.

4. Don't operate electrical equipment in an explosive atmosphere.

5. Always ground test equipment and the circuit under test.

6. Consult service manuals.

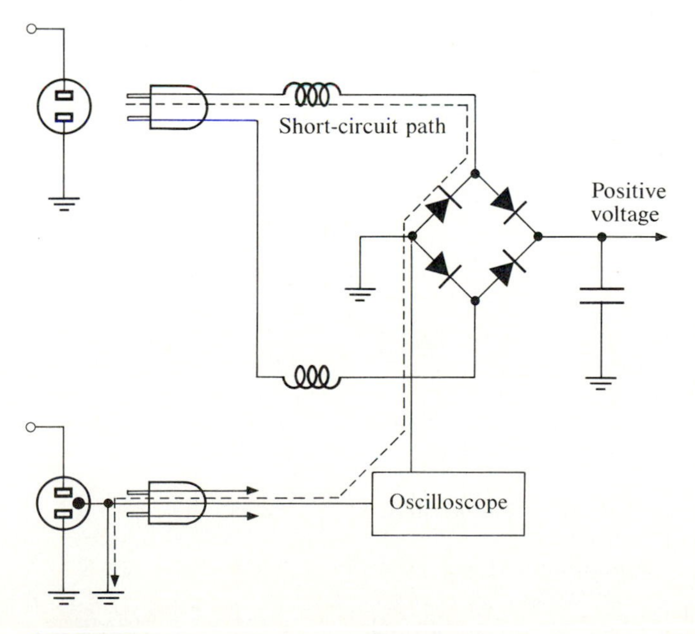

FIGURE 1–7
Floating ground using an oscilloscope

7. Check power cords frequently for condition and don't use a cord that is cracked, broken, or has a missing ground pin.

8. Use proper fuses to avoid fire hazard.

9. Always read instructions.

10. Retain all reference material.

11. Unplug equipment before cleaning its case.

12. Use recommended cleaning products only. Avoid using aerosol spray cleaners on the exterior of equipment cases since they can penetrate equipment and may damage components on boards inside.

13. Don't use attachments not recommended by the manufacturer.

14. Follow the assembly and mounting instructions that accompany equipment. Don't use shortcuts.

15. Make sure ventilation slots are unimpeded. The slots permit cooling of electric and electronic circuitry and should never be blocked or placed in front of a heat source such as a radiator or heat register. Don't encase equipment in bookcases or cabinets if this action will prevent proper ventilation of the equipment. Leads and power cords should be routed so that they cannot be stepped on, tripped over, or pinched by items placed upon or against them. Pay particular attention to cords at plugs, powerline receptacles, and the point where they exit the equipment they power. Unplug equipment that is not in use to prevent possible damage through powerline surges. Replace components with those specified by the manufacturer. Improper substitutions can cause damage.

LAB PROCEDURES

The following lab procedures will help you avoid delays and become a more efficient worker.

1. Keep a logbook. In it list the date, kind of equipment, model number, specifications and/or schematic numbers, symptoms encountered, tests performed, results, servicing done, and results of servicing. This information may be used as a reference for future work. See Appendix F for an example of a logbook entry.

2. Tube-driven test equipment requires warm-up time, which is the time necessary for the heater filaments to reach a temperature sufficient to produce the necessary emission for proper and stable operation. Turn on the equipment at least 15 minutes before using it to take measurements, and don't turn it off between tests.

3. Static electricity can destroy many types of solid-state components. Simply walking across a rug or swiveling on a lab stool can cause static to build up on the human body. Whenever possible, carpeting should be removed from the floor of the work area. If removal is not feasible, a spray chemical can be used to reduce static, or sheets of antistatic plastic can be installed over the carpeting.

4. Electrostatic discharge (ESD), also known as static electricity, is a problem that affects electronic components today. ESD, in the form of voltage and current, causes damage that may range from total component failure to performance degradation, which may be impossible to detect. It is estimated that ESD causes catastrophic damage only 10% of the time; 90% of the time the damage is intermittent and may not appear until the unit is in service.

When two nonconductive materials are rubbed together, electrons from one material are transferred to the other. The accumulation of electricity on the surface of a material is called a *triboelectric charge,* or an electrostatic charge. The amount of charge is a function of the separation of the materials as located on Table 1–2. (The combination of human hair and polyester is worse than paper and wood.) Conductive materials (such as the human body) can be drained of static charge by connecting them to earth ground. Nonconducting materials cannot be drained of accumulated static charge. Using a static wrist or ankle strap can reduce the danger of ESD from the body or from tools handled during service. Because styrofoam cups and plastic report covers are not electrical conductors, they cannot be discharged in this manner and should not be allowed in the service area. Additional factors in charge generation are the intimacy of contact, rate of separation, and humidity. Table 1–3 shows the relative susceptibility of various devices to electrostatic voltages.

Of the several factors that contribute to the generation of ESD, reduced humidity creates the

TABLE 1–2
Electrostatic (triboelectric) series

Material	
Air	↑
Human skin	
Glass	More
Human hair	positive
Wool	
Fur	
Paper	
Cotton	Neutral
Wood	
Hard rubber	
Acetate rayon	
Polyester	More
Polyurethane	negative
Polyvinylchloride (PVC)	
Teflon®	↓

TABLE 1–3
Component susceptibility to ESD

Device	Range of Susceptibility
Power MOSFET	100–200V
JFET	140 V–10 kV
CMOS	250 V–2 kV
Schottky diodes, TTL	300 V–2.5 kV
Bipolar transistors	380 V–7 kV
ECL	500 V
SCR	680 V–1 kV

TABLE 1–4
Typical ESD voltages

	Voltage	
	Relative Humidity	
Source	*(10–20%)*	*(65–90%)*
Walking across a carpet	35 kV	1.5 kV
Picking up a polyethylene bag	20 kV	1.2 kV
Sitting on a chair padded with polyurethane foam	18 kV	1.5 kV
Walking across a vinyl floor	12 kV	250 V
Handling a vinyl envelope	7 kV	600 V
Standing at a workbench	6 kV	250 V

most severe results. More than 300 times more electrostatic charge is created at 20% relative humidity than at 65% relative humidity. Table 1–4 shows the typical electrostatic voltage levels generated by various actions at different humidity levels. From Tables 1–2, 1–3, and 1–4 it is apparent that sensitive electronic components can easily be damaged or destroyed if proper precautions are not taken.

The following steps can be taken to avoid ESD problems:

a. Use a workbench with a grounded metal top or a grounded sheet of metal for a work surface.
b. Ground all soldering irons or solder/desolder guns.
c. Use a grounded wrist or ankle strap in contact with the skin.
d. Place all MOS devices on a grounded bench surface *prior* to handling—one can be electrostatically charged with respect to the bench surface.
e. Never insert or remove an MOS device from a circuit with the power applied.
f. Use antistatic bags for storing or transporting MOS devices.
g. Keep the workbench free of objects such as paper, cigarette ashes, styrofoam cups, and plastic report sheets.

5. LCDs are susceptible to environmental changes. If an LCD is exposed to cold weather, its action will slow dramatically; the time required for a change from one reading to another could take seconds or even minutes. When the LCD is allowed to warm up, it returns to normal operating speed. If the device is exposed to high temperatures or direct sunlight, it may become dark, making the display difficult to read. Returning the LCD to room temperature will restore its usefulness.

6. Capacitors should be discharged before being measured or tested. This should be done in a nonexplosive atmosphere. A 10-kΩ resistor to suppress sparking can be placed across both leads of the capacitor simultaneously. This procedure should be repeated to ensure that any charge stored in the capacitor has been reduced to zero. Be sure not to exceed the working voltage dc (WVDC) rating of the capacitor when wiring circuits.

7. Measuring a capacitor will yield a value of capacitance. However, this does not tell if a capacitor will operate properly under load. Leakage can be detected using equipment that performs tests while the capacitor has a dc working voltage applied to it. This will show if the capacitor can operate under "pressure."

8. When an ohmmeter is turned on, it should read infinity or overload if the probes are not touching. When the probe tips are touched together, enough current should flow through the ohmmeter circuitry to move the needle or the digital (LED or LCD) display to zero ohms, indicating a short circuit. If this does not happen, you may have to calibrate the instrument. Calibration is normally accomplished with a potentiometer adjustment called *zero adjust.* If the instrument cannot be "zeroed," check the batteries. Weak or dead batteries should be replaced. If the results are still not satisfactory, check the probes for continuity. Beyond this, equipment should be tested at the manufacturer, where test setups for troubleshooting specific equipment are available.

9. Any probe used with a piece of test equipment should be designed for that piece of equipment. Frequency compensation networks and other coupling networks are frequently built into probes. Proper probes with sufficient protection should be used for measuring high voltages. Make sure the probe is properly and securely connected to the test equipment and to the circuit under test. Usually a continuity test between the appropriate ends of the test probe will indicate an open circuit if one exists. When a two-channel oscilloscope is used, probes may be checked by swapping probes or by switching the probe in question to another channel.

10. If you suspect that a component is overheating, spray a coolant on it. Reducing the temperature of a component may cause the circuit to function properly for a short time, helping to isolate the problem. Many noncorrosive and nonconducting coolant sprays are available.

11. An avalanche effect is not uncommon. Components sometimes burn out because other components are not functioning properly. For example, a shorted capacitor may cause a transistor to burn out by allowing excessive current or voltage to affect the transistor.

TROUBLESHOOTING

Troubleshooting may be highly individualized. Most technicians eventually develop their own technique. However, many basic troubleshooting procedures are common for most equipment. It makes no difference whether the device being checked is a television set, computer terminal, or electronic depth finder. The following are common troubleshooting techniques and suggestions for good work habits.

1. Prepare your work area.

2. Select the tools and test equipment you think you'll need.

3. Check the simple and obvious problems first.

4. Look over the equipment to be tested. Careful examination of the equipment or circuit to be tested will often expose the problem.

5. Find out what symptoms indicate a specific malfunction. It may be possible to localize a problem by turning on the equipment and observing it. If possible, talk to someone who is familiar with the equipment.

6. Observe the condition of any indicator lights (e.g., power on).

7. Check to see that the equipment is plugged in if it requires powerline voltage for operation. Check its batteries if it is battery powered.

8. Make sure there is power at the service outlet into which the equipment is plugged.

9. Make sure the powerline cords are securely plugged into the equipment they service.

10. Check fuses. Replace fuses with the exact value of fuse required by the manufacturer. "Smoke tests" are dangerous and can be expensive mistakes.

11. Make sure the equipment is turned on.

12. Visually inspect the equipment for foreign materials or burned, missing, or broken components that could cause the malfunction. This inspection can be done using a good light source and a common paintbrush. Brushing equipment cleans dust (a potential problem) and may dislodge broken or extraneous parts, which could be the cause of the trouble.

13. Check ground and voltage levels.

14. Sometimes one section of the circuitry causes a problem by "loading down" other sections because of a short or open circuit. If possible, isolate sections of the circuit by disconnecting them one at a time. If no apparent change occurs, reconnect the section and disconnect another until you find the problem section.

15. Using several pieces of test equipment at the same time may cause loading problems through a common powerline. Use isolation transformers to prevent this undesired interaction.

16. Specialized equipment is available for the isolation of problems, including logic probes, current tracers, logic clips, logic pulsers, tube extenders, and extender boards.

Two common methods of troubleshooting active circuits are signal tracing and signal injection.

Signal Tracing

If the circuit in Figure 1–8 is to work properly, it must have a positive dc voltage. This voltage can be measured at points *B, C, D,* and *E* with a dc voltmeter. If the dc voltage is absent at any of these points, the resistor between the test point and the power supply is probably burned open. An appropriate signal is then injected at point *A*. Assume that this circuit is biased to operate as a cascaded class-A amplifier. In this case, the same signal should be observed at point *B*. An oscilloscope can be used to view the signals. An amplified, inverted signal should be "seen" at point *C*. The same signal seen at point *C* should be seen at point *D*. An amplified (again) and inverted (again) signal should be seen at point *E* and point *F*. If a signal is seen at one point but not at the adjacent point, the faulty component has probably been identified. A capacitor identified as open can be shunted with a "good" capacitor without disturbing the circuit under test. If the capacitor is shorted, it will likely cause the following transistor to go into saturation. Since the purpose of this amplifier is to amplify, the amplitude of the signal injected at point *A* must be kept small enough that the amplifier is not driven into saturation. Figure 1–9 shows the signal to be in-

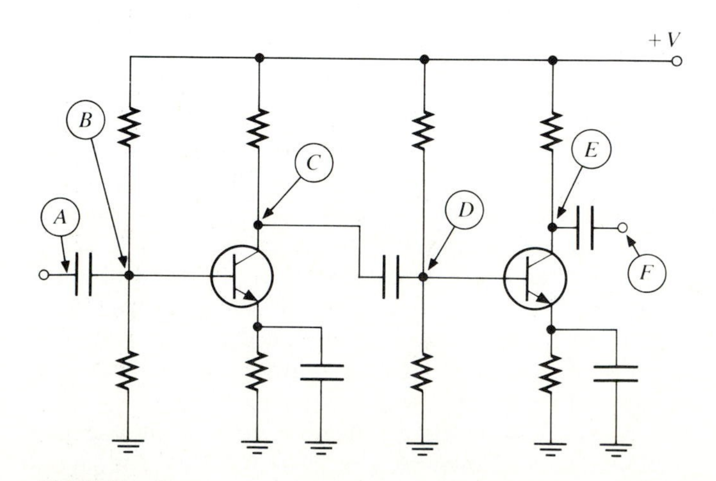

FIGURE 1–8
Signal-tracing a two-stage class-A amplifier

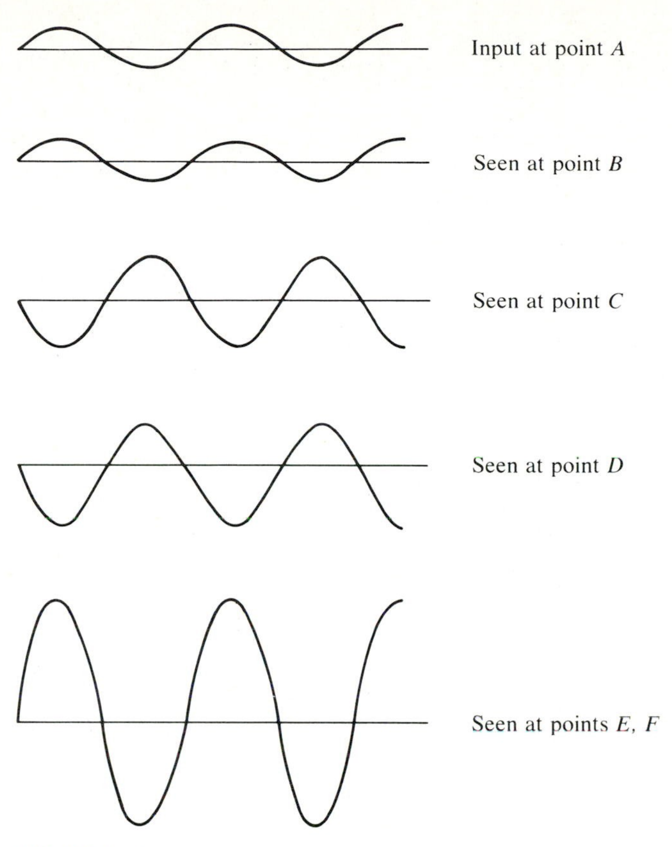

FIGURE 1–9
Signals for troubleshooting the circuit in Figure 1-8

jected into the amplifier and the signals that would be seen at points *B* through *F*.

The signal-tracing procedure can also be illustrated using Figure 1–10. This is a block diagram of a conventional superheterodyne (superhet) receiver found in most AM radios. If an RF generator set at about 800 kHz with a modulating signal of 400 Hz is attached to point *A* and a 400-Hz hum is heard at the speaker, the antenna is faulty. If an oscilloscope is used to observe the signal at point *K*, a 400-Hz signal should be seen. If the probe is moved to point *J*, a variable-amplitude (as the volume control is adjusted) 400-Hz signal should be seen. As the probe is moved through the point sequence *I, H, G, . . . , A,* appropriate signals should be observed. If a signal is seen at one point but not at the preceding point, the faulty circuit has been isolated. Figure 1–11 shows the signal at point *A* and the signals that should be seen at the other points.

Signal Injection

With reference again to Figure 1–8, a detector (voltmeter or oscilloscope) is connected to point *F*. By injecting the appropriate signal at points *E, D, C, B,* and *A* and observing the output, a defective component can be isolated. This method is a little more cumbersome than signal tracing since the signal source must be changed to provide the appropriate signal that would appear at each injection point if the amplifier were operating properly. Figure 1–12 illustrates the signal that would be seen and the signals that would be injected at the various points. Remember that the dc measurements must still be made.

The block diagram of Figure 1–10 can also be used to demonstrate signal injection. If a detector (oscilloscope) is attached to point *K* and the appro-

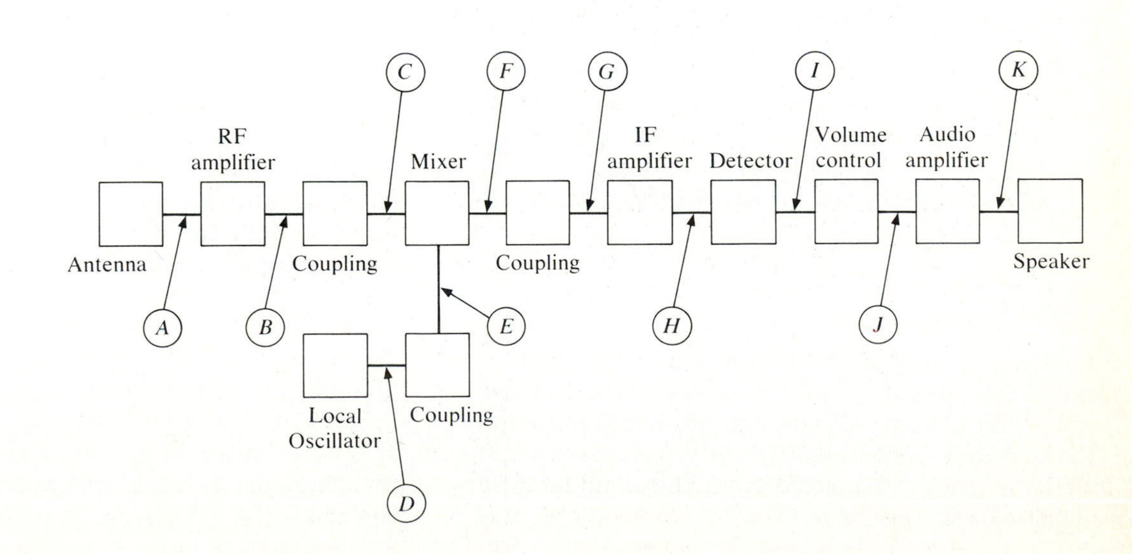

FIGURE 1–10
Troubleshooting block diagram for a superheterodyne receiver

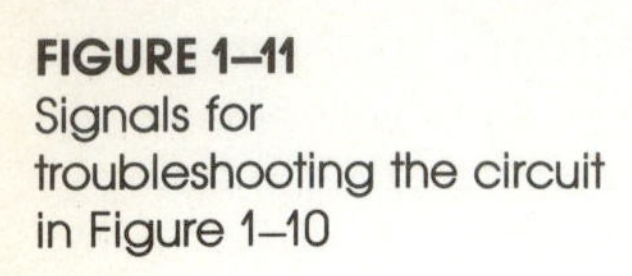

FIGURE 1–11
Signals for troubleshooting the circuit in Figure 1–10

Signal	Signal tracing	Signal injection
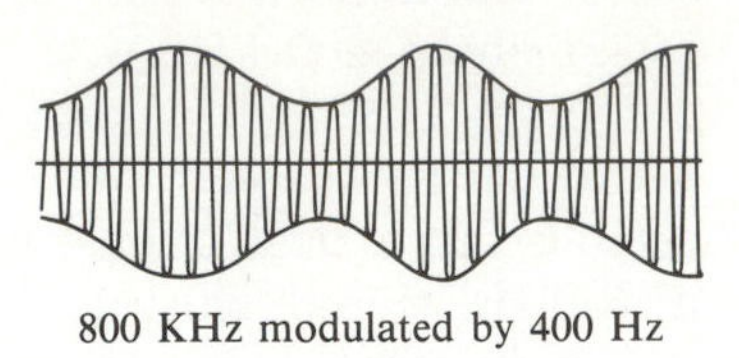 800 KHz modulated by 400 Hz	Applied at *A*	Injected at *A, B, C*
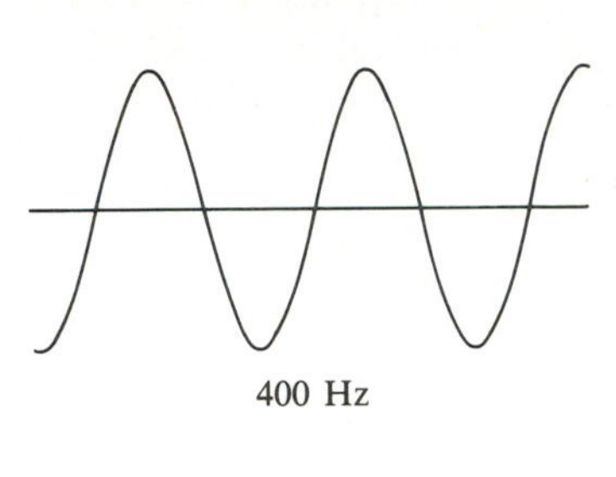 400 Hz	Seen at *K*	Injected at *K*
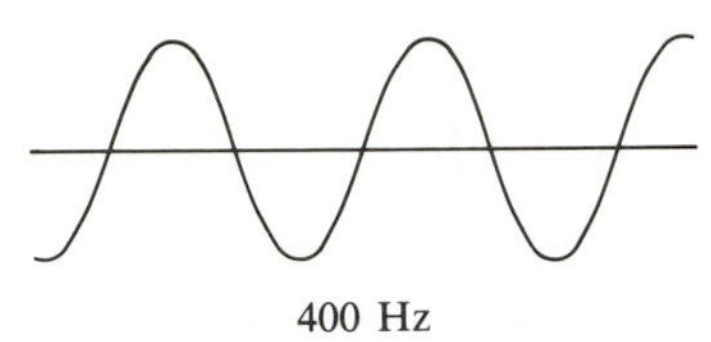 400 Hz	Seen at *I, J*	Injected at *I, J*
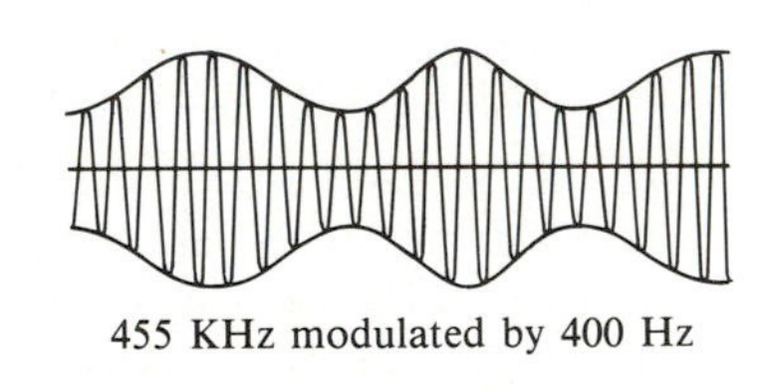 455 KHz modulated by 400 Hz	Seen at *F, G, H*	Injected at *C, F, G, H* (If there is an output when this signal is injected at point C, the mixer/local oscillator is inoperative)
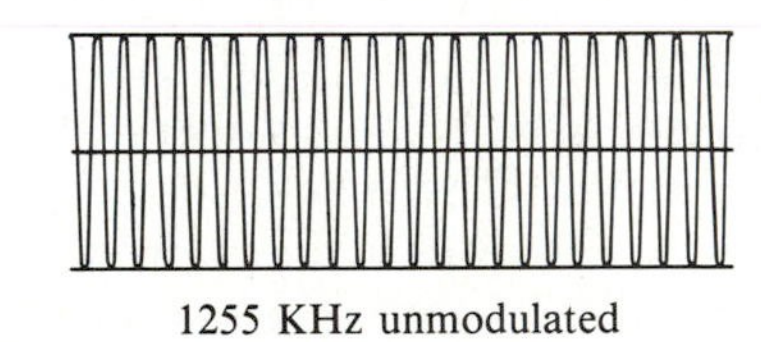 1255 KHz unmodulated	Seen at *D, E*	Injected at *D, E*

priate signals are injected at points *A, B, C, . . . , J*, a faulty circuit can be isolated. Because it is important that the appropriate signal be injected at each point, the technician must know what the circuits do and what to expect from them. Figure 1–11 shows the expected signal from point *K* if the appropriate signals are injected at the points listed.

REFERENCES

The test technician should have a variety of reference books and materials handy. Service and operating manuals for all equipment should be on file. These manuals come with newly purchased equipment or can be requested from the manufacturers at little or no cost.

Data books contain specific information about discrete components and integrated circuits along with electrical and physical data that can prove useful. Substitution manuals, distributed by most large component manufacturers, contain cross-referenced listings. If an exact substitute cannot be obtained, an equivalent component produced by another manufacturer, can be substituted. Component and equipment catalogs, usually available at no cost, can be useful. Trade newspapers such as *Mass High Tech*, published biweekly by High Tech Times, Inc., 755 Mount Auburn Street, Watertown, MA 02172, present news about industries, stock market analyses, new developments in equipment, and many possible employment contacts in industry. They are regional but available in all parts of the country.

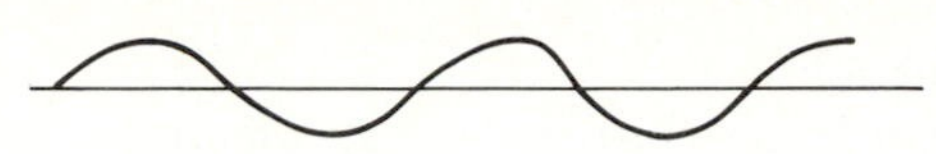

Small audio signal injected at points *A*, *B*

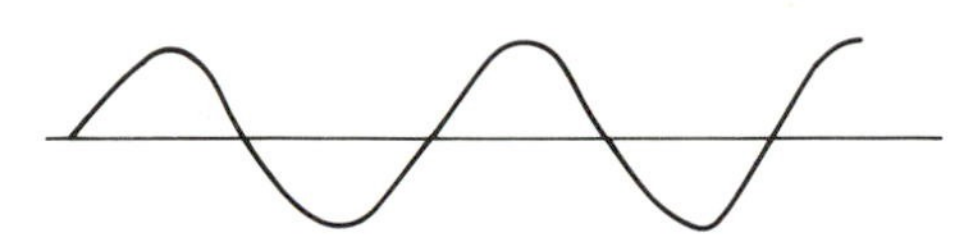

Larger audio signal injected at points *C*, *D*

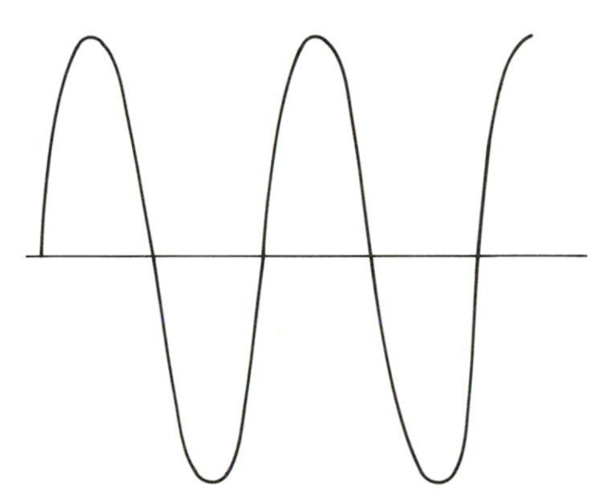

Larger audio signal injected at point *E* and seen at point *F*

FIGURE 1–12
Signals for signal injection troubleshooting the circuit in Figure 1–8. Phase changes are not shown. If a two-channel oscilloscope is used, the phase changes will be observed.

Many magazines can be had for the asking. Periodicals such as *Computer/Electronic Service News*, published by CESN Publications, 20 Grove Street, P. O. Box 428, Peterborough, New Hampshire 03458 and *Test & Measurement World* can be obtained by writing the publisher at P. O. Box 5341, Denver, Colorado 80217.

Logbooks will prove to be an important tool in troubleshooting equipment. They list a history of symptoms and solutions that the technician has observed and used.

Another valuable reference source is the local technical college or university. Most staff members will answer questions if they have time. Most college libraries are available for general use if the materials are used on-site. And their copying services will enable you to take information with you.

Fellow technicians can also be helpful in answering questions and offering suggestions.

SUMMARY

The preceding discussion of safety precautions, lab procedures, troubleshooting tips, and reference materials should provide useful information to the technician who is using test equipment. The discussion is not complete but should give ideas about how to proceed and where to get important information.

EXERCISES

1. What are some reasons for learning safety precautions?
2. Why do most accidents in an electronics laboratory occur?
3. Why is electrostatic discharge of concern to the service technician?
4. Which generates a greater static charge, brushing a cat's fur with a hard rubber comb or combing your own hair with a Teflon® comb? Why?
5. Describe the effect of humidity on electrostatic charge buildup.
6. How can static be minimized?
7. Many different types of electronics equipment are available. Why is it possible to present troubleshooting techniques in a relatively few pages of text?
8. Why is a clean, orderly workbench important?
9. Describe a floating ground and how to prevent one.
10. How can electronics equipment be tested for the presence of a floating ground?
11. What can be done to eliminate a floating ground?
12. Why keep one hand in your pocket, if possible, while testing electronic equipment?
13. Why approach a potentially hot component with the back of your fingers rather than the front?

14. Describe the entries in a useful logbook.
15. List 10 safety precautions that should be observed in an electronics testing situation.
16. Describe the reaction of LCDs to temperature variation.
17. What is the difference between measuring capacitance and testing a capacitor?
18. Describe the procedures you would follow if you turned on an ohmmeter and couldn't zero it.
19. What are the first steps in troubleshooting any kind of electronic equipment?
20. Describe signal tracing.
21. Describe signal injection.
22. List several resources the technician should have access to in an electronics testing laboratory.

2 Measurement and Errors

OUTLINE

OBJECTIVES
MEASUREMENT
- ☐ The National Bureau of Standards
- ☐ Calibration Curves
- ☐ Accuracy versus Precision
- ☐ Standards
- ☐ Technique
- ☐ Interpreting Data

ERRORS
- ☐ Gross Errors
- ☐ Systematic Errors
- ☐ Random Errors

HANDLING GROSS ERRORS
- ☐ Parallax
- ☐ Interpolation
- ☐ Significant Figures
- ☐ Rounding Off

STATISTICAL ANALYSIS
- ☐ Deviation and Standard Deviation
- ☐ Probable Error
- ☐ Handling Probable Errors

SUMMARY
EXERCISES
PROGRAMMING PROBLEM 2–1—Statistical Analysis

OBJECTIVES

After completing this chapter, you will be able to:

- ☐ Define measurement.
- ☐ Differentiate between the terms "measure" and "indicate."
- ☐ Explain the difference between primary and secondary standards.
- ☐ Explain the purpose of standards.
- ☐ Describe the relationship between accuracy and error.
- ☐ Explain the importance of technique in measurement.
- ☐ Explain the importance of correct interpretation of data.
- ☐ Differentiate among gross, systematic, and random errors and describe how to minimize the effects of each.
- ☐ Explain the difference between a linear and a nonlinear scale.
- ☐ Apply a technique for rounding off numbers.
- ☐ Apply basic statistical methods for handling errors.
- ☐ Explain standard deviation and probable error.
- ☐ Apply a simple method of handling calculation errors.

This chapter introduces some concepts of measurement, error, data handling, and error handling. The technician must be able to work comfortably with these concepts. The technician should also be aware of the drawbacks of some techniques and the implications of probable errors. A greater appreciation of the tolerance of a range of readings rather than an "exact" reading is also important.

MEASUREMENT

Measurement is the comparison of a value obtained from an instrument and an accepted standard. The standard could be the frequency transmitted by the government on radio station WWV or a reliable instrument considered to be a standard. In most labs, the instrument is used to *indicate* a value rather than *measure* the value. If a highly accurate value is required, the device being used must be calibrated using an accepted standard. *Primary standards* are unique and in the hands of the government (National Bureau of Standards). *Secondary standards* are calibrated against the primary standards and are available for purchase along with documentation attesting to their accuracy, but these are very expensive.

The National Bureau of Standards

The National Bureau of Standards (NBS) is a nonregulatory agency of the U.S. Department of Commerce. Established in 1901, the NBS conducts research and provides the national system of measurements and various technical and informational services to commerce, manufacturing, educational, and government institutions. The National Measurement Laboratory, one of the NBS's four major centers, has facilities in Gaithersburg, Maryland, and Boulder, Colorado, where primary standards are housed and maintained. Secondary standards are produced commercially using techniques traceable to and approved by the NBS. These standards provide the basis for the calibration of most electronic test equipment.

Calibration Curves

Most devices that can be adjusted for calibration purposes come with instructions outlining the procedure. A piece of equipment that cannot be calibrated in any prescribed manner may come with a calibration curve. If it does not, the technician can develop one.

For example, the technician may choose to construct and use a graph of equipment values versus standard values. A series of comparison values is taken, first with the meter to be calibrated and then with the meter accepted as a standard:

Meter to Be Calibrated (V)	*Standard Meter* (V)
0.95	1.0
1.90	2.0
2.85	3.0
3.83	4.0
4.83	5.0
5.82	6.0
6.80	7.0
7.75	8.0

These data are then graphed using the horizontal axis for the reading taken from the meter to be calibrated and the vertical axis for the standard meter values (see Figure 2–1). Curve *A* is an ideal calibration curve; that is, it is the curve that would result if the meter to be calibrated were perfect. Curve *B* applies to the actual voltmeter. A reading of 3.5 V on the actual meter corresponds to the true value of 3.7 V. Although this is not an exact approach, it allows the technician to use a less accurate instrument with reasonable accuracy and confidence. Remember that once the calibration curve is developed, it can be used as a reference, but only for that particular piece of equipment!

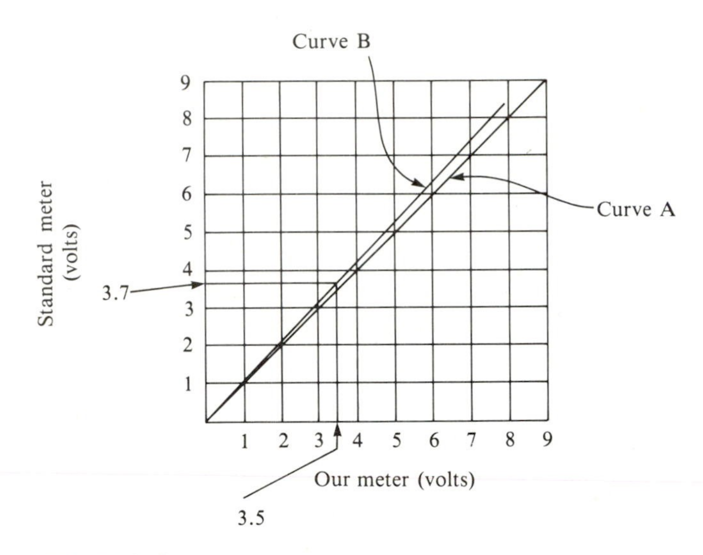

FIGURE 2–1
Calibration curve

Accuracy versus Precision

Accuracy is the closeness of a value to the true value. *Error* is the lack of accuracy. The expression "95% accuracy" indicates that a measured value is within 5% of the true value. This also indicates a 5% deviation (error) from the true value. In other words,

$$\text{Error} = 100\% - \text{accuracy} \qquad (2\text{–}1)$$

and

$$\text{Accuracy} = 100\% - \text{error} \qquad (2\text{–}2)$$

Precision refers to the reproducibility of measurements. An experiment results in a set of data. Can the results be repeated? If they can be repeated, the technique is said to show precision.

Standards

Since only primary standards can be accepted as 100% accurate, the concepts of error, probable error, tolerance, and significant figures must be dealt with.

After working with a set of test equipment for some time, the technician learns which instruments are most reliable. Usually an expensive multimeter or frequency meter is the standard in a general lab. Manufacturing facilities have more stringent requirements, however, and as a result, better standards are used there. In most field service situations, any instrument that will give an indication of failure is sufficient.

Technique

The technique used in taking a measurement is as important as the accuracy of the standard or instrument. In some instances, if the proper technique is employed, a relatively poor piece of equipment can be used to take reasonably accurate measurements. Likewise, an expensive piece of equipment can yield poor results if the technique is poor. Technique will be covered in some detail throughout the text.

Interpreting Data

The technician must know the value of the data or they are useless. When measurements are taken, it is important to relate the data obtained to the problem and recognize the implications of differences between the predicted and measured results. If the results are unexpected, the instrument could be malfunctioning, interpretation of the data could be poor, knowledge of the parameters to be measured could be lacking, or *a new discovery could have been made!* The technician must recognize these possibilities and work with them.

ERRORS

There are three general categories of errors in any measurement process: gross errors, systematic errors, and random errors. One function of this book is to demonstrate how to eliminate gross errors, minimize systematic errors, and account for and handle random errors.

Gross Errors

Gross errors are caused by the misuse of equipment, the proper use of incorrect or inadequate equipment, or the misrepresentation of data obtained. They can also be termed "human errors." The technician should learn the strengths and limitations of various measuring devices in order to be able to select the best "tool" for each job. Applying this knowledge eliminates a large amount of error and a great deal of frustration when measurements must be taken. Learning how to use a piece of equipment properly may eliminate many errors; a $2000 device could be better used as a paperweight if not used correctly. The technician should know the magnitude, units, and significance of the data obtained or the data will be of no value.

One way to acquire these skills is by trial and error, that is, by reading the operation manual and experimenting. This approach is time consuming, and time is money. It is imperative that the operation manual be read before any instrument is used. Another way to acquire these skills is by following through a course or text (such as this) in conjunction with a lab experience. The more a piece of equipment is used, the more familiar the technician becomes with it, and the more consistently valid the resulting data will be.

Systematic Errors

Systematic errors result from mechanical weaknesses of an instrument. Worn bearings on meter movements and nonlinear sweeps on oscilloscopes are the fault of the instrument, not the technician. With an adequate understanding of these problems, the technician can diagnose the situation correctly and repair or replace the defective equipment. In some cases, recalibration or compensation for the error may be a solution.

Random Errors

Random errors are errors whose cause cannot be directly established because they appear to be random variations in the electrical parameters of the measuring system or the device under test. The good technician will acknowledge their existence and compensate for them by statistical methods. A subsequent section will deal with statistical analysis at a basic level in an attempt to "normalize" random errors. Much of the remainder of the text focuses on eliminating the systematic and gross errors that pop up whenever they are least expected.

HANDLING GROSS ERRORS

As stated before, gross errors are essentially human errors. Examples are errors caused by parallax, interpolation, loading effects, and ignorance of frequency limitations of instruments. The loading effect of instruments on circuits and the loading effect of circuits on instruments result when one circuit draws enough current from another circuit to change significantly the electrical characteristics of either the test instrument or the circuit under test. Frequency considerations and limitations result when the frequency response or bandwidth of an instrument is too limited to allow proper processing of incoming signals. Errors due to parallax and interpolation are the subject of the next two sections.

Parallax

Parallax describes the situation in which only one point is available to determine a straight line from the eye to a scale, and that point is the needle or pointer of a meter. Depending on the point of view, the specific reading in Figure 2–2A would be 20 mA, and that in Figure 2–2B would be about 26 mA. The difference in readings is not caused by the instrument but by the positioning of the eye with respect to the meter's needle. Many analog meter movements use a mirror behind the needle so that the eye can align the needle with its reflected image and thus get a much more accurate result, nearly eliminating parallax errors (Figure 2–2C).

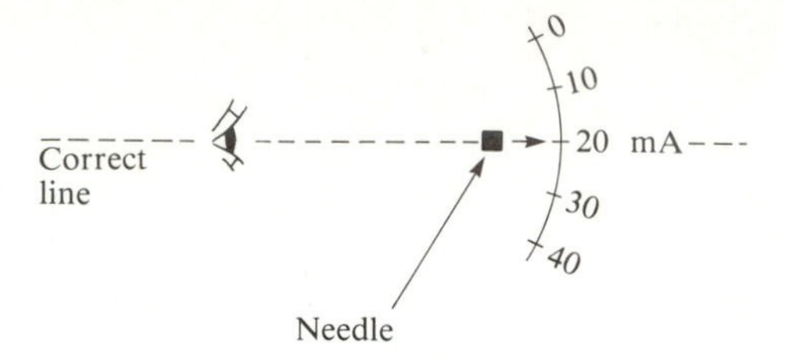

A.

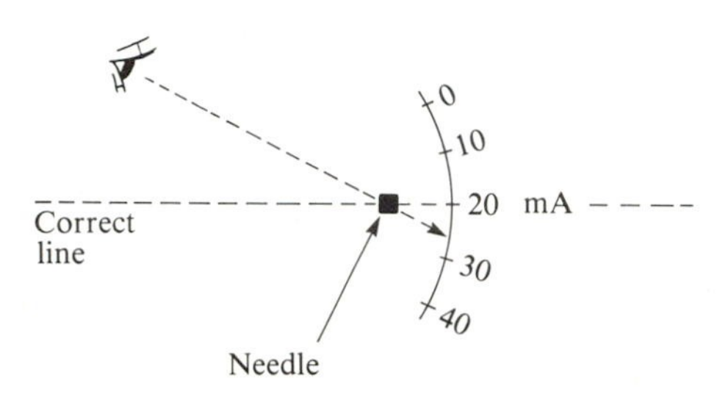

B.

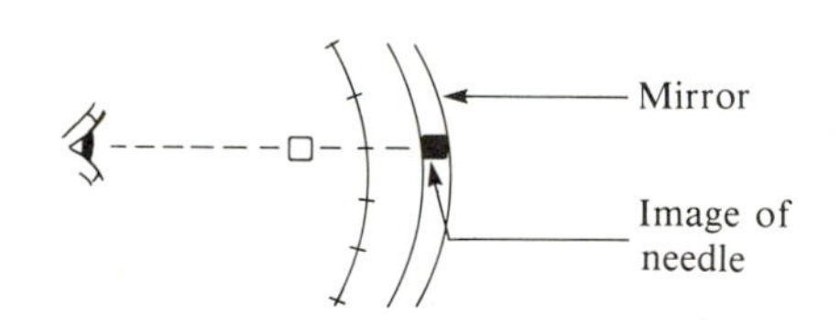

C.

FIGURE 2–2
Parallax

Interpolation

Another gross error occurs when the needle of an analog meter comes to rest between two values. It becomes necessary to interpolate (make an intelligent guess!) where the needle is pointing. From Figure 2–3A, one may guess that the value is 1320 Ω, but since the scale is nonlinear, some error in interpolating is guaranteed. Nonlinear scales have different values of parameters for equal amounts of deflection. Figure 2–3B shows a linear scale, and Figure 2–3C shows a nonlinear scale. The use of digital readout meters reduces this type of error.

Significant Figures

The number 279 indicates a value between 278.5 and 279.5. The figure 3000 indicates a value between 2500 and 3500. Significant figures can be thought of as nonzero figures that occur leftmost in a number and can be counted from left to right. If no decimal point occurs in a number, significant figures are counted from left to right until all other numbers are zero. If a number contains a decimal point, all figures approaching the decimal point from the left are considered to be significant. The number 35,000. indicates a value between 34,999.5 and 35,000.5, whereas the number 35,000 indicates a value between 34,500 and 35,500. In the first example, the 3, 5, and three 0s are considered to be significant figures, whereas in the second example, the 3 and the 5 are the only significant figures. A constant number such as the 1 in $1/R$ is considered to be perfect—1 ± 0% error. Significant figures are important since they govern the accuracy of the work.

Consider a 5.1-kΩ resistor in parallel with a 3.3-kΩ resistor. The result on the calculator should appear as 2003.5714 Ω:

$$
\begin{aligned}
R_t &= \frac{R_1 R_2}{R_1 + R_2} \\
&= \frac{(5.1\ \text{k}\Omega)\ (3.3\ \text{k}\Omega)}{5.1\ \text{k}\Omega + 3.3\ \text{k}\Omega} \qquad \text{(2–3)} \\
&= 2003.5714\ \Omega
\end{aligned}
$$

FIGURE 2–3
Linear versus nonlinear scales

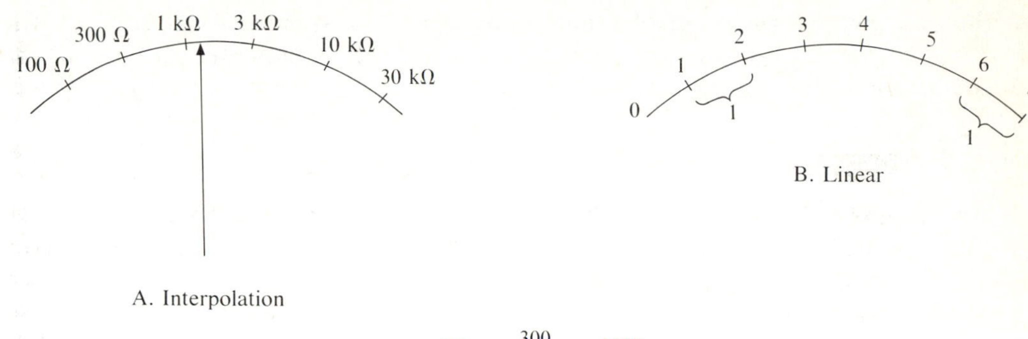

Since the most accurate original data has two significant figures (5100 and 3300), the best accuracy that can be expected is two significant figures (2003.5714), and the answer should therefore be expressed as 2000 Ω, or 2.0 kΩ. Using π accurate to nine significant figures (3.14159265...) does *not* improve the accuracy of numbers with fewer significant figures. Another viewpoint is that it is ridiculous to worry about eight significant figures if the meter being used can yield only two or three!

Rounding Off

But what happens to the extra numerals generated in the calculation? The answer is, they are rounded off. Many techniques exist for rounding off extra numerals. The method applied in this book is a simple one, having two parts. First, if a number following the last significant figure to be retained is 0, 1, 2, 3, or 4, change it and all figures that lie to the right of it to zero. For example, to round 7563.12 off to three significant figures, look at the 3, since it follows the 6, which is the last of the three figures to be retained:

Last figure to be retained

$$75\underline{6}3.12 \tag{2–4}$$

The 3 will be changed to zero, as will all numbers to the right of it, yielding 7560.00, or simply 7560. Second, if the figure following the last numeral to be retained is 5, 6, 7, 8, or 9, add one to the preceding figure and change all numbers to the right of the last significant figure to be retained to zeros. For example, to round 7563.12 off to *two* significant figures, look at the 5 as the last significant figure to be retained.

Last figure to be retained

$$7\underline{5}63.12 \tag{2–5}$$

The 6 and all other numerals to the right of it will be changed to zero, and one will be added to the preceding figure, yielding 7600.00, or 7600.

STATISTICAL ANALYSIS

Assume that a voltage must be measured using 12 different meters. The following measurements are taken:

Voltage (V)		
48	55	48
53	49	57
49	47	51
45	55	46

If one value must result, the arithmetic mean, or average, should be calculated. To do this the 12 entries are added and the sum is divided by 12:

$$48 + 53 + 49 + 45 + 55 + 49 + 47 + 55 + 48 + 57 + 51 + 46 = 603 \tag{2–6}$$

$$\frac{603\text{ V}}{12} = 50.25\text{ V}$$

Note that this figure is not even one of the original measurements. One way to handle this situation is to calculate the probable error and express the value as the average plus or minus this error. To do this, we have to look at three new concepts: deviation, standard deviation, and probable error.

Deviation and Standard Deviation

The *deviation* (d) is the difference between the arithmetic mean and the data entry. For example, for the first meter measurement in our example,

$$d = 48 - 50.25 = -2.25$$

The negative value will be of no consequence here since the next step will be to square the deviation. The resulting deviations are as follows:

	d
48 − 50.25 =	−2.25
53 − 50.25 =	2.75
49 − 50.25 =	−1.25
45 − 50.25 =	−5.25
55 − 50.25 =	4.75
49 − 50.25 =	−1.25
47 − 50.25 =	−3.25
55 − 50.25 =	4.75
48 − 50.25 =	−2.25
57 − 50.25 =	6.75
51 − 50.25 =	0.75
46 − 50.25 =	−4.25

Next, we will square the deviations:

d	d^2
−2.25	5.0625
2.75	7.5625
−1.25	1.5625
−5.25	27.5625
4.75	22.5625
−1.25	1.5625
−3.25	10.5625
4.75	22.5625
−2.25	5.0625
6.75	45.5625
0.75	0.5625
−4.25	18.0625

Then we add these squares and get 168.2500. We do this in order to calculate the *standard deviation* (σ). The standard deviation for a finite number of readings is defined as

$$\sigma = \sqrt{\frac{\sum d^2}{n - 1}} \qquad \textbf{(2–7)}$$

where $\sum$ is a mathematical symbol meaning "the sum of," $\sum d^2$ denotes the sum of the d^2 terms, and n is the number of entries used. (If the number of readings is greater than 30, use n in place of $n - 1$.) Substituting values from the example,

$$\sigma = \sqrt{\frac{168.2500}{12 - 1}} = 3.9109 \qquad \textbf{(2–8)}$$

Now let's look at the meaning of the standard deviation and the reason for calculating it.

If n readings are taken and plotted as to frequency of occurrence, they will usually follow a Gaussian distribution curve, resembling a bell (Figure 2–4). As shown, the mean value occurs most often if a large number of readings is taken, and values away from the mean occur less often. The curve demonstrates that the probability of a random reading occurring at the average value is greatest and that the probability of a random reading occurring above or below the average is less. It also shows that the greater the distance from the average, the lower the probability that any random reading will occur there. A standard deviation is a constant value for each set of readings. It is an indication of probability. This indication shows the probability any random reading has of falling within a range defined by the average plus or minus one standard deviation.

Consider the data in Table 2–1. The left column shows the number of standard deviations from the mean being considered. The area under the dis-

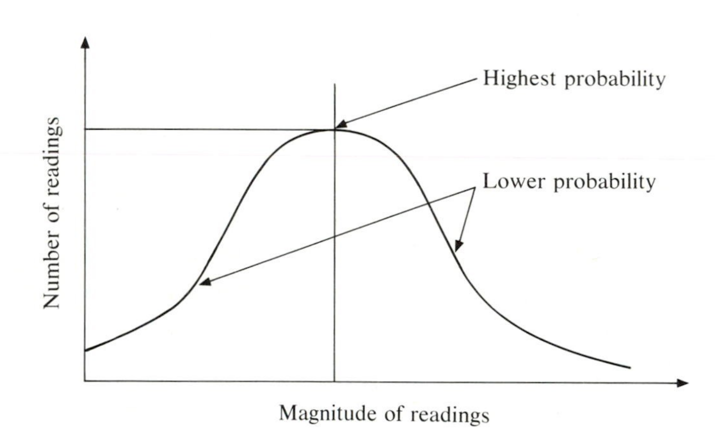

FIGURE 2–4
Gaussian distribution curve

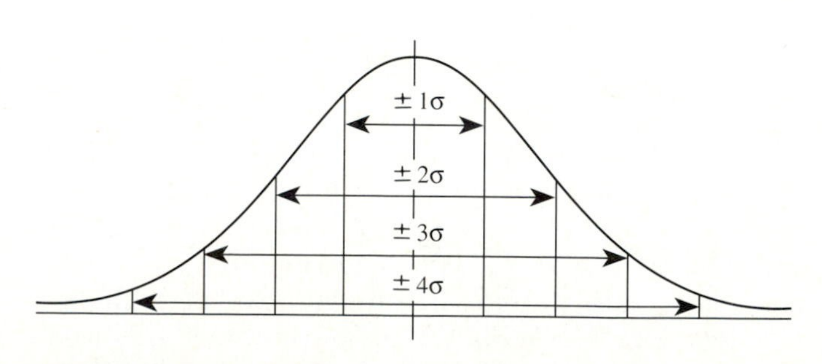

FIGURE 2–5
Area under the distribution curve related to standard deviations

TABLE 2–1
Probability related to standard deviations

No. of Standard Deviations from Mean	Percentage Probability of Reading Occurring within No. of Standard Deviations	Odds against Reading Occurring outside No. of Standard Deviations
0.6745	50	1:1
1	68.27	2.15:1
2	95.45	21:1
3	99.73	370:1
4	99.99366	15,772:1
5	99.99995	1,744,000:1

Source: Standard Mathematical Tables, Chemical Rubber Company, 1963, p. 236.

tribution curve can be thought of as the number of readings that will occur within the given tolerance (see Figure 2–5). The probability that any random reading will fall within the indicated number of standard deviations is given in the middle column of the table. The right column gives the odds against the random reading falling outside the number of standard deviations. As can be seen, the more standard deviations from the mean, the higher the probability a random reading will fall within the area under the curve representing that number of standard deviations. Also, the farther away from the mean that the reading occurs, the greater are the odds that the reading will fall within the area under the curve. Very little area is left under the curve.

Probable Error

In general, if a reading is five or more standard deviations from the mean, the probability of its being valid is extremely small. In some laboratories, any reading falling five or more standard deviations from the mean can be thrown out as being erroneous.

Notice that the odds are 50% that any random reading will fall within 0.6745σ of the mean. The 0.6745σ is referred to as *probable error*. Using this value to express the original 12 readings as one yields

$$50.25 \pm (0.6745)(3.9109) \quad \text{(2–9)}$$

or

$$50.25 \text{ V} \pm 2.638 \text{ V} \quad \text{(2–10)}$$

Expressed as a percentage, this is written

$$50.25 \pm \frac{2.638}{50.25} \quad \text{(2–11)}$$

or

$$50.25 \text{ V} \pm 5.25\% \quad \text{(2–12)}$$

The 50.25 V is the nominal value, and the ±5.25% is the probable error.

Handling Probable Errors

The random deviations that occur in taking a series of measurements can be expressed as a mean plus or minus a probable error. Most electronic components are labeled as to their tolerance. Resistors are usually ±5% or ±10%, and capacitors might be ±10%. How can these probable errors be taken into account when used as the values in simple calculations?

For example, consider the problem of adding two resistances in series to get an equivalent value of resistance. Suppose we have the following:

$$\begin{array}{r} 5.6 \text{ k}\Omega \pm 5\% \\ \underline{+2.2 \text{ k}\Omega \pm 10\%} \end{array}$$

The expected sum is 7.8 kΩ. Looking at worst-case conditions (i.e., the 5.6-kΩ resistance is +5% and the 2.2-kΩ resistance is −10%), we obtain

$$\begin{aligned} 5.6 \text{ k}\Omega + 5\% &= 5600\ \Omega + 280\ \Omega = 5880\ \Omega \\ 5.6 \text{ k}\Omega - 5\% &= 5600\ \Omega - 280\ \Omega = 5320\ \Omega \end{aligned} \quad \text{(2–13)}$$

The manufacturer guarantees that the 5.6 kΩ ± 5% resistance lies between 5320 Ω and 5880 Ω. And we also get

$$\begin{aligned} 2.2 \text{ k}\Omega + 10\% &= 2200\ \Omega + 220\ \Omega = 2420\ \Omega \\ 2.2 \text{ k}\Omega - 10\% &= 2200\ \Omega - 220\ \Omega = 1980\ \Omega \end{aligned} \quad \text{(2–14)}$$

The manufacturer guarantees that the 2.2 kΩ ± 10% resistance lies between 1980 Ω and 2420 Ω. Therefore, the sum of these resistance combinations is

$$\begin{array}{ccc} 5320\ \Omega & 5600\ \Omega & 5880\ \Omega \\ 1980\ \Omega & 2200\ \Omega & 2420\ \Omega \\ \hline 7300\ \Omega & 7800\ \Omega & 8300\ \Omega \end{array}$$

If both resistances were found to be at the lower end of their tolerance range, their sum would be 7300 Ω. If both resistances were found to be at the upper limit of their tolerance range, their sum would be 8300 Ω. The probable error found in adding these values would be 7800 Ω ± 500 Ω, or 7800 Ω ± 6.4%. In general, it would be safe to use the poorest tolerance when expressing the sum of probable errors. *Calculations cannot improve tolerance or probable errors.*

Multiplication, for example $V = IR$, also involves probable error. Assume that current can be measured as 3 mA ± 3% and resistance as 2 kΩ ± 5%. The expected result would be

$$(2\ \text{k}\Omega)(3\ \text{mA}) = 6\ \text{V} \tag{2–15}$$

Worst-case conditions yield the following values:

$$\begin{matrix} 1900\ \Omega & 2000\ \Omega & 2100\ \Omega \\ 0.00291\ \text{A} & 0.003\ \text{A} & 0.00309\ \text{A} \end{matrix} \tag{2–16}$$

The largest value of resistance multiplied by the largest value of current will yield

$$(2100\ \Omega)(0.00309\ \text{A}) = 6.489\ \text{V} \tag{2–17}$$

Multiplying the current reading that happened to be at the upper limit of its tolerance by a resistance value that happened to be at the upper limit of its tolerance leads to an answer approximately 8% from the expected answer. Generally, with multiplication of numbers with expressed probable error, the resulting probable error can be approximated by adding the individual probable errors. Division can be handled the same way since it is simply the inverse of multiplication.

EXAMPLE 2–1

For two resistances in parallel,

$$R_t = \frac{R_1 R_2}{R_1 + R_2} \tag{2–18}$$

With

$R_1 = 10\ \text{k}\Omega \pm 10\%$ and $R_2 = 5\ \text{k}\Omega \pm 5\%$,

$$\begin{aligned} R_t &= 3.3\ \text{k}\Omega \pm (10\% + 5\% + 10\%) \\ &= 3.3\ \text{k}\Omega \pm 25\% \end{aligned} \tag{2–19}$$

The probable error tends to add up very quickly. Using another form of the equation to calculate the equivalent of the two parallel resistances, consider

$$\frac{1}{R_t} = \frac{1}{R_1} + \frac{1}{R_2} \tag{2–20}$$

The number 1 is exact (1 ± 0%). With the values from Example 2–1,

$$\begin{aligned} \frac{1}{10\ \text{k}\Omega \pm 10\%} + \frac{1}{5\ \text{k}\Omega \pm 5\%} &= 0.0001\ \text{S} \pm 10\% \\ &\quad + 0.0002\ \text{S} \pm 5\% \\ &= 0.0003\ \text{S} \pm 10\% \end{aligned} \tag{2–21}$$

$$\frac{1}{0.0003\ \text{S} \pm 10\%} = 3.3\ \text{k}\Omega \pm 10\%$$

The probable error has been reduced by reducing the number of mathematical operations. As a general rule, use the least number of mathematical operations to reach a solution since doing so will reduce the built-in probable error.

Now consider subtraction. Assume that the voltage on two sides of a differential amplifier is measured with a meter guaranteed to yield a maximum error of ±1%, with the following results:

$$74.2\ \text{V} \pm 1\% \quad \text{and} \quad 72.8\ \text{V} \pm 1\% \tag{2–22}$$

The expected potential difference would be

$$74.2\ \text{V} - 72.8\ \text{V} = 1.4\ \text{V} \tag{2–23}$$

Consider the worst-case conditions:

$$\begin{aligned} 74.2\ \text{V} + 1\% &= 74.2\ \text{V} + 0.742\ \text{V} \\ &= 74.942\ \text{V} \\ 72.8\ \text{V} - 1\% &= 72.8\ \text{V} - 0.728\ \text{V} \\ &= 72.072\ \text{V} \end{aligned} \tag{2–24}$$

This combination of values yields the largest possible error with both readings still taken within guaranteed tolerance limits:

$$74.942\ \text{V} - 72.072\ \text{V} = 2.87\ \text{V} \tag{2–25}$$

The measured value is more than 100% from the calculated or expected value

$$\begin{aligned} &2.87\ \text{V} - 1.4\ \text{V} = 1.47\ \text{V} \\ &\frac{1.47\ \text{V}}{1.4\ \text{V}} \times 100\% = 105\%! \end{aligned} \tag{2–26}$$

Remember that we started with 1% values! From this example, it is obvious that subtraction should be avoided because it can lead to a very large error.

SUMMARY

The student working in a carefully controlled laboratory environment is rarely concerned with random and loading errors. The technician in industry, however, must be aware of these problems and know how to deal with them. This chapter serves as an introduction to many of these situations so that the student will be more comfortable facing them later on.

EXERCISES

1. To what accuracy does a 15% error correspond?
2. Draw a calibration curve for the following data:

Meter to Be Calibrated (Hz)	*Standard Frequency (Hz)*
110	100
215	200
423	400
840	800
1263	1200

   What true value would be indicated if the meter reading corresponded to 635 Hz? What meter reading would you have to see to know that you are measuring a true frequency of 1000 Hz?
3. Given the following data, calculate the arithmetic mean, the standard deviation, and the probable error: 5802, 5790, 5810, 5821, 5762, 5781, 5809, 5798, 5774, 5811.
4. Repeat Exercise 3 for the following data: 5.25, 4.69, 5.00, 4.92, 5.20, 4.89, 4.78, 5.05, 5.10, 4.92, 5.10, 5.10, 4.88, 4.95.
5. Repeat Exercise 3 for the following data: 31.2, 31.0, 30.8, 30.4, 31.1, 30.5, 30.0, 31.0, 31.4, 30.8, 30.9, 31.3, 30.6, 11.2.
6. In Exercise 5, are we justified in dropping the 11.2 reading as probably being in error? Explain.
7. How many significant figures are there in each of the following numbers?
   a. 25,700 b. 0.00035 c. 4000. d. 4000 e. 93,000,000
8. Express the answer to each of the following using the greatest valid number of significant figures:
   a. 35.2 + 19.7 + 24.36 + 14. + 39.253
   b. (3500)(8100)
   c. (2)(π)(62) (π = 3.14159265 . . .)
   d. $\dfrac{1.01892}{6100.}$
9. Round off each of the following to two significant figures:
   a. 652.14 b. 593,200 c. 1.9935 d. 2085 e. 1
10. Round off each of the following to three significant figures:
    a. 3.14159265 . . . b. 355.555 c. 0.2146 d. 5919 e. 4.944
11. Power is to be calculated from current and resistance values using the formula $P = I^2R$. The current is measured at 20 mA on a meter rated at $\pm 2\%$, and the resistance is measured at 2200 Ω on a meter with a probable error of $\pm 1\%$. What is the probable error of the calculated power?
12. Give the probable error in calculating the series equivalent of two resistors if R_1 = 270 Ω ± 1% and R_2 = 540 Ω ± 10%.
13. If the two resistors in Exercise 12 are wired in parallel, give their equivalent value plus or minus the probable error.

14. Show any difference in the calculation in Exercise 13 if the calculation is performed using:

a. $\frac{1}{R_t} = \frac{1}{R_1} + \frac{1}{R_2}$ b. $R_t = \frac{R_1 R_2}{R_1 + R_2}$

PROGRAMMING PROBLEM 2–1 Statistical Analysis

Object

To write and execute a BASIC program that will perform a statistical analysis on data read during an experiment, and to print out the nominal result (arithmetic mean) and probable error.

Procedure

You are measuring resistors. Your task is to have the computer list the individual values of resistance (input) in ohms and then print out:

1. The arithmetic mean
2. The standard deviation
3. The arithmetic mean plus or minus the probable error

The resistors measure as follows:

Resistance (Ω)

5802	5762	5774
5790	5781	5811
5810	5809	
5821	5798	

Use the program to evaluate the following voltage readings in the same way:

Voltage (V)

5.25	5.20	5.10	4.88
4.69	4.89	4.92	4.95
5.00	4.78	5.10	
4.92	5.05	5.10	

Use the program to analyze the following data:

Current (mA)

31.2	31.1	31.4	30.6
31.0	30.5	30.8	11.2
30.8	30.0	30.9	
30.4	31.0	31.3	

If you recalculate the analysis in the last problem eliminating the 11.2-mA reading as probably erroneous, what data do you get? Are you justified in eliminating this reading under the specifications given? Explain.

3 Operational Amplifiers

OUTLINE

OBJECTIVES
HISTORY OF THE OPERATIONAL AMPLIFIER
USES OF OP-AMPS
CHARACTERISTICS OF A TYPICAL OP-AMP
OPEN-LOOP AMPLIFICATION
NONINVERTING OP-AMP WITH A FEEDBACK LOOP
INVERTING OP-AMP WITH A FEEDBACK LOOP
THE OP-AMP AS A BUFFER AMPLIFIER
OP-AMPS IN MATHEMATICAL OPERATIONS
☐ Multiplication
☐ Division
☐ Addition
☐ Subtraction
☐ Solving Systems of Equations
☐ Differentiation
☐ Integration
SUMMARY
EXERCISES
LAB PROBLEM 3–1—Operational Amplifiers

OBJECTIVES

After completing this chapter, you will be able to:

☐ Outline the development of operational amplifiers.
☐ List the characteristics of a typical op-amp.
☐ List the uses of the op-amp as an amplifier without feedback (open loop).
☐ List the uses of the op-amp as an amplifier with feedback (closed loop).
☐ List the uses of the op-amp as a buffer amplifier.
☐ Describe how an op-amp can be used to perform multiplication, division, addition, and subtraction.
☐ Demonstrate how an op-amp can be used to invert a signal.
☐ Show how op-amps can be used to solve systems of algebraic equations.
☐ Show how an op-amp can be used to take a derivative.
☐ Demonstrate how an op-amp can be used to take an integral.
☐ Describe the use of the op-amp in signal processing.

This chapter introduces some fundamentals of the operational amplifier (op-amp). It begins with the historical development of the device and then considers the op-amp's general characteristics. Use of the op-amp in circuits is extensive and vital to modern-day instrumentation. Op-amps used as analog computing devices can perform arithmetic and calculus operations on input signals. The text discussion gives a brief overview of these capabilities.

HISTORY OF THE OPERATIONAL AMPLIFIER

In 1948, at Bell Laboratories, semiconductor devices called transistors were developed. The year 1958 saw the first integrated circuit (IC) conceived and constructed. The first solid-state operational amplifier (op-amp) was introduced in 1962.

Fairchild Semiconductor produced the first commercial op-amp, the μA702, in 1963. The device had low voltage gain and used −6-V and +12-V dc power supply voltages. Although it burned out easily when temporarily shorted, the μA702 was the best device of its time.

The μA709, also produced by Fairchild Semiconductor, had higher gain, a larger signal range, lower input current, and a ±15-V power supply voltage.

In 1967, National Semiconductor introduced the LM101, which had a gain of up to 160,000. Equipped with short-circuit protection, the LM101 also had simplified frequency compensation through the use of an externally wired capacitor. This arrangement eliminated internally generated oscillations in the direct-coupled amplifier network. Fairchild Semiconductor introduced its μA741 in 1968. Similar to the LM101, it had internal frequency compensation.

Chips containing multiple op-amps were introduced in 1974, when Raytheon Semiconductor developed the RC4558. The RCA CA3130 introduced FET input op-amps.

Many types and configurations of op-amps have been introduced since that time.

USES OF OP-AMPS

The characteristics and behavior of the operational amplifier that lead to its many useful applications are of particular interest to the technician involved in electronic instrumentation and measurements. The op-amp is a powerful device that can be used as a linear amplifier, a signal-processing device, and a device for performing mathematical operations through electronic circuitry applications. It can also be used for the electronic isolation of in-circuit components.

Linear amplifiers usually operate as class-A amplifiers with an output signal that is an exact reproduction of the input signal. The op-amp can be used to sharpen leading and trailing edges of pulses and to establish voltage thresholds for shaping signals. Because of its unique characteristics, it can be used in relatively simple circuits to perform addition, subtraction, multiplication, and division of input voltages to give results that reflect mathematical operations. Differentiation and integration are performed with similar ease. The result is analog computer circuits with relatively few components capable of solving sophisticated mathematical problems. Because of its high impedance, the op-amp can be used to isolate devices in active circuits without any significant loading effects. Op-amps are also used in logic and audio circuits.

CHARACTERISTICS OF A TYPICAL OP-AMP

The μA741 is a typical op-amp. Figure 3–1 shows the internal circuit of the μA741. Figure 3–2 illustrates different package configurations of op-amps.

In Figure 3–1, transistors Q_1 and Q_2 are, respectively, noninverting and inverting inputs to a differential amplifier configuration. Transistors Q_{16} and Q_{17} constitute a high-gain Darlington driver. Transistors Q_{14} and Q_{20} form a class-AB complementary symmetry output stage. A 10-kΩ potentiometer is between pins 1 and 5 with the wiper tied to $-V_{cc}$, providing offset null adjustments. (Offset null adjustments are used to compensate for small irregularities in the manufacturing of the op-amp.) Tiny differences in the characteristics of transistors Q_1 and Q_2 sometimes mean that one transistor will conduct a little better than the other. When this slight imbalance is amplified many thousands of times, an output may occur even without any input. The offset null is used to 'zero' the op-amp so that with zero input there will be zero output.

An ideal amplifier has the following characteristics:

1. Infinite gain
2. Infinite input impedance
3. Zero output impedance
4. Infinite bandwidth
5. Zero output voltage with no input voltage
6. Extremely small input current (bias current)

The μA741 is not an ideal amplifier, but its characteristics are impressive:

1. Gain of up to 200,000
2. Input impedance between 1 MΩ and 50 MΩ
3. Output impedance generally less than 75 Ω
4. Bandwidth between 1 MHz and 5 MHz

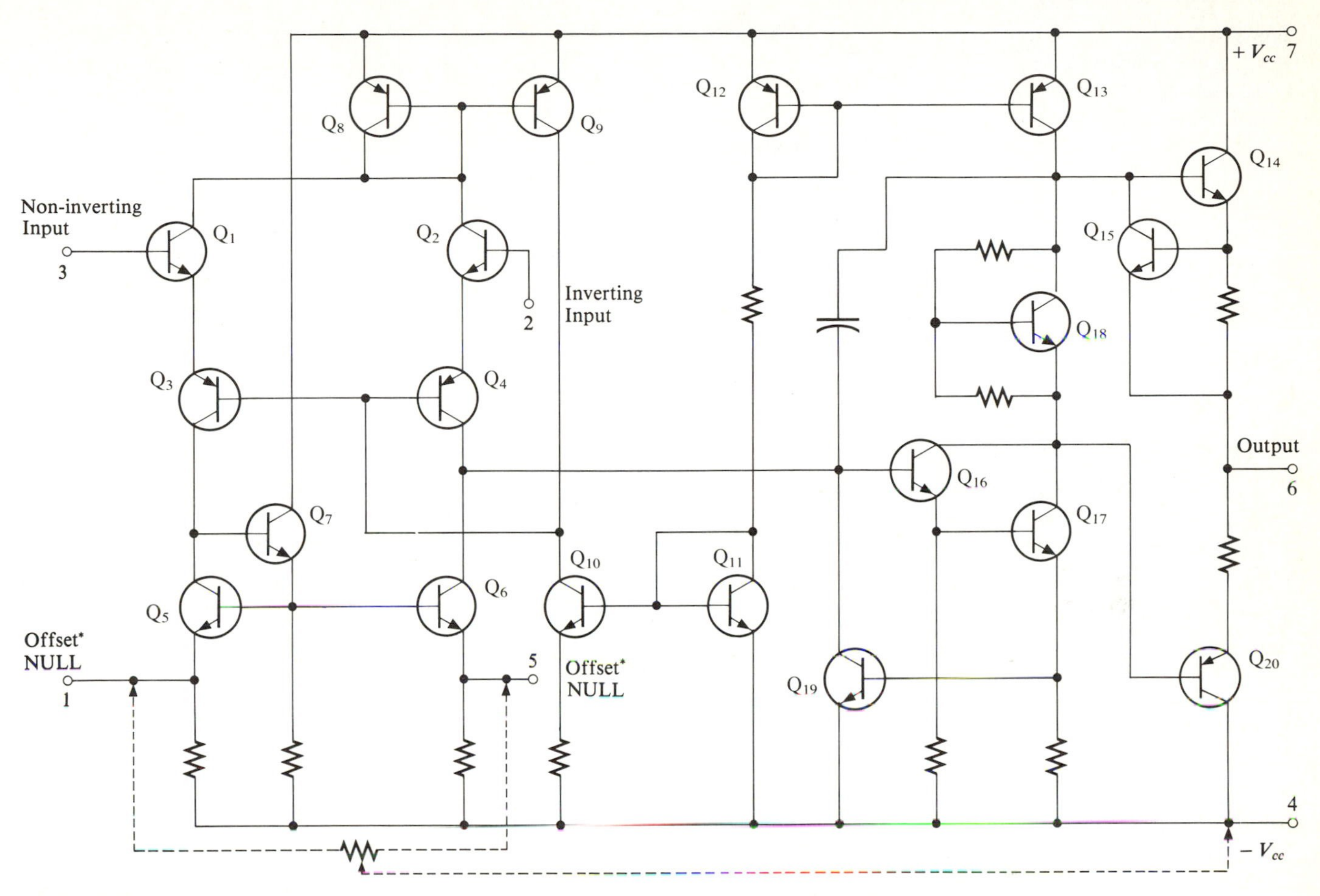

FIGURE 3–1
Internal circuitry of a μA741 op-amp

5. Zero output voltage with zero input voltage when the offset null adjustment is used
6. Very small input current because of the large input impedance

OPEN-LOOP AMPLIFICATION

Open loop means without a feedback loop (see Figure 3–3). Open-loop gain (A) is typically around 200,000 and can be expressed as

$$A = \frac{V_{out}}{V_{in}} \qquad \textbf{(3–1)}$$

The output voltage range across a 75-Ω resistive load is ±14 V. The short-circuit output current is 25 mA maximum with typical values of ≤ 5 mA.

Consider a typical μA741 with the following characteristics:

Gain—200,000
Supply voltage—±18 V
Input voltage—±15 V
Output voltage—±14 V
Output short-circuit current—
25 mA maximum, ≤ 5 mA typical
Input impedance—2 MΩ
Output impedance—75 Ω

With a possible output voltage swing of ±14 V, the maximum input voltage swing for class-A amplification would be

$$V_{in} = \frac{V_{out}}{A} = \frac{\pm 14\ \text{V}}{200{,}000} = \pm 70\ \mu\text{V} \qquad \textbf{(3–2)}$$

Any input voltage swing greater than 70 μV will

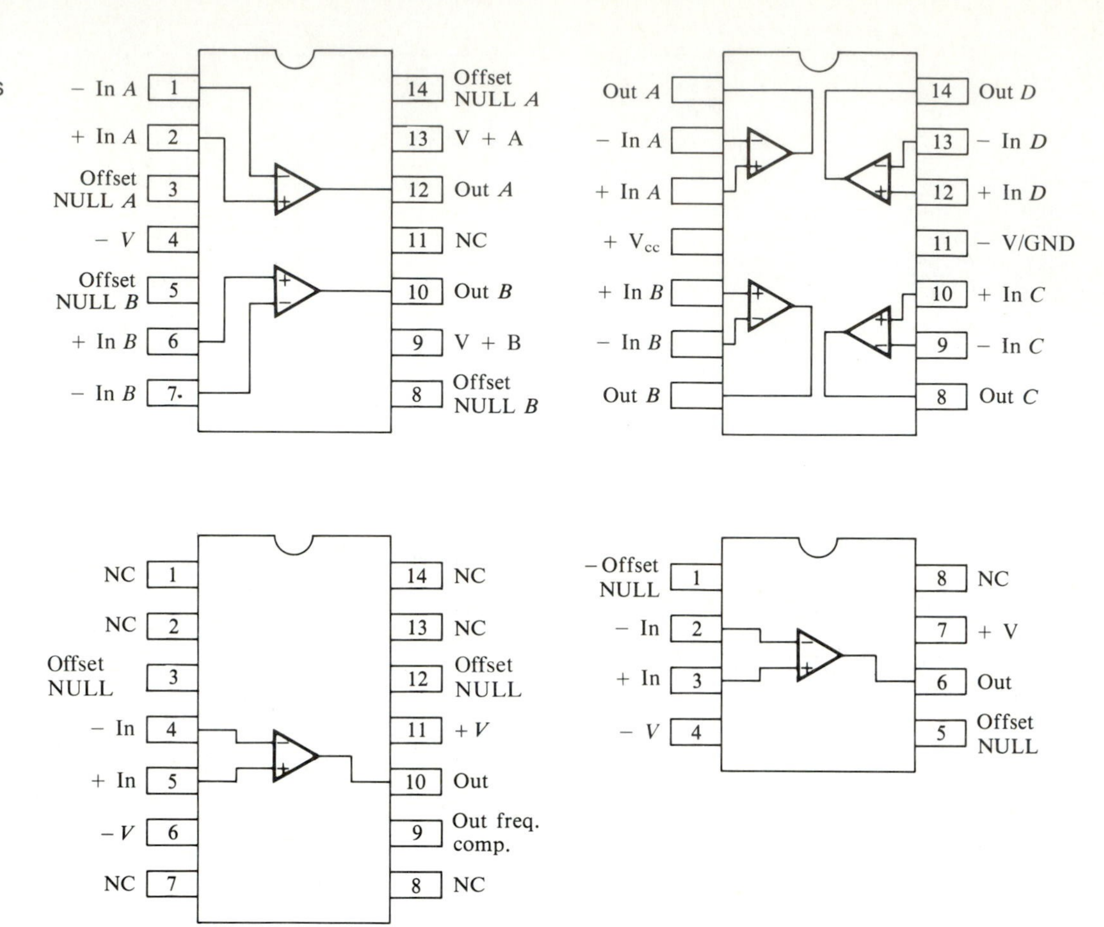

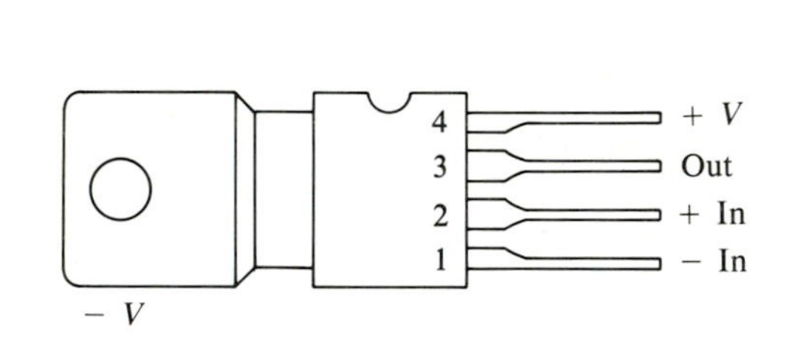

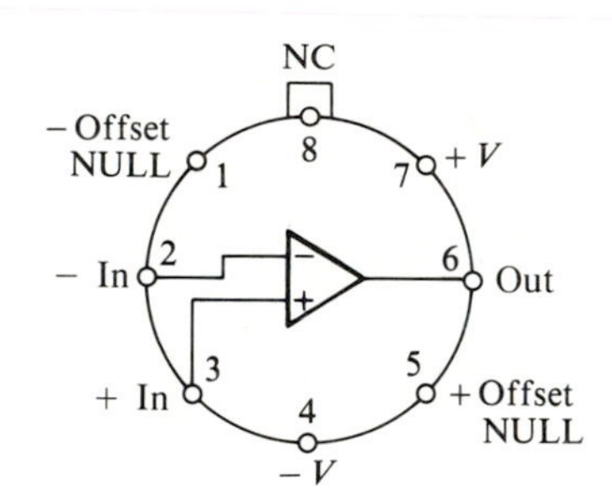

FIGURE 3–2
Package configurations of several different op-amps

drive the op-amp into saturation. A 70-μV signal across the 2-MΩ input impedance gives an input current of

$$\begin{aligned} I_{\text{in}} &= \frac{V_{\text{in}}}{R_{\text{in}}} \\ &= \frac{\pm 70 \times 10^{-6}\ \text{V}}{2 \times 10^{6}\ \Omega} \\ &= \pm 35 \times 10^{-12}\ \text{A} \\ &= \pm 35\ \text{pA} \end{aligned} \tag{3–3}$$

This open-loop op-amp circuit has most of the characteristics of the ideal amplifier. It has serious limitations for use as a class-A amplifier, however, since few practical applications involve input volt-

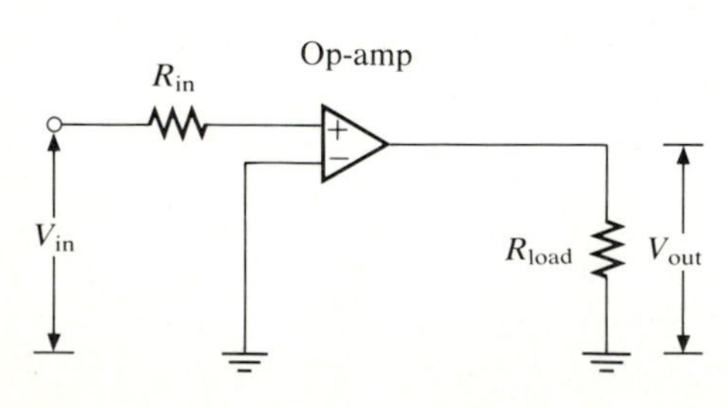

FIGURE 3–3
The op-amp as an open-loop amplifier

ages of $\pm 70\ \mu V$. Used as an open-loop device, the op-amp can clip signals or generate square-wave pulses. When the input from any signal exceeds 70 μV, the op-amp is driven into saturation and the output will rise or fall 14 V from zero volts. Whether the output rises or falls depends on whether the inverting or noninverting input is used. When the input signal drops from 70 μV to below $-70\ \mu V$, the output will change by 28 V to the opposite polarity. This output is just an amplification of the change from $+70\ \mu V$ to $-70\ \mu V$ by a factor of 200,000. The result will be a pulse with a very sharp leading or trailing edge.

NONINVERTING OP-AMP WITH A FEEDBACK LOOP

Since the gain of an op-amp is so high, the amplification of even small signals is impossible. The way to reduce and control the gain of an op-amp is through the use of feedback loop. The feedback loop returns a portion of the output to the input 180° out of phase, resulting in the reduction of the effective input to the op-amp and thus to the apparent overall gain. The closed-loop amplification is presented through two techniques, one using voltages and the other using currents.

Figure 3–4 shows an op-amp with a feedback loop through resistance R_f. The op-amp is wired as a noninverting amplifier (the input is to the positive input terminal). Voltage V_{in} is the input signal, R_f is the feedback resistance, R_L is the load resistance ($R_L \gg R_1$), V_{out} is the output voltage, and V_1 and V_2 are the voltages at the input terminals of the op-amp relative to ground.

The gain *with* feedback (A_f) is the ratio of the output signal voltage to the input signal voltage. The gain *without* feedback is

$$A = \frac{V_{out}}{V_{in}} \tag{3–4}$$

The input signal is

$$V_{in} = V_1 - V_2 \tag{3–5}$$

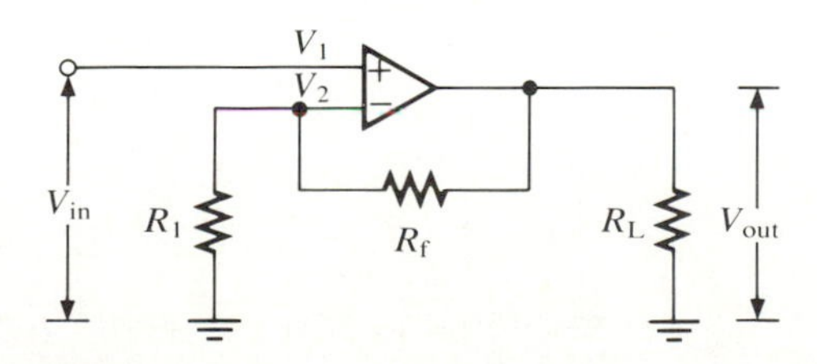

FIGURE 3–4
The op-amp as a closed-loop amplifier

Therefore,

$$V_{out} = A(V_1 - V_2) \tag{3–6}$$

where A is the open-loop gain of the amplifier and $V_2 = V_f$ (the feedback voltage). Using voltage division,

$$V_2 = V_{out}\left(\frac{R_1}{R_1 + R_f}\right) \tag{3–7}$$

$$V_1 = V_{in} \tag{3–8}$$

Therefore,

$$V_{out} = A\left(V_{in} - \frac{R_1 V_{out}}{R_1 + R_f}\right) \tag{3–9}$$

Combining fractions gives (3–10)

$$V_{out} = \frac{AV_{in}(R_1 + R_f) - AR_1V_{out}}{R_1 + R_f}$$

$$V_{out}R_1 + V_{out}R_f = AV_{in}(R_1 + R_f) - AR_1V_{out}$$

$$V_{out}R_1 + V_{out}R_f + AR_1V_{out} = AV_{in}(R_1 + R_f)$$

$$V_{out}(R_1 + R_f + AR_1) = AV_{in}(R_1 + R_f)$$

$$\frac{V_{out}}{V_{in}} = \frac{A(R_1 + R_f)}{R_1 + R_f + AR_1}$$

Since

$$AR_1 \gg R_1 + R_f$$

then

$$R_1 + R_f + AR_1 \cong AR_1$$

and

$$\frac{V_{out}}{V_{in}} = A_f \tag{3–11}$$

Therefore,

$$\frac{A(R_1 + R_f)}{AR_1} = \frac{R_1 + R_f}{R_1} = A_f \tag{3–12}$$

Thus, the gain with feedback is

$$A_f = 1 + \frac{R_f}{R_1} \tag{3–13}$$

This relationship illustrates three important characteristics of the op-amp circuit used. First, the results are positive, indicating a 0° (in phase) relationship between the input and the output. Second, the gain with feedback depends on the feedback and input resistances and is independent of the gain of the op-amp. Third, because the gain with feedback does not depend on the gain of a specific op-amp, field service substitution is simplified. The technician does not have to provide a perfectly matched op-amp for replacement.

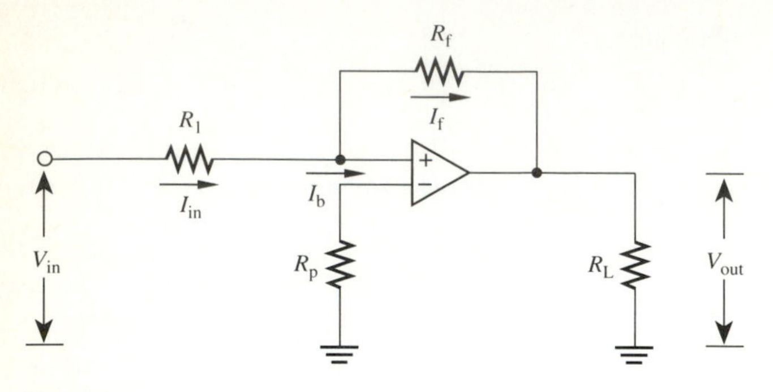

FIGURE 3–5
Noninverting op-amp with a feedback loop

The noninverting amplifier is used in logic circuits and audio amplifiers. The following example illustrates the development of the "gain with feedback" relationship of a noninverting op-amp circuit using Kirchhoff's current law.

Resistor R_p in Figure 3–5 is a *pull-up* resistor. It is typically between 1 kΩ and 3 kΩ. Since the input impedance of the op-amp is so large, the current through R_p is nearly zero. A very small voltage drop occurs across R_p. The input to the op-amp is very close to ground potential and is said to be at *virtual ground.*

By Kirchhoff's current law,

$$I_{in} = I_f + I_b \tag{3–14}$$

where I_{in} is the total input current to the amplifier, I_f is the current in the feedback loop, and I_b is the current into the op-amp. Since I_b is nearly zero amperes, it can be assumed that

$$I_{in} \cong I_f \tag{3–15}$$

Ohm's law gives

$$V_{out} = I_{in}R_1 + R_f I_{in} \tag{3–16}$$

Since

$$V_{in} = V_2$$

then

$$A_f = \frac{V_{out}}{V_{in}} = \frac{(R_f + R_1)I_{in}}{R_1 I_{in}} = \frac{R_f + R_1}{R_1} = \frac{R_f}{R_1} + 1 \tag{3–17}$$

INVERTING OP-AMP WITH A FEEDBACK LOOP

Figure 3–6 shows an op-amp wired as an inverting amplifier with a feedback loop. The input is placed across the negative input terminal of the op-amp. By Kirchhoff's current law,

$$I_{in} = I_b + I_f \tag{3–18}$$

But

$$I_b \cong 0$$

Therefore,

$$I_{in} = I_f \tag{3–19}$$

and by Ohm's law,

$$\frac{V_{in} - V_2}{R_1} = \frac{V_2 - V_{out}}{R_f} \tag{3–20}$$

The gain without feedback is

$$A = \frac{V_{out}}{V_{in}} \tag{3–21}$$

Since

$$V_{in} = V_1 - V_2$$

then

$$A = \frac{V_{out}}{V_1 - V_2} \tag{3–22}$$

and

$$V_1 - V_2 = \frac{V_{out}}{A} \tag{3–23}$$

But

$$V_1 \cong 0$$

Therefore,

$$V_2 = -\frac{V_{out}}{A} \tag{3–24}$$

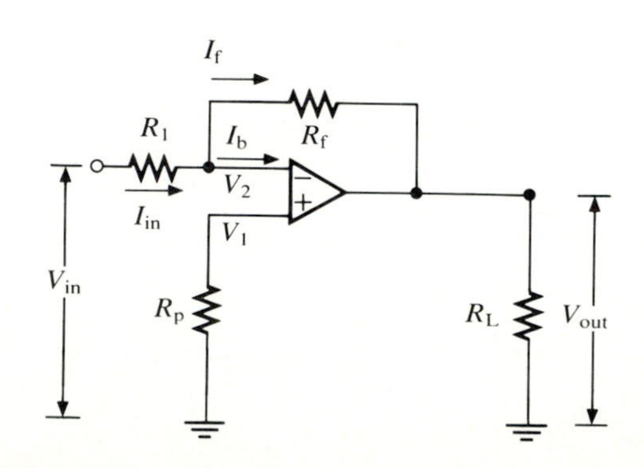

FIGURE 3–6
Inverting op-amp with a feedback loop

Substitution gives

$$\frac{V_{in} - (-V_{out}/A)}{R_1} = \frac{(-V_{out}/A) - V_{out}}{R_f}$$

$$\frac{V_{in} + V_{out}/A}{R_1} = \frac{-V_{out}/A - V_{out}}{R_f} \qquad (3\text{–}25)$$

$$\frac{AV_{in}}{AR_1} + \frac{V_{out}}{AR_1} = \frac{-V_{out}}{AR_f} - \frac{AV_{out}}{AR_f}$$

$$\frac{V_{out}}{R_1} + \frac{V_{out}}{R_f} + \frac{AV_{out}}{R_f} = \frac{-AV_{in}}{R_1}$$

Factoring gives

$$V_{out}\left(\frac{1}{R_1} + \frac{1}{R_f} + \frac{A}{R_f}\right) = V_{out}\left(\frac{R_f + R_1 + AR_1}{R_1 R_f}\right)$$

$$= V_{in}\frac{-A}{R_1} \qquad (3\text{–}26)$$

$$\frac{V_{out}}{V_{in}} = -\frac{AR_f}{R_1 + R_f + AR_1}$$

But

$$AR_1 \gg R_1 + R_f$$

Therefore,

$$R_1 + R_f + AR_1 = AR_1 \qquad (3\text{–}27)$$

and

$$\frac{V_{out}}{V_{in}} = -\frac{AR_f}{AR_1} = -\frac{R_f}{R_1} \quad \text{(ideal)} \qquad (3\text{–}28)$$

The ideal gain of an inverting amplifier with a feedback loop is the ratio of the feedback resistance to the input resistance. The minus sign indicates that the output signal is an inverted version of the input signal (180° phase difference).

In summary, the gain of an amplifier using a feedback loop is

$$A = 1 + \frac{R_f}{R_1} \qquad (3\text{–}29)$$

for a noninverting configuration, and

$$A = -\frac{R_f}{R_1} \qquad (3\text{–}30)$$

for an inverting configuration. This amplifier circuit operates independently of the gain of the individual op-amp, making replacement simple.

THE OP-AMP AS A BUFFER AMPLIFIER

If the feedback loop in an amplifier configuration is a short circuit (R_f = 0 ohms), the gain of the amplifier is 1 and the amplifier circuit can be used as a buffer amplifier (voltage follower) (Figure 3–7). When an op-amp buffer is connected to the output of another circuit, the large input resistance of the buffer does not load down the preceding circuit. In addition, the low output resistance of the buffer does not load down any circuit connected to the buffer's output (Figure 3–8).

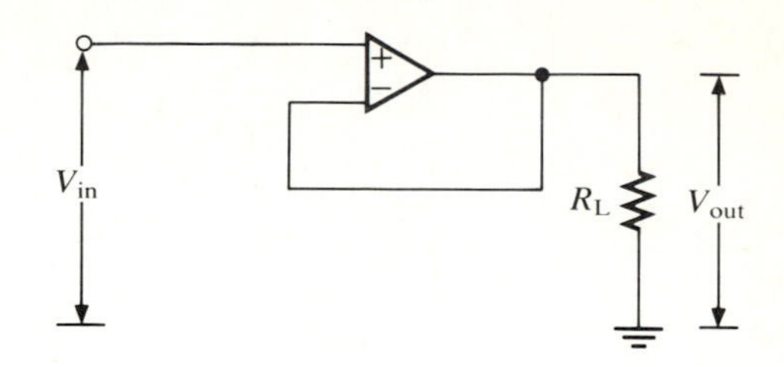

FIGURE 3–7
The op-amp as a buffer amplifier (voltage follower)

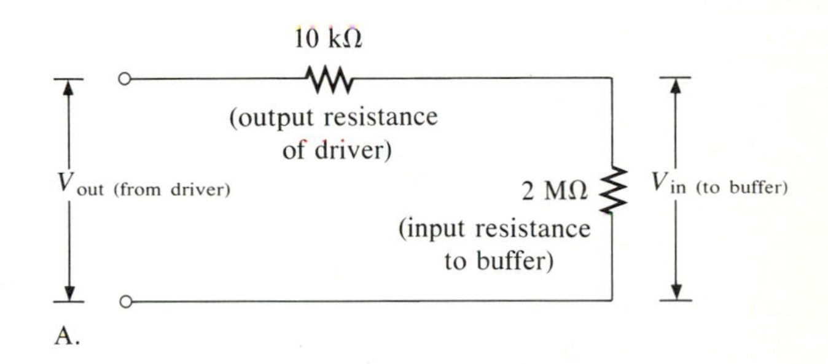

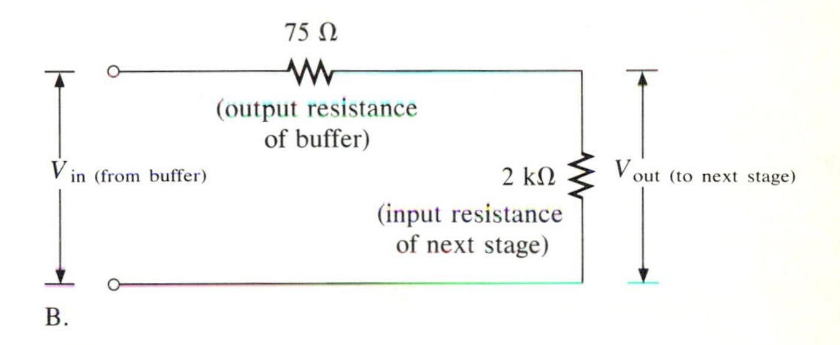

FIGURE 3–8
Operation of a buffer amplifier

In Figure 3–8A, the voltage across the input to the buffer amplifier, V_{out}, is

$$V_{out(to\ buffer)} = V_{in}\frac{2{,}000{,}000\ \Omega}{2{,}010{,}000\Omega} \qquad (3\text{–}31)$$

$$= 0.995V_{in}$$

In Figure 3–8B, the input to the next stage, $V_{out(to\ next\ stage)}$, is

$$V_{out(to\ next\ stage)} = V_{in(from\ buffer)}\frac{2000\ \Omega}{2075\ \Omega} \qquad (3\text{–}32)$$

$$V_{out(to\ next\ stage)} = 0.964V_{in(from\ buffer)}$$

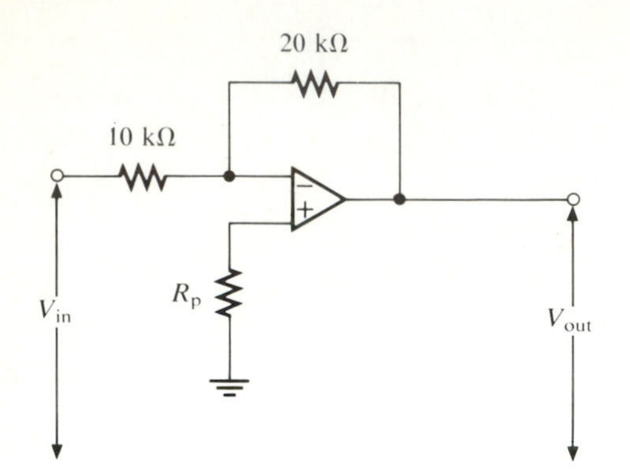

FIGURE 3–9
The op-amp used to multiply

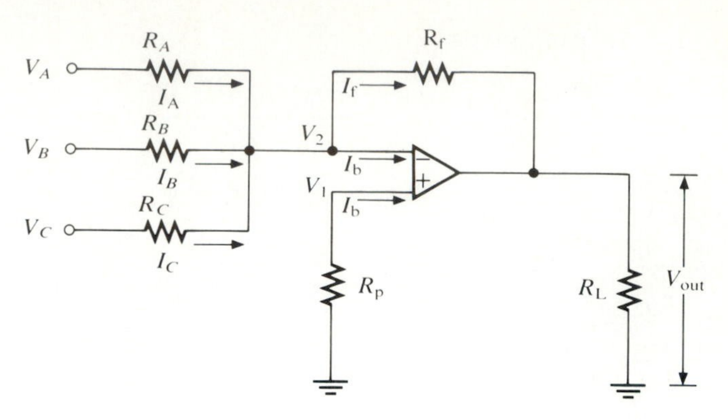

FIGURE 3–11
The op-amp as an adder or summing amplifier

OP-AMPS IN MATHEMATICAL OPERATIONS

Multiplication

Since the gain of an inverting amplifier is $-R_f/R_{in}$, an op-amp can be used to multiply the magnitude of an input signal voltage by a constant. Using the information in Figure 3–9 gives

$$\begin{aligned} V_{out} &= -V_{in}\left(\frac{20\ \text{k}\Omega}{10\ \text{k}\Omega}\right) \\ &= -2V_{in} \end{aligned} \qquad \textbf{(3–33)}$$

Division

Division (multiplication by a fraction) is accomplished in a similar fashion by using a resistance value in the feedback loop that is smaller than the input resistance. Referring to Figure 3–10,

$$\begin{aligned} V_{out} &= -V_{in}\left(\frac{5\ \text{k}\Omega}{20\ \text{k}\Omega}\right) \\ &= -\frac{1}{4}V_{in} \end{aligned} \qquad \textbf{(3–34)}$$

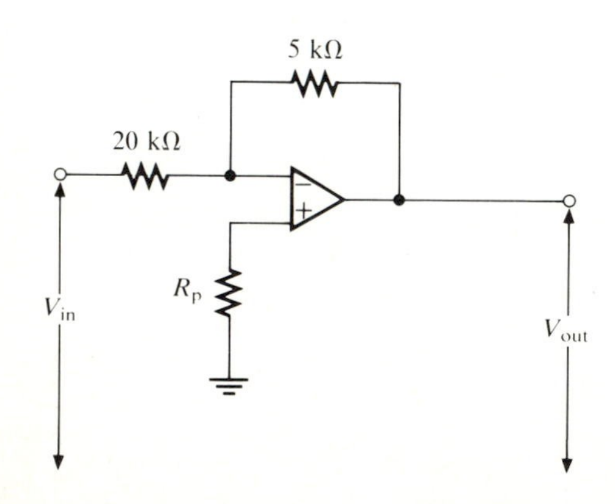

FIGURE 3–10
The op-amp used to divide

Addition

Op-amps can be used to perform the common mathematical algorithm addition (see Figure 3–11). The value of the pull-up resistor, R_p, can be calculated using

$$R_p = R_A \parallel R_B \parallel R_C \parallel R_f \qquad \textbf{(3–35)}$$

By Kirchhoff's current law,

$$I_A + I_B + I_C = I_b + I_f \qquad \textbf{(3–36)}$$

but

$$V_1 = 0 \quad \text{and} \quad I_b = 0$$

Substituting from Ohm's law,

$$\frac{V_A}{R_A} + \frac{V_B}{R_B} + \frac{V_C}{R_C} = \frac{-V_{out}}{R_f} \qquad \textbf{(3–37)}$$

The output can be expressed as

$$V_{out} = -\left(\frac{R_f}{R_A}V_A + \frac{R_f}{R_B}V_B + \frac{R_f}{R_C}V_C\right) \qquad \textbf{(3–38)}$$

This op-amp circuit can be used either in summing, by having

$$\begin{aligned} V_A &= V_B = V_C = V \\ V_{out} &= -VR_f\left(\frac{1}{R_A} + \frac{1}{R_B} + \frac{1}{R_C}\right) \ \textit{(summing amplifier)} \end{aligned} \qquad \textbf{(3–39)}$$

or in averaging, by having

$$\begin{aligned} R_A &= R_B = R_C = R \\ V_{out} &= -\frac{R_f}{R}\left(V_A + V_B + V_C\right) \ \textit{(averaging amplifier)} \end{aligned} \qquad \textbf{(3–40)}$$

This type of configuration can be used to add input voltages, for example as a *mixer* circuit that provides a single output from several input sources.

Subtraction

Figure 3–12 shows an op-amp wired as a differential amplifier. The output from the op-amp circuit is

$$V_{out} = -\frac{R}{R}(V_A - V_B) = V_B - V_A \quad \textbf{(3–41)}$$

This is straightforward subtraction.

Since an inverting amplifier can be wired to give a gain of -1 by having the feedback resistance equal the input resistance, it can be used as an inverting circuit. The inverting amplifier simply creates a 180° phase shift between the input and output and, with the values in Figure 3–13, gives

$$V_{out} = -V_{in}\left(\frac{10\ \text{k}\Omega}{10\ \text{k}\Omega}\right) = -V_{in} \quad \textbf{(3–42)}$$

Inverters are used to make a quantity negative. Addition of a negative quantity gives the same result as subtraction. These methods of subtraction can be used interchangeably.

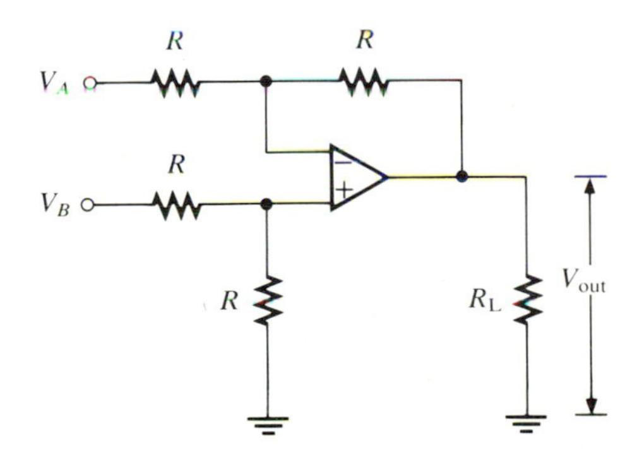

FIGURE 3–12
The op-amp as a differential amplifier

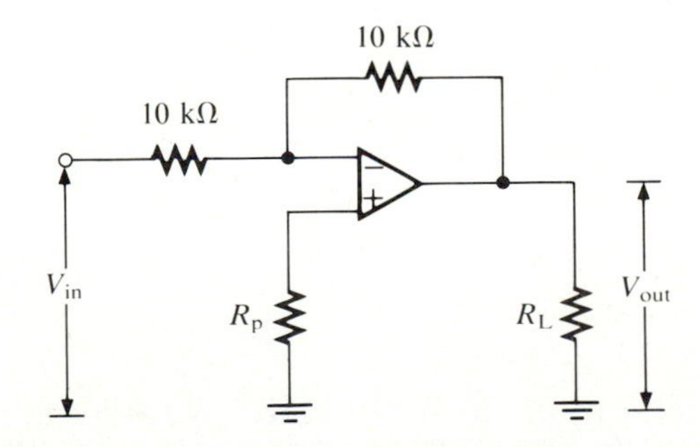

FIGURE 3–13
The op-amp as an inverter

Solving Systems of Equations

Op-amp circuits can be used together to perform calculations and solve equations involving changing input voltages.

EXAMPLE 3–1
Solve the equation

$$2x - 3y = z$$

where x and y are variable input voltages at points X and Y, respectively (see Figure 3–14).

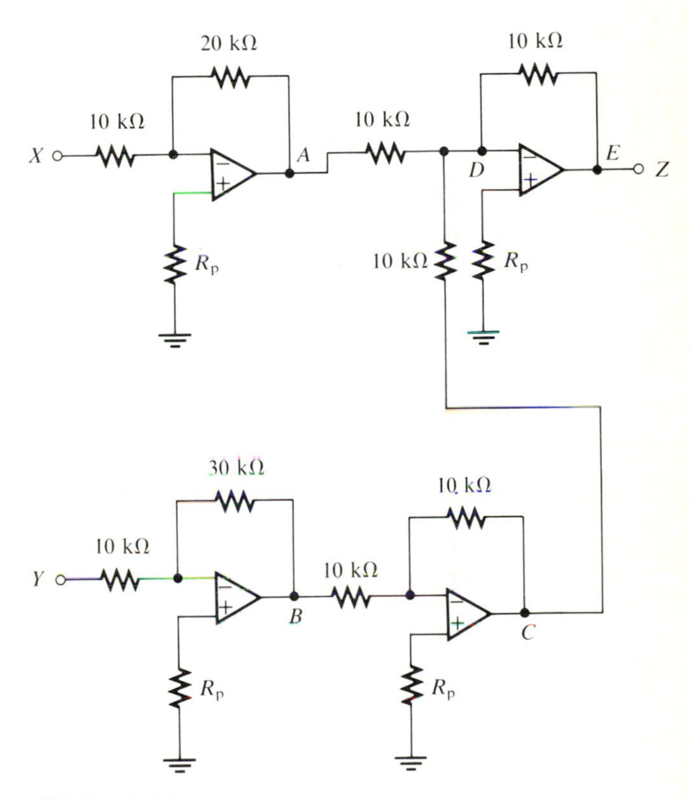

FIGURE 3–14
Op-amp configuration for solving the equation $2x - 3y = z$

Solution:
The voltage at point A is $-2x$. The voltage at point B is $-3y$; this voltage is inverted at point C to give $+3y$. The voltage at point C and that from point A are added from point D to point E to give the negative of their sum:

$$z = -(-2x + 3y) = 2x - 3y \quad \textbf{(3–43)}$$

A meter across point Z (to ground) will give the value of $2x - 3y$ for whatever values of x and y are

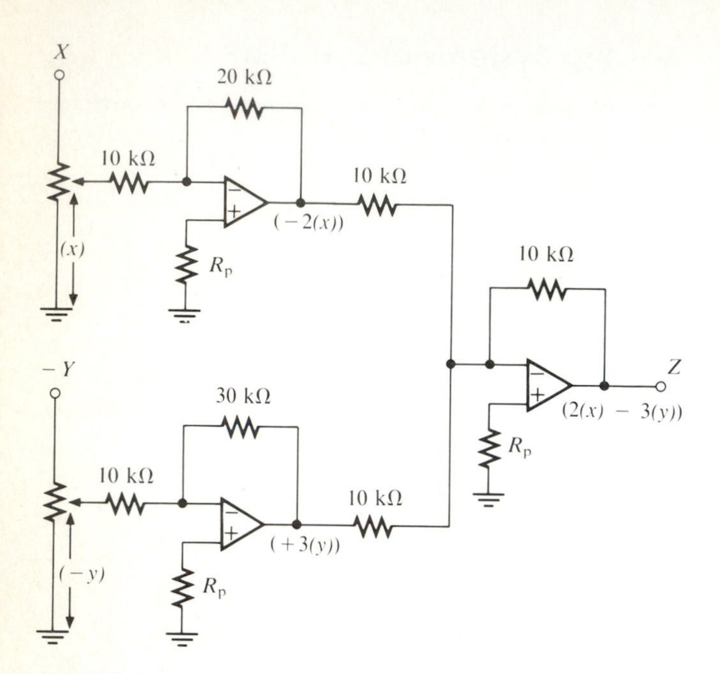

FIGURE 3–15
Another way of solving the equation $2x - 3y = z$

input to the system. Figure 3–15 shows another circuit that will accomplish this. The only difference is that a negative y voltage is used. This latter arrangement reduces the number of op-amps required.

EXAMPLE 3-2

Show how op-amps can be used to solve the following two equations with two unknowns (see Figure 3–16):

$$2x + 3y = 28 \quad (3\text{–}44)$$
$$2x - y = 4 \quad (3\text{–}45)$$

Solution:
First, solve equation (3–44) for x:

$$2x = -3y + 28 \quad (3\text{–}46)$$
$$x = -3/2y + 14$$

Then, solve equation (3–45) for y:

$$-y = -2x + 4 \quad (3\text{–}47)$$
$$y = 2x - 4$$

Next, develop the circuit in section A to give the output for equation (3–45). The y input will be developed in section B. Then develop section B to give the output from equation (3–44). This output provides the y input to section A, and the x output of section A provides the x input to section B. A meter

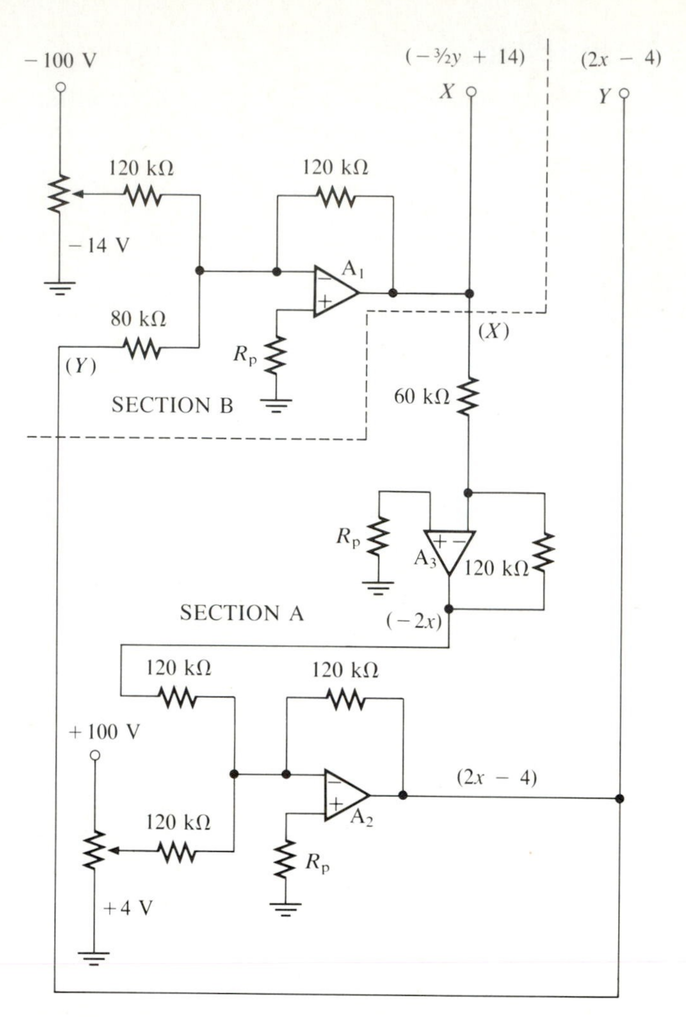

FIGURE 3–16
Configuration for solving two equations with two unknowns

at point X will read the solution for x, and a meter at point Y will read the solution for y. An algebraic solution using substitution yields the answers expected in this example:

$$\begin{aligned} y &= 2x - 4 \\ y &= -3y + 28 - 4 \\ 4y &= 24 \\ y &= 6 \end{aligned} \quad (3\text{–}48)$$

$$\begin{aligned} x &= -\frac{3}{2}(6) + 14 \\ &= -9 + 14 \\ &= 5 \end{aligned} \quad (3\text{–}49)$$

The meter at point X will read 5 V, and the meter at point Y will read 6 V, which are the solutions for this system of equations.

EXAMPLE 3–3
Solve the following system of equations using an op-amp analog circuit:

$$2x - y + 3z = 14 \qquad \textbf{(3–50)}$$
$$3x - 2y - z = 5 \qquad \textbf{(3–51)}$$
$$4x + 3y - 2z = -1 \qquad \textbf{(3–52)}$$

Solution:
Solve equation (3–50) for x:

$$x = 7 + \frac{1}{2}y - \frac{3}{2}z \qquad \textbf{(3–53)}$$

Solve equation (3–51) for z:

$$z = -5 + 3x - 2y \qquad \textbf{(3–54)}$$

Solve equation (3–52) for y:

$$y = -\frac{1}{3} + \frac{2}{3}z - \frac{4}{3}x \qquad \textbf{(3–55)}$$

Draw a circuit that will yield x, y, and z from this information as shown in equations (3–53), (3–54), and (3–55). Remember that the output for each part can become the input for another. Meters wired at points X, Y, and Z, respectively, will indicate the values $x = 2$ V, $y = -1$ V, and $z = 3$ V (Figure 3–17).

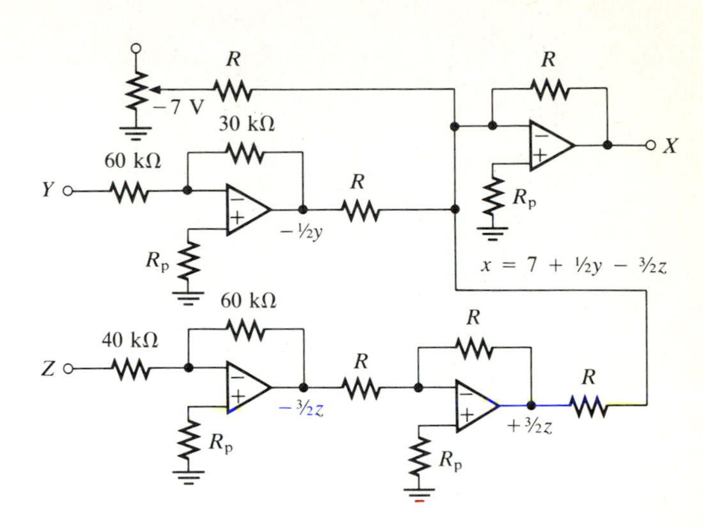

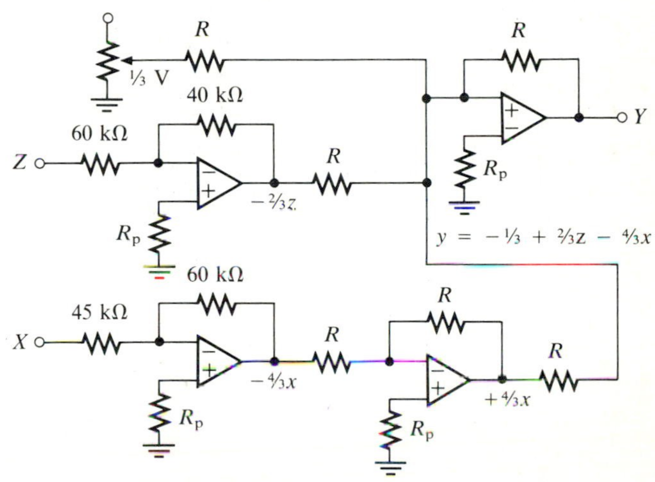

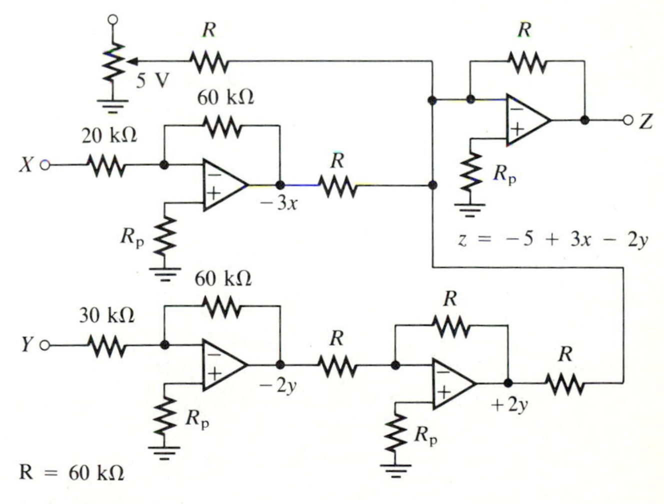

FIGURE 3–17
Configuration for solving three equations with three unknowns. All points labeled X are wired together, as are all points labeled Y and all points labeled Z. This arrangement provides the inputs for each part of the circuit.

One advantage of this type of solution is that it can give immediate results for systems with time-varying inputs. The inputs can be from transducers representing either varying voltages or resistances. Since this "computer" has no memory, it cannot remember instructions. If the problem is changed, the circuitry must be altered. Circuit changes are accomplished by a patch panel on the front of the computer. Accuracy of the solutions is also a problem since it depends on the accuracy of the components and output devices. Generally, this kind of circuitry is used in conjunction with digital circuitry. The resulting device is called a *hybrid* computer. Because nature is analog, this computer provides a useful interface between the time-varying events in the physical world and the high-speed calculations of the digital computer.

Differentiation

Differentiation is a process in calculus whereby the slope of the graph of a function can be found by

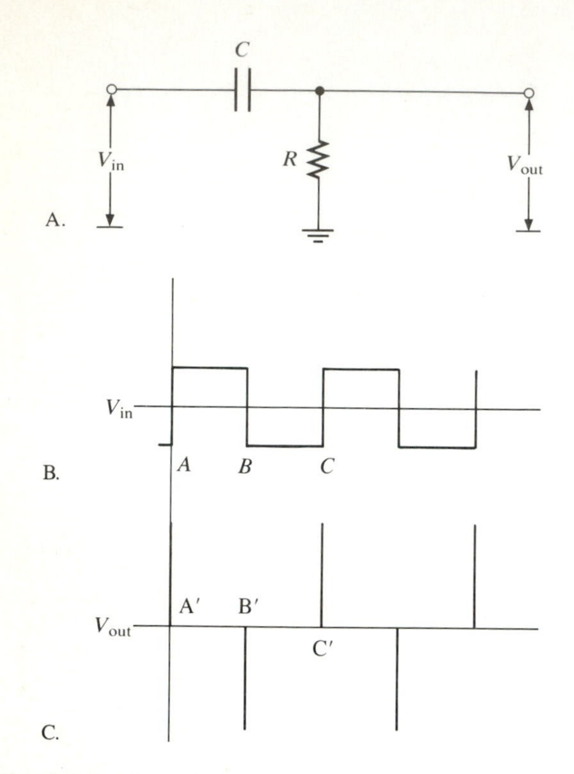

FIGURE 3–18
A passive differentiator

mathematical means. The same result can be obtained using the circuit in Figure 3–18A.

The RC time constant must be very small so that only the changes in input voltage will be "seen" through the capacitor. The capacitor charges and discharges very quickly. Consider the square-wave input signal in Figure 3–18B. At point A, the input voltage undergoes a large positive change. The output voltage follows the input, as shown at point A' in Figure 3–18C. The output voltage quickly returns to zero (from point A' to point B'). Since no change occurs in the input voltage, the capacitor discharges through R to zero volts. A large negative change in the input voltage occurs at point B, reflected in the output by the sharp negative spike at point B'. The output graph represents the slope of the input signal, which can be described mathematically as

$$V_{out} = \frac{d(V_{in})}{dt} \tag{3–56}$$

The passive circuit in Figure 3–18A consumes energy. The output signal power is always smaller than the input signal power. By using an op-amp, as shown in Figure 3–19, the same mathematical operation is performed. Because the op-amp is an

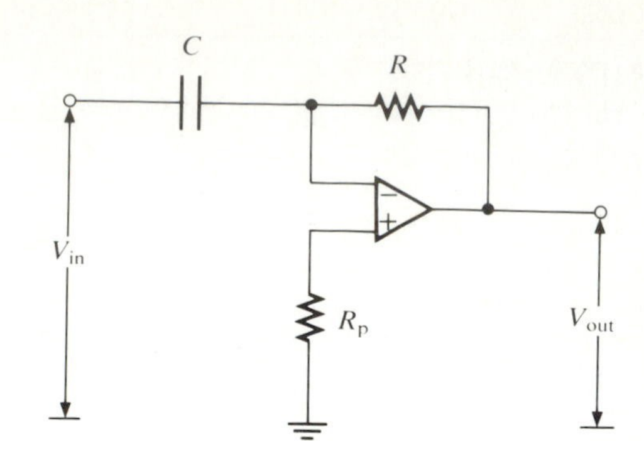

FIGURE 3–19
The op-amp as an active differentiator

active device, the power level of the input signal can be maintained, or even increased if necessary.

Integration

Integration is a mathematical operation that can be considered the inverse of differentiation. It can be used to find the area under the graph of a mathematical function. Integration can also be accomplished using the circuit in Figure 3–20A.

If the RC time constant of the circuit is very large, integration is achieved. Consider the input pulse V_{in} in Figure 3–20B. As the voltage rises to

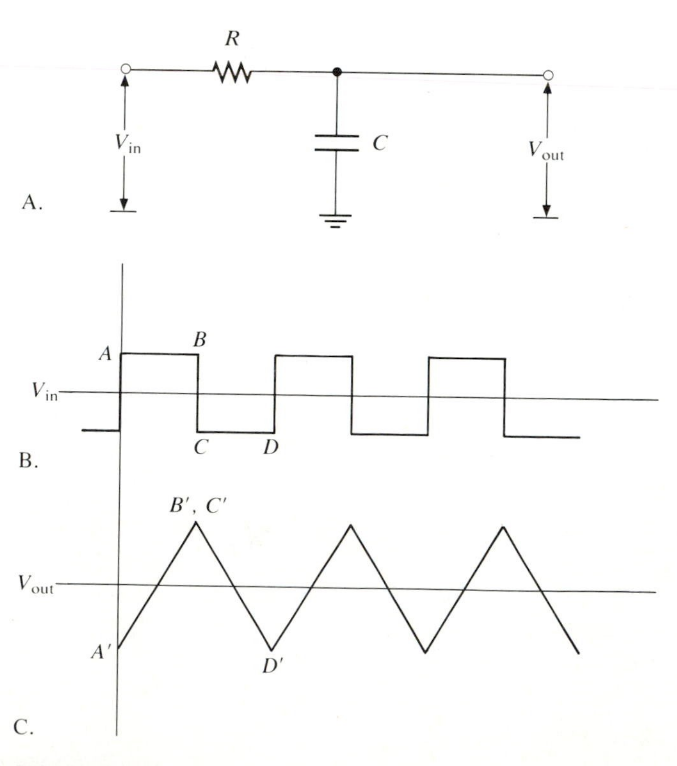

FIGURE 3–20
Passive integrator

its peak and remains there (from point A to point B), the area under the graph increases as the time increases. This effect is shown between points A' and B' on the graph of V_{out} in Figure 3–20C. When the voltage changes from the level at point B to the level at point C, a "negative" area develops. At point D, the sum of the areas under the curves has reached a negative value, as shown by the negative peak at point D'. The net result is that the output voltage, V_{out}, is the integral of the input voltage, V_{in}.

Figure 3–21 illustrates an active integrator circuit. In this circuit,

$$V_{out} = -\frac{1}{RC}\int_0^t V_{in}\,dt \qquad (3\text{–}57)$$

The active circuit maintains or increases the power level of the signal passing through it. The passive integrator in Figure 3–20 consumes some of the power, making it less desirable in most applications.

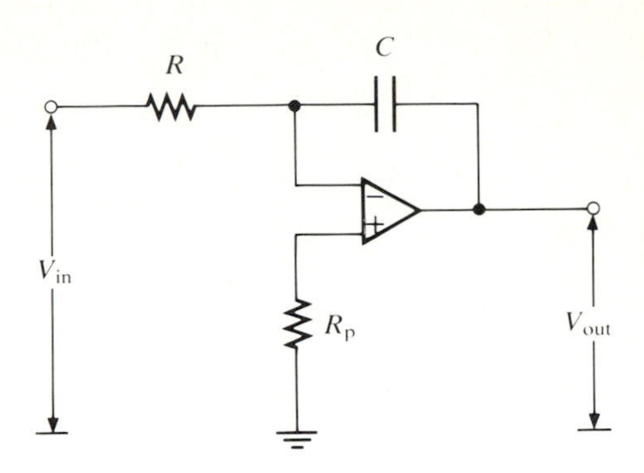

FIGURE 3–21
The op-amp as an active integrator

SUMMARY

The preceding presentation of analog mathematical circuits is intended to expose the technician to these signal-processing networks. The circuits can be used together to obtain outputs that represent solutions to sophisticated problems involving inputs comprising time-varying voltages or resistances. Other op-amp circuits can be used to develop logarithms and other mathematical quantities. The op-amps, capacitors, resistors, diodes, and power supplies needed to construct these circuits are found on the control panel of an analog computer.

The special characteristics of op-amps that make them useful in isolating components in active circuits will be discussed in Chapter 15.

EXERCISES

1. How do the typical op-amp and the ideal amplifier differ?
2. How can an op-amp with an open-loop gain of 150,000 be used to amplify a ±0.1-V signal?
3. What is meant by a closed-loop amplifier gain of −10?
4. What is the output voltage of the circuit in Figure 3–22?

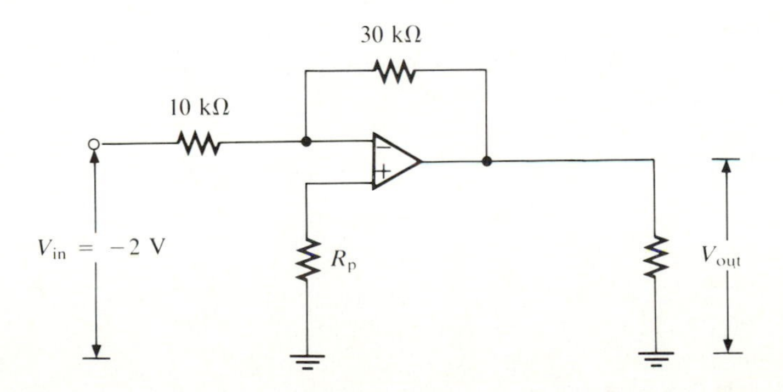

FIGURE 3–22
Circuit for Exercise 4

5. What is the output voltage of the circuit in Figure 3–23?

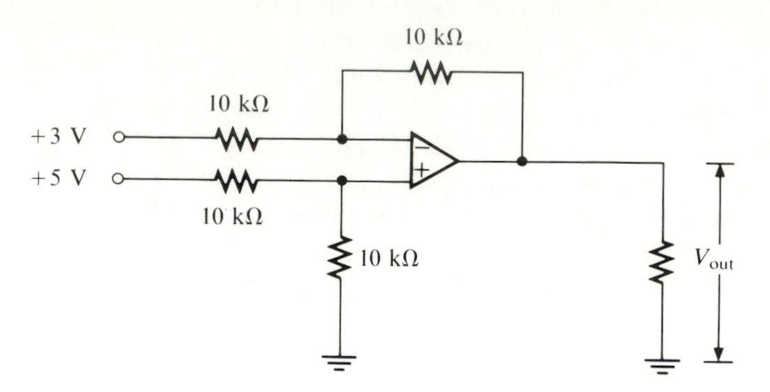

FIGURE 3–23
Circuit for Exercise 5

6. Graph the output of the circuit in Figure 3–24 if the input signal is 3 V peak-peak.

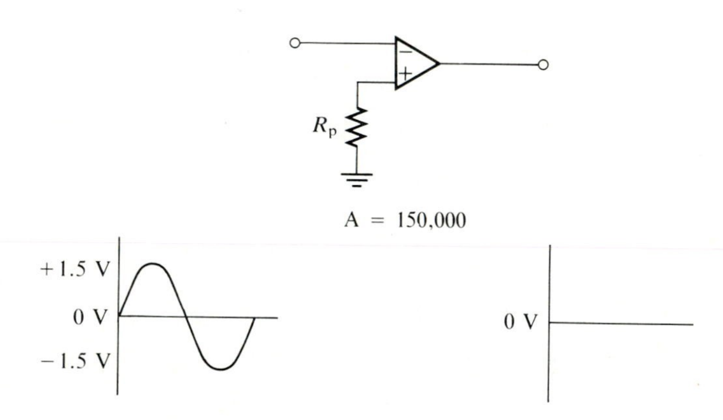

FIGURE 3–24
Circuit for Exercise 6

7. What is the output voltage of the circuit in Figure 3–25?

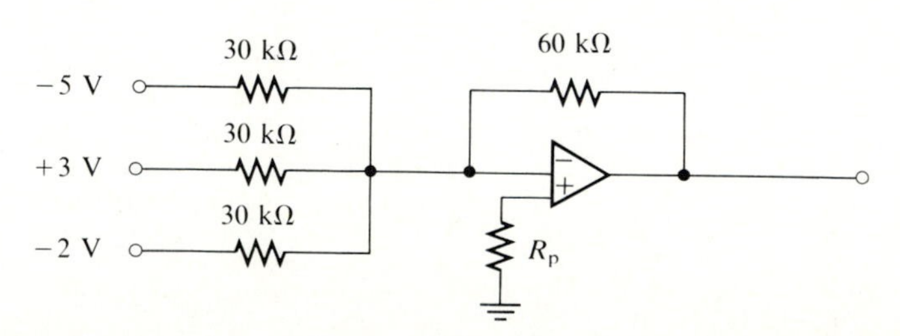

FIGURE 3–25
Circuit for Exercise 7

8. What is the output voltage of the circuit in Figure 3–26?

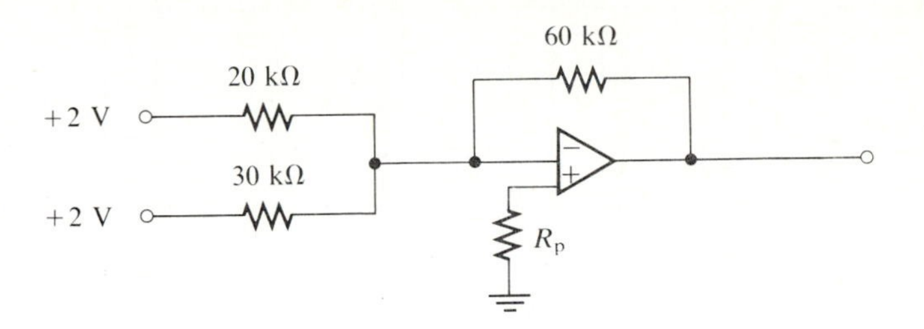

FIGURE 3–26
Circuit for Exercise 8

9. Design a circuit that will solve the following system of equations:

$$\begin{aligned} 2x + 3y &= 4 \\ 3x - y &= -5 \end{aligned}$$

10. Design a circuit that will solve the following system of equations:

$$\begin{aligned} x + 2y + z &= 12 \\ 2x + 3y - 2z &= 10 \\ 3x - y - 3z &= 4 \end{aligned}$$

LAB PROBLEM 3–1
Operational Amplifiers

Equipment
Low-voltage power supply; DVM; signal generator; 2-channel oscilloscope; selection of ¼, W resistors; 2 = 2k, potentiometers; several μA741 op-amps.

Object 1
To use an op-amp with various types of feedback control to yield linear voltage amplifiers.

Procedure
Refer to the circuit of Figure 3–27.

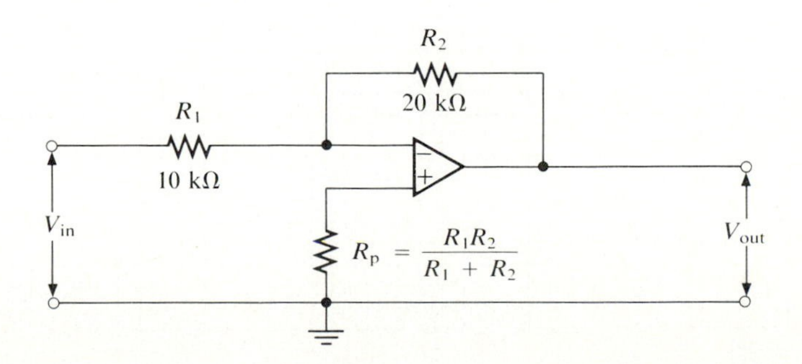

FIGURE 3–27
Op-amp with feedback control

FIGURE 3–28
Graph of input and output waveforms

Use a DVM to measure the true values of the resistors. Remember to apply $+V_{cc}$ and $-V_{cc}$ to the op-amp. V_{cc} must exceed the predicted V_{out} voltages or clipping of the output will result. Record these data and construct the op-amp circuit as shown. For the input, use a signal generator to produce a 1-kHz sinewave of 1 V peak-peak amplitude with zero volts of dc offset voltage. Use an oscilloscope to monitor V_{in} and V_{out} simultaneously. Sketch the input and output waveforms on Figure 3–28.

Take ac measurements of the amplitude of V_{in} and V_{out}. Next, with R_1 unchanged, calculate the value of R_2 needed to yield a gain of -5. Select a resistor for this new value of R_2. Replace R_2 in the circuit with the new value and repeat the procedure. Now adjust the dc offset until severe clipping occurs. Try this with both positive and negative dc offset. Record the results.

Construct the circuit of Figure 3–29, which illustrates a noninverting op-amp.

Repeat the steps used in the first part of this procedure with the inverting op-amp circuit.

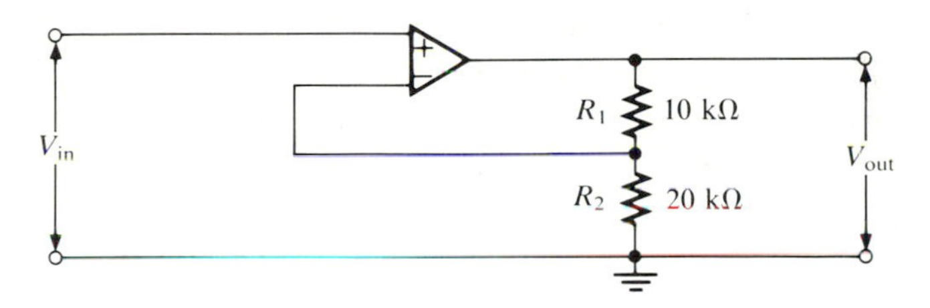

FIGURE 3–29
Noninverting op-amp

Object 2

To investigate the behavior of an op-amp used as a comparator. A *comparator* is a device that compares two voltage levels. The state of its output depends on which input is higher.

Procedure

Construct the circuit of Figure 3–30.

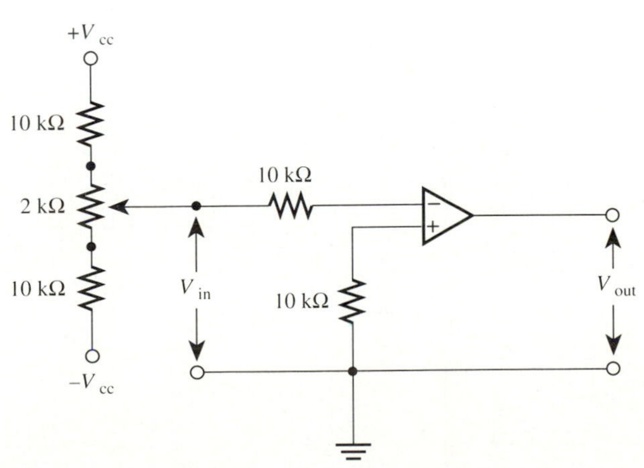

FIGURE 3–30
The op-amp as a comparator

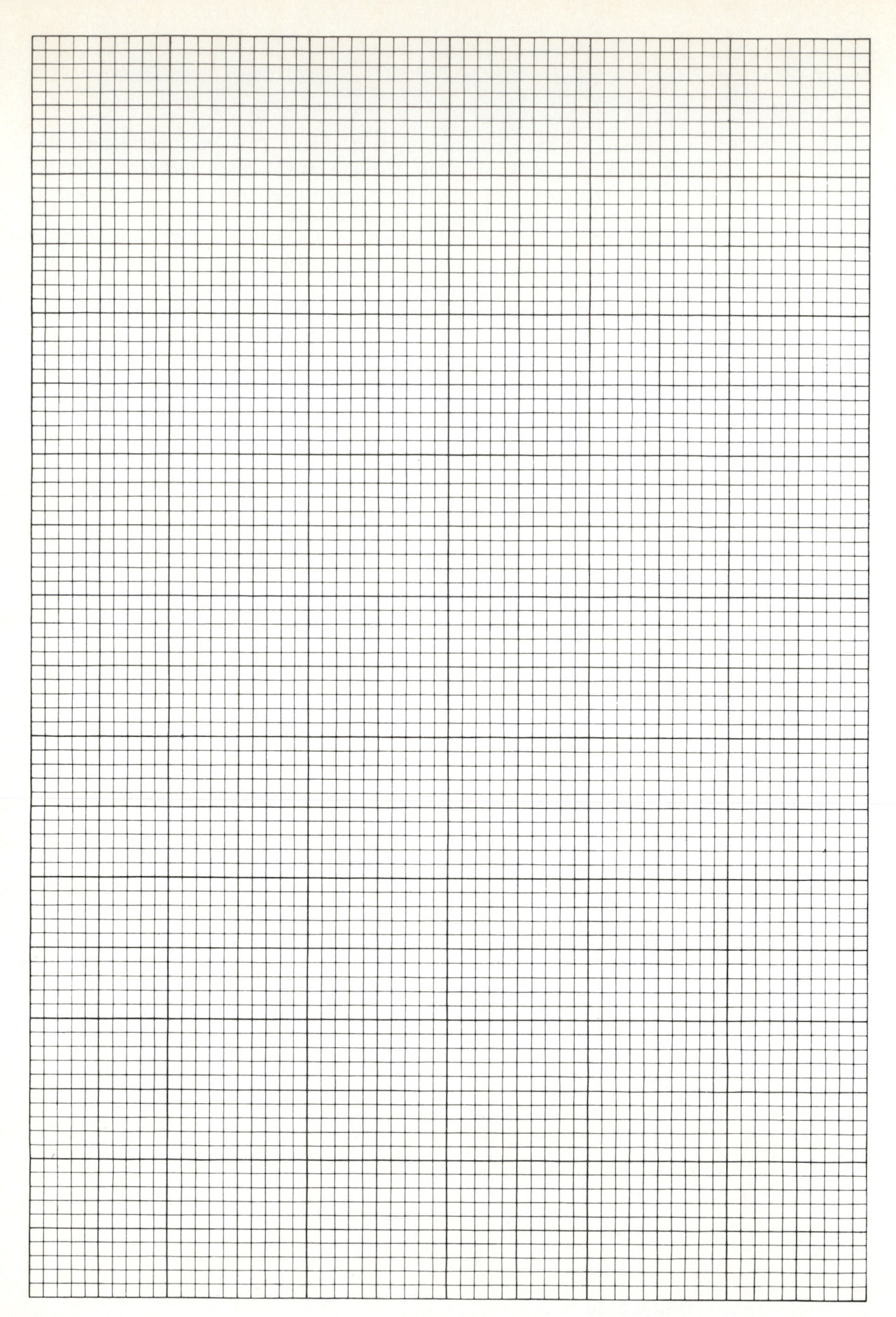

FIGURE 3–31
Graph illustrating the effect of offset voltage

Use the oscilloscope to monitor V_{out}. Connect the DVM across point A and ground to monitor V_{in}. Set V_{in} at $+2$ V. Then decrease V_{in} slowly while observing V_{out} on the oscilloscope. Record the value of V_{in} that causes V_{out} to switch from $-V_{cc}$ to $+V_{cc}$. Repeat the experiment using a starting V_{in} of -2 V and increasing the voltage.

Replace the potentiometer input with a signal generator. Set the signal generator to produce a 4-V peak-peak sinewave at 1 kHz with zero volts offset. Record V_{in} and V_{out}. Increase the signal generator dc offset to $+1$ V; then change the dc offset to $+2$ V and above and record the results for each setting on Figure 3–31.

Object 3

To investigate the behavior of an op-amp used as a Schmitt trigger.

Procedure

Construct the circuit of Figure 3–32.

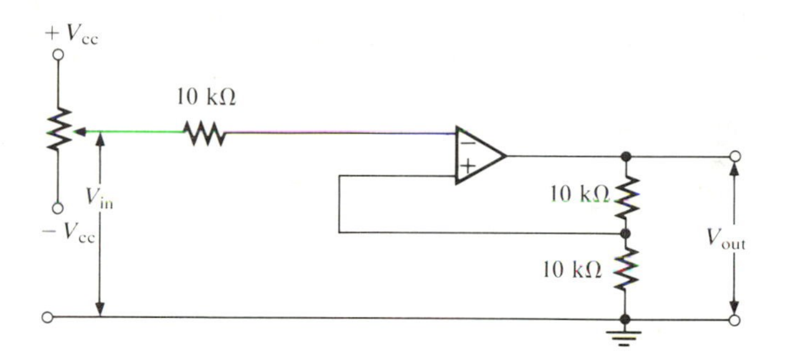

FIGURE 3–32
The op-amp as a Schmitt trigger

Repeat the procedure used in Object 2 of this lab. Use $+10$ V and -10 V as the starting points for V_{in}. Replace the potentiometer with a signal generator, and the lower 10-kΩ resistor with a 1.8-kΩ resistor. Set the generator to produce an 8-V peak-peak sinewave at 1 kHz with zero volts dc offset, $+1$ V and $+3$ V and above. Record the results on Figure 3–33.

Object 4

To wire a basic analog circuit that will solve two equations simultaneously, and to investigate the effect of changing the electrical parameters of the circuit.

Procedure

Wire the circuit shown in Figure 3–34.

This circuit should give for values at points X and Y the solutions for the system of equations

$$x = 5 - 2y$$
$$y = 1 + x$$

If the potentiometer providing a constant voltage to op-amp A_1 is set at -5 V and the potentiometer providing the constant voltage to op-amp A_3 is set to deliver -1 V, the voltages at X and Y should correspond to the mathematical solutions of the system of equations.

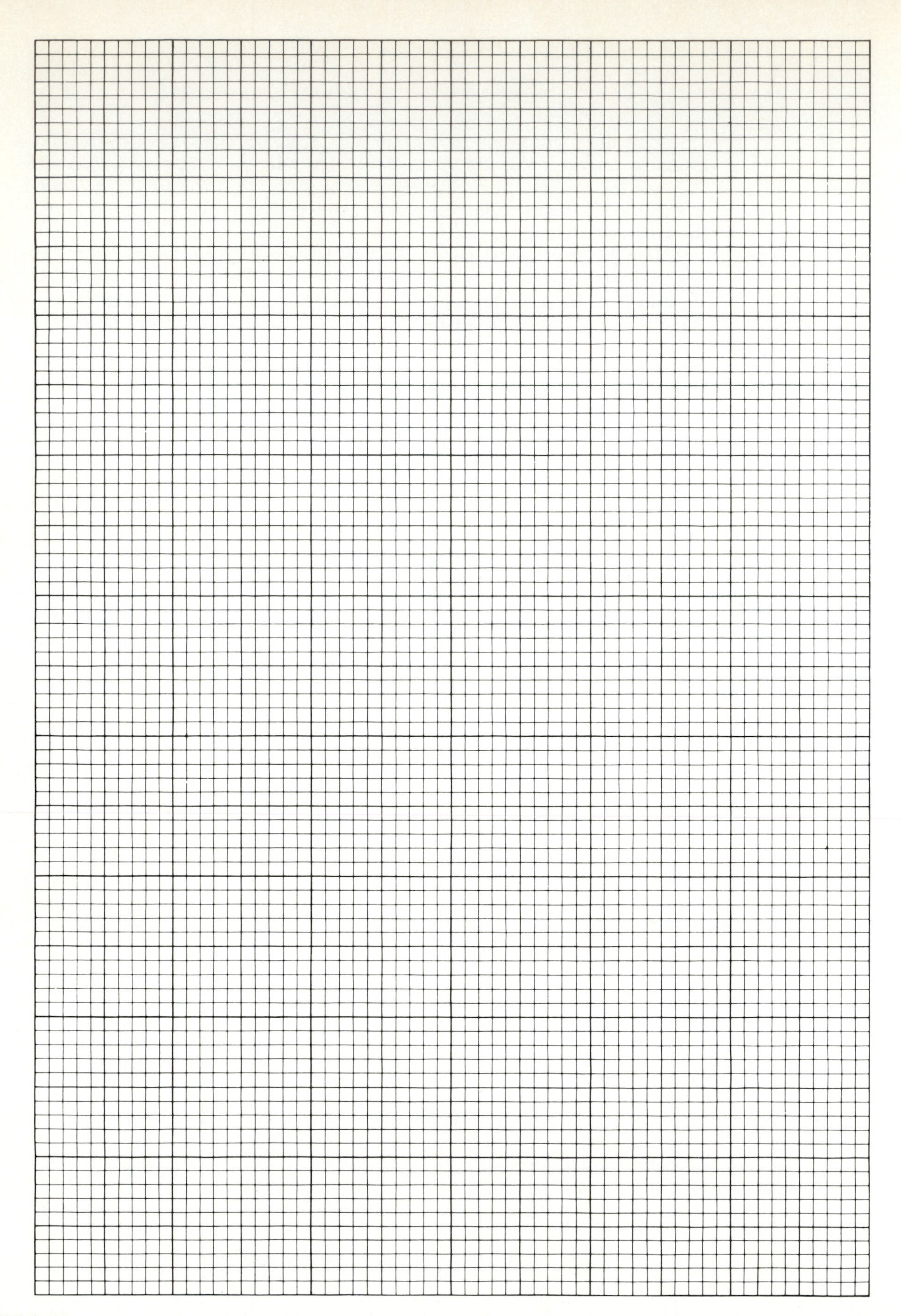

FIGURE 3–33
Graph of results of a Schmitt trigger circuit

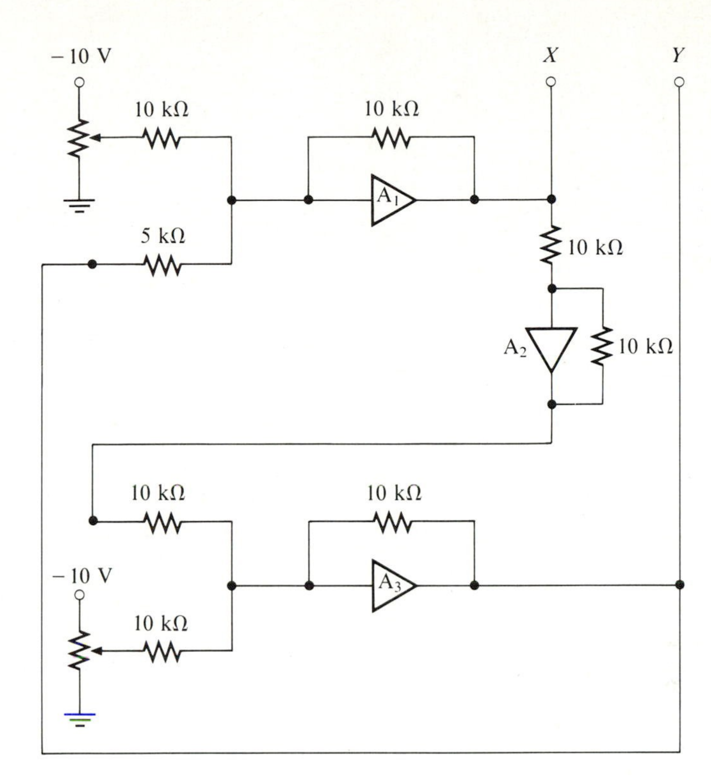

FIGURE 3–34
Circuit for solving two equations with two unknowns

With the constant voltages changed to -9 V and -3 V, respectively, the circuit will yield the solutions for the system

$$x = 9 - 2y$$
$$y = 3 + x$$

A change in the ratio of feedback resistors will change the circuit to represent the system

$$x = 5 - y$$
$$y = 3x + 1$$

Try changing the 5-kΩ resistor to 10 kΩ and the feedback resistor at A_3 to 30 kΩ. Compare the theoretical values to the measured results.

4 dc Analog Meters

OUTLINE

OBJECTIVES

THE BASIC METER MOVEMENT

INTERNAL RESISTANCE OF THE METER MOVEMENT

FULL-SCALE DEFLECTION

SHUNTING

MULTIRANGE AMMETER

AYRTON SHUNT

VOLTMETERS

MULTIRANGE VOLTMETERS

SENSITIVITY

LOADING PROBLEMS

SERIES-TYPE OHMMETER

SHUNT-TYPE OHMMETER

SUMMARY

EXERCISES

LAB PROBLEM 4–1—Measuring Internal Resistance

LAB PROBLEM 4–2—Ayrton Shunt

PROGRAMMING PROBLEM 4–1—Two-Range Ayrton Shunt

PROGRAMMING PROBLEM 4–2—Three-Range Ayrton Shunt

OBJECTIVES

After completing this chapter, you will be able to:

- ☐ Describe the operation of the basic meter movement.
- ☐ Discuss the internal resistance of a meter movement.
- ☐ Describe full-scale deflection and discuss its significance.
- ☐ Design single- and multirange meter shunts.
- ☐ Discuss the significance of make-before-break contact switches.
- ☐ Design an Ayrton shunt.
- ☐ Design multirange voltmeters.
- ☐ Discuss meter sensitivity.
- ☐ Explain what is meant by loading problems.
- ☐ Discuss the basic operation of a series-type ohmmeter.
- ☐ Discuss the basic operation of a shunt-type ohmmeter.
- ☐ Measure the internal resistance of a meter.

The dc analog meter movement can be made to measure dc current and voltage; ac current and voltage; and resistance. This chapter introduces the basic dc meter movement. With appropriate circuitry, dc meter movements can be adapted to indicate various ranges of dc or ac current and voltage or resistance. Limitations of this type of instrument are presented. Sensitivity of voltmeters and the loading effects of meters are also discussed.

THE BASIC METER MOVEMENT

The basic meter movement consists of a coil of wire suspended in the field of a permanent or fixed (field) magnet (Figure 4–1). When a current flows into the wire coil comprising the meter movement (electromagnet), a magnetic field is built up around the coil. If the current flow is in the proper direction, the magnetic field will be established, as shown in Figure 4–2.

In this case, the two north poles will repel each other while the north pole of the coil will be attracted to the south pole of the field magnet, producing a clockwise torque (force). Likewise, the two south poles will repel each other and the south pole of the coil will be attracted to the north pole of the field magnet, again producing a clockwise torque. If the coil is pivoted (suspended) in the middle, the two torques will result in a clockwise rotational motion.

If the current being forced through the electromagnet is in the opposite direction, the torque will likewise be in the opposite direction and will force the meter movement backward, possibly damaging or destroying it. The correct signal condition should always be observed when using an analog meter movement.

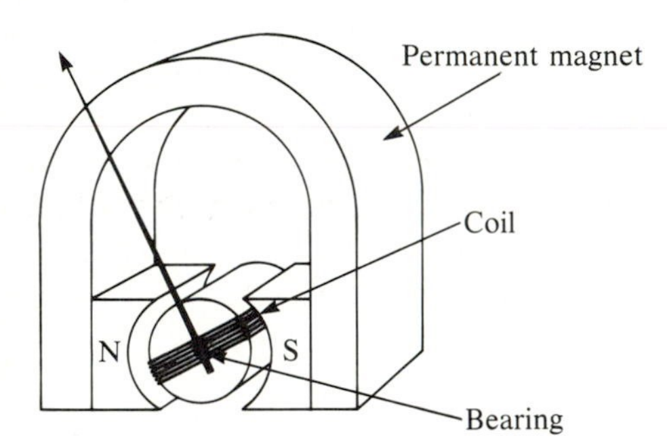

FIGURE 4–1
Basic permanent magnet moving coil meter movement

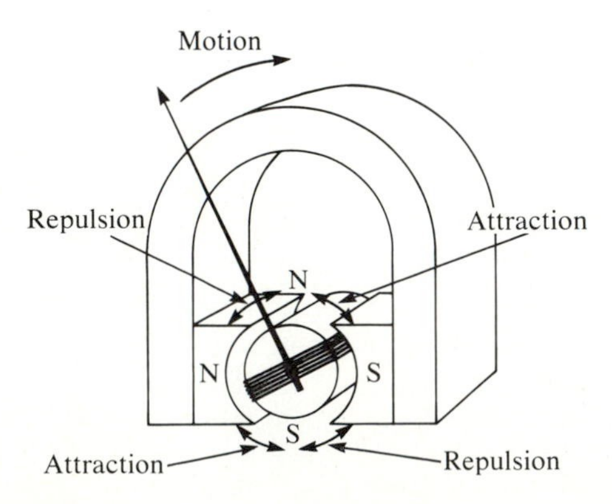

FIGURE 4–2
Operation of a basic meter movement

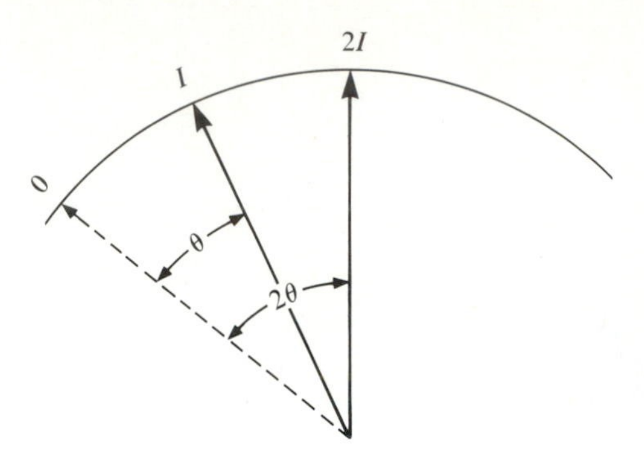

FIGURE 4–3
Proportional currents cause proportional deflection angles.

The motion of the coil's angular displacement is directly proportional to the amount of current flowing in the coil (Figure 4–3). If a certain amount of current, I, results in the angular deflection θ of the needle, twice that amount of current ($2I$) will result in twice the angular deflection (2θ) of the needle. This action produces a linear displacement scale that is much easier to use than a nonlinear scale.

INTERNAL RESISTANCE OF THE METER MOVEMENT

Since the meter movement is a coil of very fine wire with little mass, it has a high resistance (R_m) and, as a result, little torque (a small current) will deflect it. If the meter movement is measuring heavier current, it must be larger, with greater mass and smaller resistance (Figure 4–4). (Remember that

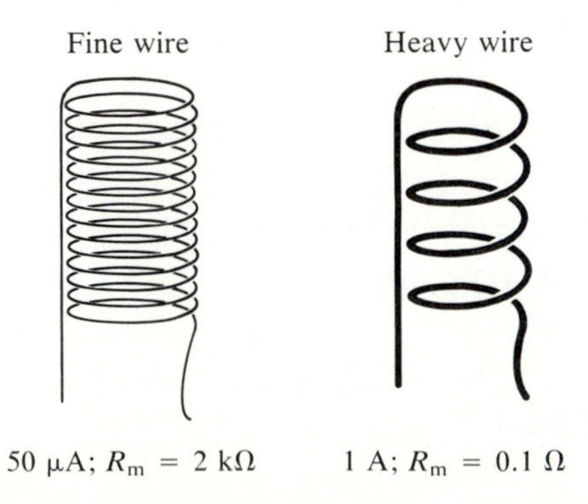

FIGURE 4–4
Size versus resistance of meter movement coils

resistance is inversely proportional to the square of the diameter of the conductor.)

A meter measuring microamperes will most probably be used in a circuit with a resistance of 100 kΩ or larger. A meter measuring milliamperes will most likely be used in a circuit with a resistance of the order of 1000 Ω. In both cases, the relative meter resistance is tolerable. It is small with respect to the circuit resistance.

Placing a meter in a circuit changes the circuit and affects its operation. This effect, called *loading,* will be considered in some detail later in the chapter.

Although the meter resistance is an integral part of the meter movement, the meter movement can be thought of as having two separate parts: the ideal meter (Figure 4–5A), with zero resistance, and the meter resistance (Figure 4–5B). Figure 4–5C shows the resulting composite circuit diagram.

The following is the procedure for measuring the internal resistance of a panel meter (see Figure 4–6). Adjust R_s to give full-scale deflection. Then adjust R_s to give half-scale deflection. Based on this resistance and Ohm's and Kirchhoff's laws, it should be possible to determine R_m. Remember that R_s should be measured when it is *not* in the circuit.

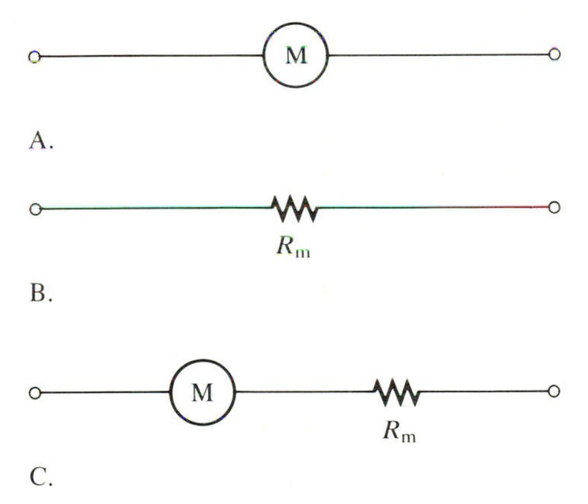

FIGURE 4–5
The meter movement as two components. A. Ideal meter. B. Meter resistance. C. Composite circuit.

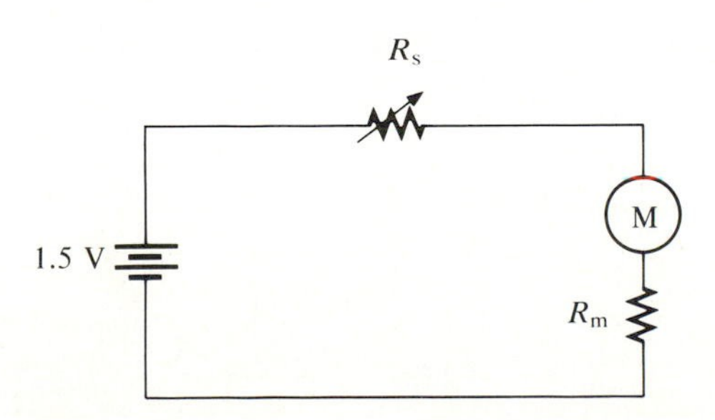

FIGURE 4–6
Circuit for measuring R_m

TABLE 4–1
Approximate internal resistance of panel-mounted current meters

Meter Range	Resistance
0–10 μA	17,000 Ω
0–20 μA	4,750 Ω
0–50 μA	2,000 Ω
0–100 μA	825 Ω
0–200 μA	360 Ω
0–500 μA	156 Ω
0–1 mA	55 Ω
0–1.5 mA	30 Ω
0–3 mA	14 Ω
0–5 mA	8.5 Ω
0–10 mA	3.1 Ω
0–15 mA	6.6 Ω
0–25 mA	4.0 Ω
0–50 mA	2.0 Ω
0–75 mA	1.3 Ω
0–100 mA	1.0 Ω
0–200 mA	0.5 Ω
0–500 mA	0.2 Ω
0–1 A	0.1 Ω

Table 4–1 is a general listing of common panel meter resistances. These values are approximations. Each time a meter movement is to be used to construct a meter, its internal resistance should be measured to ensure accuracy.

FULL-SCALE DEFLECTION

Full-scale deflection is important for two major reasons. First, it represents the highest reading a meter can indicate without suffering physical damage. Second, it is related to the maximum percentage of systematic error guaranteed by the manufacturer. If a meter is guaranteed to yield a maximum of 3% systematic error, that means 3% of full-scale deflection.

Suppose that a current of 97 mA is to be measured on a meter with 100-mA and 1000-mA ranges full-scale ± 3%. If the meter range is set at 100 mA, the reading is 97 mA ±3 mA or 97 mA ±3.1%. The true value could be anywhere between 94 mA and 100 mA. If the meter range is switched to 1000 mA, the needle points to 97 mA. This represents 97 mA ± 30 mA (3% of 1000 mA is 30 mA). The reading is now 97 mA ± 31%. The true value could be anywhere between 67 mA and 127 mA.

Therefore, the meter should be used to indicate as close to full-scale deflection as possible in order to minimize the inherent systematic error.

SHUNTING

If a 50-μA meter movement with an internal resistance of 2 kΩ is to be used to measure a 500-μA current, some changes must be made. A current of 450 μA can be shunted away from the meter movement, leaving 50 μA to go through the movement (Figure 4–7). By Kirchhoff's current law, the current through the shunt equals the total current minus the current through the meter movement:

$$\begin{aligned} I_{sh} &= I_t - I_m \\ &= 500\ \mu A - 50\ \mu A \\ &= 450\ \mu A \end{aligned} \qquad \textbf{(4–1)}$$

Since the meter and the shunt are in parallel,

$$V_{sh} = V_m \qquad \textbf{(4–2)}$$

By Ohm's law,

$$I_{sh}R_{sh} = I_m R_m \qquad \textbf{(4–3)}$$

Thus,

$$R_{sh} = \frac{I_m}{I_{sh}} R_m \qquad \textbf{(4–4)}$$

and

$$\begin{aligned} R_{sh} &= \frac{50\ \mu A}{450\ \mu A}(2\ k\Omega) \\ &= 222\ \Omega \end{aligned} \qquad \textbf{(4–5)}$$

Notice that as the current-handling capability of the meter increased (50 μA to 500 μA), the meter resistance decreased. Since the new meter circuit has 222.2 Ω in parallel with 2 kΩ, the meter's effective resistance now equals 200 Ω. The resistance decreased from 2 kΩ to 200 Ω.

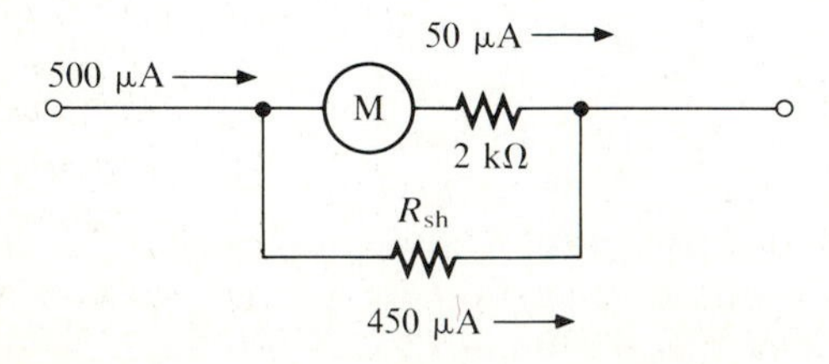

FIGURE 4–7
Single-range shunted ammeter

Of course, the meter readings must be correctly interpreted—10 μA really means 100 μA, and 40 μA really means 400 μA.

A 50-μA meter movement cannot be made more sensitive (e.g., 10 μA full-scale) without using an amplifier. In general, a meter movement can be made to read a heavier current full-scale by shunting, but it cannot be used to measure a lighter current full-scale by the same means. In other words, a meter can be made less sensitive but never more sensitive than what its original design provided.

MULTIRANGE AMMETER

A selection of shunt resistances can be used to provide a corresponding selection of current ranges, as shown in Figure 4–8.

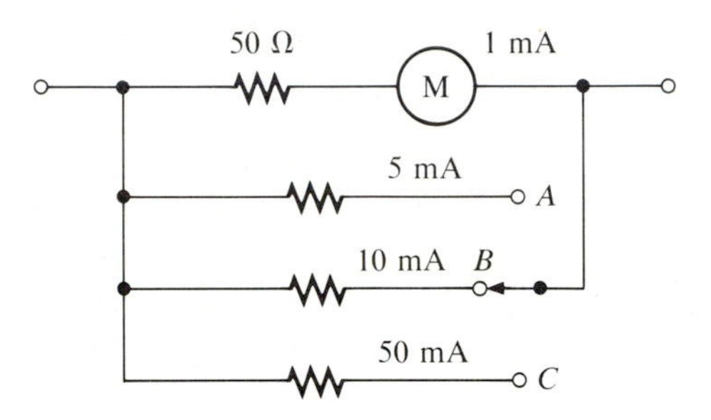

FIGURE 4–8
Multirange ammeter

EXAMPLE 4–1
Design a four-range ammeter using a 50-μA, 2-kΩ meter movement. The four ranges are to be 1 mA, 500 μA, 100 μA, and 50 μA full-scale deflection (see Figure 4–9).

Solution:
From Figure 4–9,

Switch Position	*Current Range*
A	1 mA
B	500 μA
C	100 μA
D	50 μA

When the switch is in position *A*,

$$V_{sh} = V_m \qquad \textbf{(4–6)}$$

$$R_{sh}I_{sh} = R_m I_m \qquad \textbf{(4–7)}$$

$$R_{sh} = R_m \frac{I_m}{I_{sh}} \qquad \textbf{(4–8)}$$

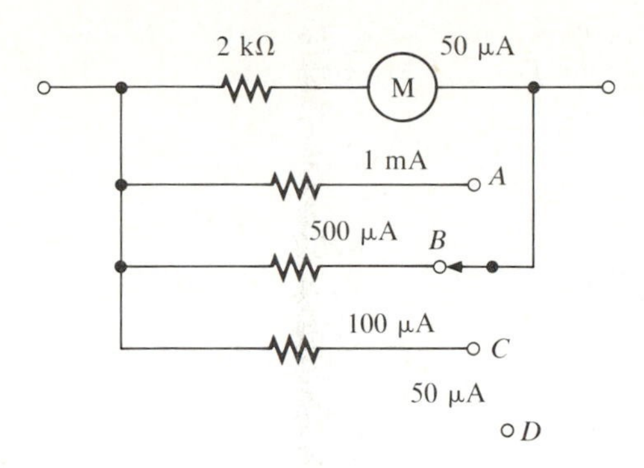

FIGURE 4–9
Four-range ammeter

With 1 mA = 1000 μA flowing through the meter movement, 950 μA will be flowing through the shunt. Therefore,

$$R_1 = 2\ \text{k}\Omega\left(\frac{50\ \mu\text{A}}{950\ \mu\text{A}}\right) = 105.26\ \Omega \quad \textbf{(4–9)}$$

When the switch is in position B, a current of 500 μA in the circuit means 50 μA through the meter movement and 450 μA through the shunt. Therefore,

$$R_2 = 2\ \text{k}\Omega\left(\frac{50\ \mu\text{A}}{450\ \mu\text{A}}\right) = 222.22\ \Omega \quad \textbf{(4–10)}$$

When the switch is in position C, a current of 100 μA in the circuit means 50 μA through the meter movement and 50 μA through the shunt. Therefore,

$$R_3 = 2\ \text{k}\Omega\left(\frac{50\ \mu\text{A}}{50\ \mu\text{A}}\right) = 2\ \text{k}\Omega \quad \textbf{(4–11)}$$

When the switch is in position D, no shunt exists. All of the current must flow through the meter movement to produce full-scale deflection. Whereas a 105.26-Ω resistor (R_1) is not normally commercially available, 105-Ω and 106-Ω resistors are easy to obtain. But which should be used? It is safer to use the smaller shunting value since this action will draw more current from the meter movement, and thus the movement will not reach full-scale deflection. If the larger value is used, the shunt will force more current through the meter movement than it was designed for. Although this excess current may not damage the meter movement, it will not be possible to determine how far the needle would have been deflected past the end of the scale. It is always safer in this case to use the smaller value of shunt resistance.

This seems to be a good solution to the problem of having one meter serve the purpose of many. Remember, however, that an ammeter is always in series with the circuit. As the switch wiper is moved from point A to point B, there is a moment when it is between contacts. At this instant, *there is no shunt protecting the meter movement!* We can eliminate this problem by using *make-before-break* contact switches. These switches are designed so that the wiper makes electrical contact with the adjacent terminal before breaking electrical contact with the present terminal (Figure 4–10).

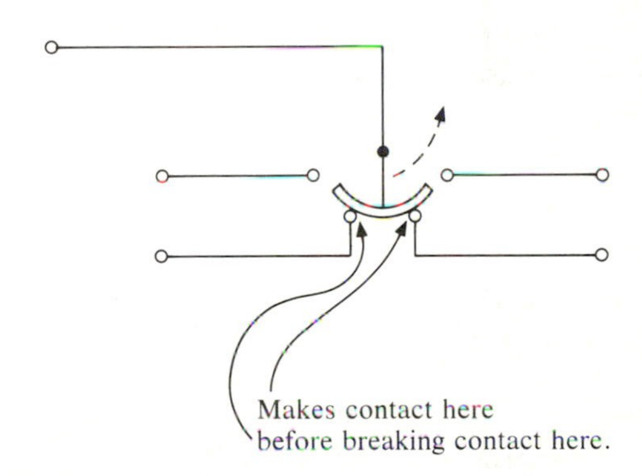

FIGURE 4–10
Make-before-break contact switch

AYRTON SHUNT

Another popular circuit arrangement used to extend the range of a meter movement is the *Ayrton shunt*. Figure 4–11 shows a three-range ammeter arrangement using an Ayrton shunt. Notice in the

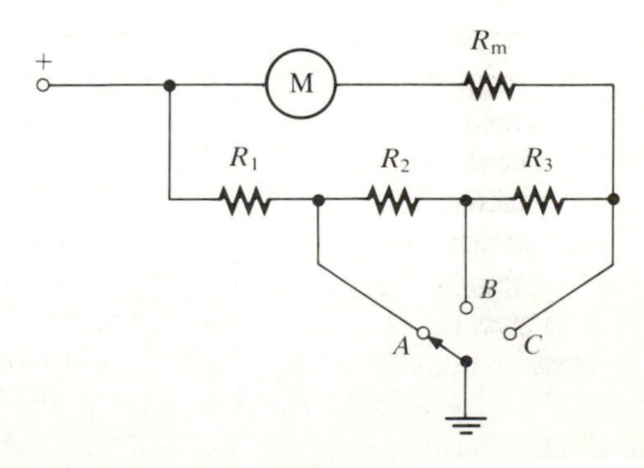

FIGURE 4–11
Three-range ammeter using an Ayrton shunt

diagram that when switch contact is between points A and B, the circuit is opened, protecting the meter movement from possible overload.

The design of a three-range meter movement using an Ayrton shunt is as follows. Assume that the ranges to be used are 1 mA, 300 μA, and 100 μA. With the switch in position A, the circuit looks like Figure 4–12A. The smallest combination of shunt resistance (R_1 alone) allows the highest current range to be served. Since

$$V_{sh} = V_m$$

then

$$I_{sh}R_{sh} = I_mR_m \quad \textbf{(4–12)}$$

and

$$(950\ \mu A)R_1 = (50\ \mu A)(2\ k\Omega + R_2 + R_3) \quad \textbf{(4–13)}$$

With the switch in position B, the circuit looks like Figure 4–12B, and

$$(250\ \mu A)(R_1 + R_2) = (50\ \mu A)(2\ k\Omega + R_3) \quad \textbf{(4–14)}$$

With the switch in position C, the circuit looks like Figure 4–12C, and

$$(50\ \mu A)(R_1 + R_2 + R_3) = (50\ \mu A)(2\ k\Omega) \quad \textbf{(4–15)}$$

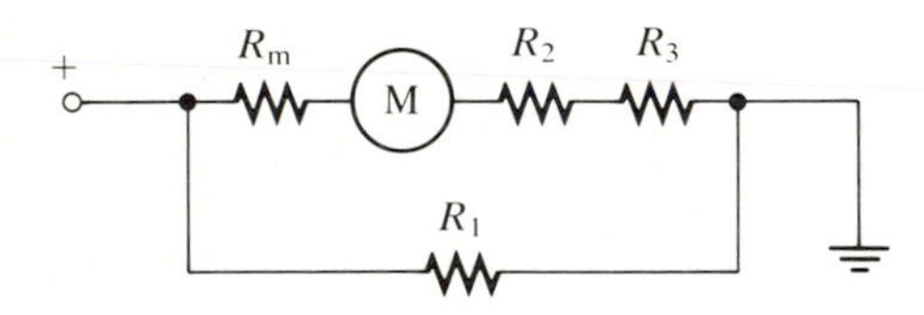

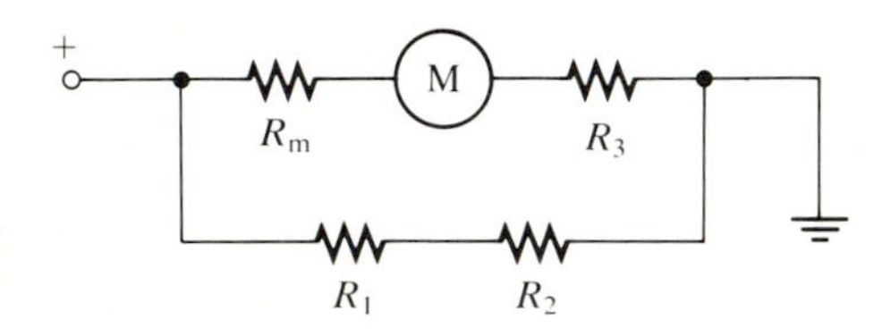

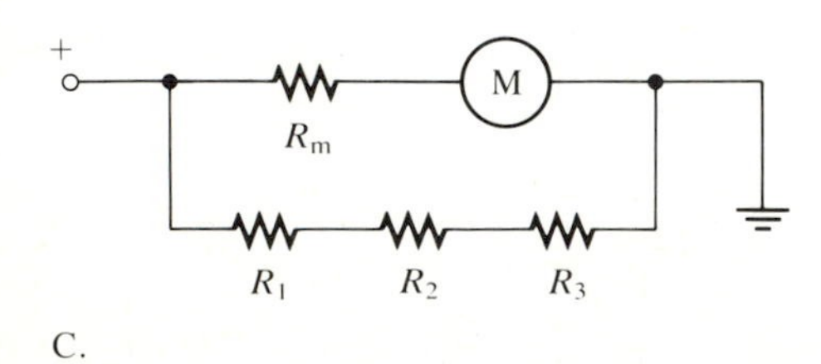

FIGURE 4–12
Circuits resulting from each switch position of a three-range Ayrton shunt

From equations (4–13) through (4–15), respectively, the following are obtained:

$$(950\ \mu A)R_1 - (50\ \mu A)R_2 - (50\ \mu A)R_3 = (50\ \mu A)(2\ k\Omega) \quad \textbf{(4–16)}$$

$$(250\ \mu A)R_1 + (250\ \mu A)R_2 - (50\ \mu A)R_3 = (50\ \mu A)(2\ k\Omega) \quad \textbf{(4–17)}$$

$$(50\ \mu A)R_1 + (50\ \mu A)R_2 + (50\ \mu A)R_3 = (50\ \mu A)(2\ k\Omega) \quad \textbf{(4–18)}$$

Dividing all equations by 50 μA yields

$$19R_1 - R_2 - R_3 = 2\ k\Omega \quad \textbf{(4–19)}$$

$$5R_1 + 5R_2 - R_3 = 2\ k\Omega \quad \textbf{(4–20)}$$

$$R_1 + R_2 + R_3 = 2\ k\Omega \quad \textbf{(4–21)}$$

Adding equations (4–19) and (4–21) gives

$$20R_1 = 4\ k\Omega \quad \textbf{(4–22)}$$

Therefore,

$$R_1 = 200\ \Omega \quad \textbf{(4–23)}$$

Substituting this value into equations (4–19) and (4–20) yields

$$19(200\ \Omega) - R_2 - R_3 = 2\ k\Omega \quad \textbf{(4–24)}$$

$$5(200\ \Omega) + 5R_2 - R_3 = 2\ k\Omega \quad \textbf{(4–25)}$$

and subtracting equation (4–25) from equation (4–24) gives

$$\begin{aligned} 14(200\ \Omega) - 6R_2 &= 0 \\ 6R_2 &= 14(200\ \Omega) \quad \textbf{(4–26)} \\ R_2 &= 466.7\ \Omega \end{aligned}$$

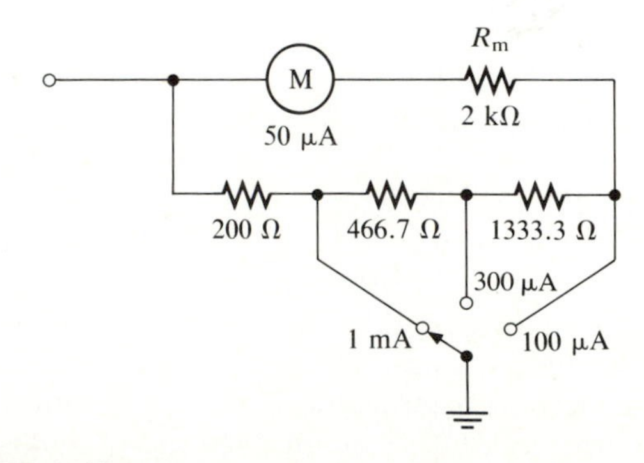

FIGURE 4–13
Results of Ayrton shunt design

By substituting these values of R_1 and R_2 into equation (4–21), we have

$$200\ \Omega + 466.7\ \Omega + R_3 = 2\ \text{k}\Omega$$
$$\begin{aligned} R_3 &= 2000\ \Omega - 200\ \Omega - 466.7\ \Omega \\ &= 1333.3\ \Omega \end{aligned} \quad \textbf{(4–27)}$$

The circuit now looks like Figure 4–13.

VOLTMETERS

At full-scale deflection, the voltage drop across a 50-μA, 2-kΩ meter movement is

$$\begin{aligned} V_m &= I_m R_m \\ &= (50\ \mu\text{A})(2\ \text{k}\Omega) \\ &= 0.1\ \text{V} \end{aligned} \quad \textbf{(4–28)}$$

If a *series dropping* resistance is placed in series with the meter movement so that it drops the appropriate amount of voltage, the meter movement will still indicate the current flow. By Ohm's law, this current is directly proportional to the voltage across the meter (see Figure 4–14).

For a 10-V meter,

$$\begin{aligned} R_s + 2\ \text{k}\Omega &= \frac{10\ \text{V}}{50\ \mu\text{A}} \\ R_s + 2\ \text{k}\Omega &= 200\ \text{k}\Omega \\ R_s &= 198\ \text{k}\Omega \end{aligned} \quad \textbf{(4–29)}$$

(see Figure 4–15). Therefore,

$$V_{R_s} = (50\ \mu\text{A})(198\ \text{k}\Omega) = 9.9\ \text{V} \quad \textbf{(4–30)}$$

Here 0.1 V remains to be dropped across the meter movement giving full-scale deflection. The meter now can be used to read 10 V full-scale (5 V half-scale).

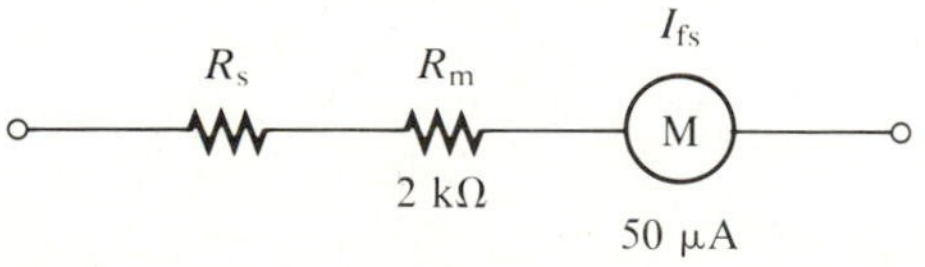

FIGURE 4–14
Conversion of an ammeter to a voltmeter

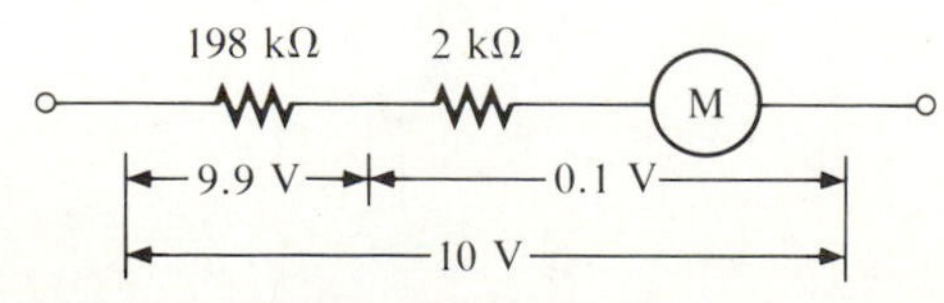

FIGURE 4–15
Results of voltmeter design

MULTIRANGE VOLTMETERS

Different voltage ranges can be provided for by using different series dropping or multiplier resistances (see Figure 4–16). Since

$$(1\ \text{mA})(50\ \Omega) = 0.05\ \text{V} \quad \textbf{(4–31)}$$

will be dropped across the meter movement at full-scale, the remaining voltage in each range must be dropped across R_1, R_2, R_3, and R_4.

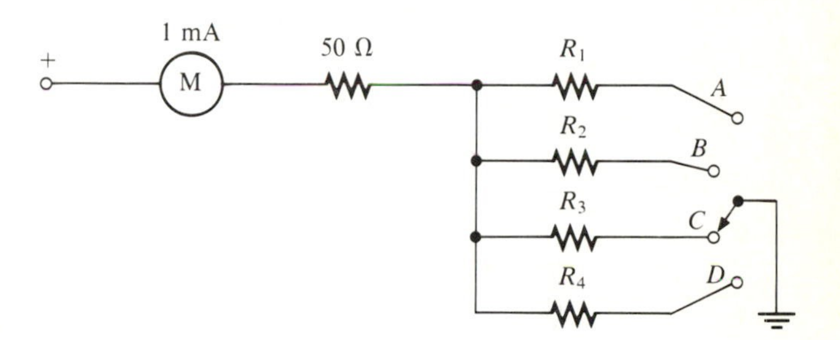

FIGURE 4–16
Multirange voltmeter

EXAMPLE 4–2
Design a meter that will measure the following:

Switch Position	*Voltage (V)*
A	1
B	5
C	10
D	50

Solution:
When the switch is in position *A*,

$$R_m + R_s = \frac{V}{I_{fs}} \quad \textbf{(4–32)}$$

where I_{fs} denotes the full-scale current. Rearranging,

$$R_s = \frac{V}{I_{fs}} - R_m \quad \textbf{(4–33)}$$

and so

$$\begin{aligned} R_1 &= \frac{1\ \text{V}}{1\ \text{mA}} - 50\ \Omega \\ &= 1\ \text{k}\Omega - 50\ \Omega \\ &= 950\ \Omega \end{aligned} \quad \textbf{(4–34)}$$

When the switch is in position B,

$$R_2 = \frac{5\text{ V}}{1\text{ mA}} - 50\ \Omega = 5\text{ k}\Omega - 50\ \Omega = 4950\ \Omega \quad (4\text{–}35)$$

When the switch is in position C,

$$R_3 = \frac{10\text{ V}}{1\text{ mA}} - 50\ \Omega = 10\text{ k}\Omega - 50\ \Omega = 9950\ \Omega \quad (4\text{–}36)$$

When the switch is in position D,

$$R_4 = \frac{50\text{ V}}{1\text{ mA}} - 50\ \Omega = 50\text{ k}\Omega - 50\ \Omega = 49{,}950\ \Omega \quad (4\text{–}37)$$

(see Figure 4–17).

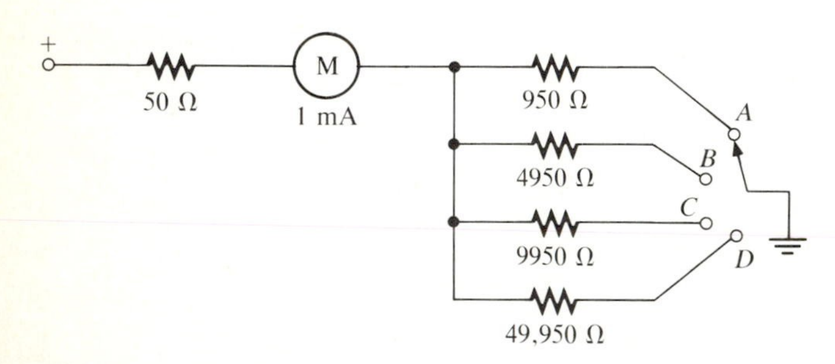

FIGURE 4–17
Results of multirange voltmeter design

If a 49-kΩ and a 50-kΩ resistance are available for R_4, which would be safer to use? The larger value will drop more than the prescribed voltage, and the meter will indicate slightly less than full-scale. If a smaller value is chosen, less voltage will be dropped across the protective resistance, and more than the prescribed voltage across the meter movement, causing the meter to deflect too far. The meter could be damaged or at least give an incorrect reading. It is safer to use a larger value of resistance to protect the meter by placing it in series with the meter movement.

Notice that when the switch is between points A and B, the circuit is open and the meter is protected.

Another arrangement for a multirange voltmeter is shown in Figure 4–18. As before,

$$R_1 + 50\ \Omega = \frac{1\text{ V}}{1\text{ mA}} = 1\text{ k}\Omega \quad (4\text{–}38)$$

$$R_1 = 950\ \Omega \quad (4\text{–}39)$$

and

$$R_1 + R_2 + 50\ \Omega = \frac{5\text{ V}}{1\text{ mA}} = 5\text{ k}\Omega \quad (4\text{–}40)$$

$$R_2 = 5\text{ k}\Omega - 50\ \Omega - 950\ \Omega = 4\text{ k}\Omega \quad (4\text{–}41)$$

and

$$R_1 + R_2 + R_3 + 50\ \Omega = \frac{10\text{ V}}{1\text{ mA}} = 10\text{ k}\Omega \quad (4\text{–}42)$$

$$R_3 = 10\text{ k}\Omega - 50\ \Omega - 950\ \Omega - 4\text{ k}\Omega = 5\text{ k}\Omega \quad (4\text{–}43)$$

and

$$R_1 + R_2 + R_3 + R_4 + 50\ \Omega = \frac{50\text{ V}}{1\text{ mA}} = 50\text{ k}\Omega \quad (4\text{–}44)$$

$$R_4 = 50\text{ k}\Omega - 5\text{ k}\Omega - 4\text{ k}\Omega - 950\ \Omega - 50\Omega = 40\text{ k}\Omega \quad (4\text{–}45)$$

These values will usually be easier to find in the supply room.

SENSITIVITY

Voltmeters are rated by their loading resistance. This rating is called *sensitivity* (S) and is expressed in ohms per volt. In the example of Figure 4–18,

$$S = \frac{1\text{ k}\Omega}{1\text{ V}} = 1\text{ k}\Omega/\text{V} \quad (4\text{–}46)$$

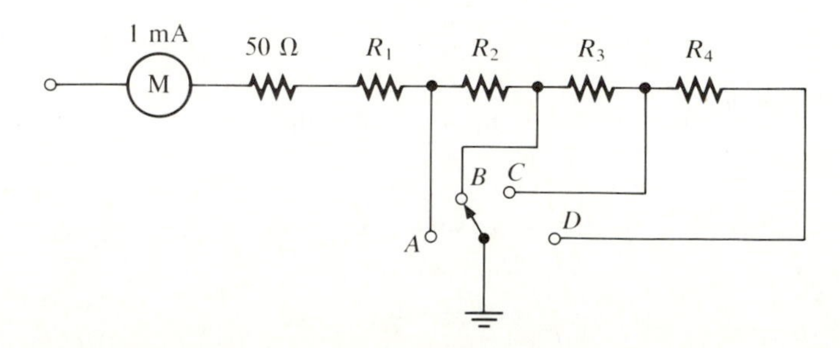

FIGURE 4–18
Another multirange voltmeter circuit arrangement

$$S = \frac{5\ \text{k}\Omega}{5\ \text{V}} = 1\ \text{k}\Omega/\text{V} \tag{4–47}$$

Sensitivity is useful in considering the *loading effect.* In dc meters, the sensitivity can be calculated as follows:

$$S = \frac{1}{I_{fs}} \tag{4–48}$$

In our voltmeter using the 1-mA meter movement, (see Figure 4–18)

$$S = \frac{1}{1\ \text{mA}} = 1\ \text{k}\Omega/\text{V} \tag{4–49}$$

A popular meter movement used in analog dc meters is the 50-μA, 2-kΩ movement, where

$$S = \frac{1}{50\ \mu\text{A}} = 20\ \text{k}\Omega/\text{V} \tag{4–50}$$

On a 5-V scale, the meter represents

$$R_m = 5\ \text{V} \times 20\ \text{k}\Omega/\text{V} = 100\ \text{k}\Omega \tag{4–51}$$

On a 100-V scale, the meter represents

$$R_m = 100\ \text{V} \times 20\ \text{k}\Omega/\text{V} = 2\ \text{M}\Omega \tag{4–52}$$

In general,

$$\begin{aligned} R_m &= \text{full-scale voltage} \times \text{sensitivity} \\ &= V_{fs} \times S \end{aligned} \tag{4–53}$$

LOADING PROBLEMS

Loading effects can be demonstrated using the circuit in Figure 4–19. Consider the effect of a multirange voltmeter with ranges of 2.5 V, 10 V, and 50 V. The meter resistance on any dc range can be calculated using the formula

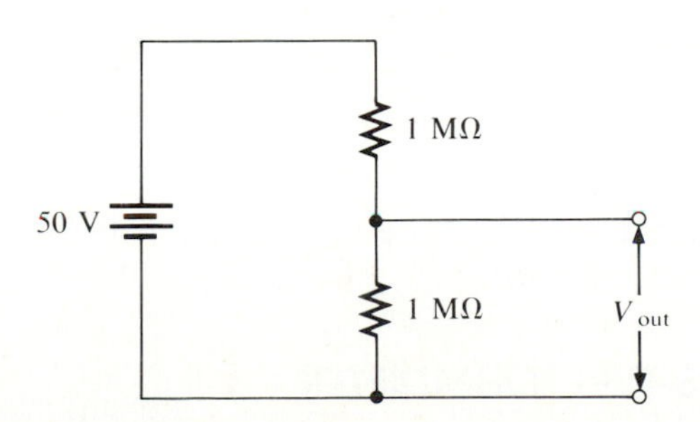

FIGURE 4–19
Circuit demonstrating loading problems

$$R_m = \text{meter range} \times \text{sensitivity} \tag{4–54}$$

Since we can assume that the voltage across the 1-MΩ resistance should be 25 V (use voltage division), let's see what we get when we set the meter for 50 V full-scale:

$$\begin{aligned} R_m &= 50\ \text{V} \times 20\ \text{k}\Omega/\text{V} \\ &= 1\ \text{M}\Omega \end{aligned} \tag{4–55}$$

Figure 4–20 illustrates the effect of using the meter to measure the voltage required.

The 1-MΩ resistance in parallel with the 1-MΩ meter resistance results in 500 kΩ, giving the results shown in Figure 4–20. By voltage division,

$$V_m = 50\ \text{V}\left(\frac{500\ \text{k}\Omega}{1500\ \text{k}\Omega}\right) = 16.67\ \text{V} \tag{4–56}$$

This corresponds to an error of (25 − 16.67)/25 = 33%!

Changing the range switch to the 10-V range gives the results shown in Figure 4–21, where

$$R_m = 10\ \text{V} \times 20\ \text{k}\Omega/\text{V} = 200\ \text{k}\Omega \tag{4–57}$$

and

$$V_m = 50\ \text{V}\left(\frac{166.7\ \text{k}\Omega}{1166.7\ \text{k}\Omega}\right) = 7.144\ \text{V} \tag{4–58}$$

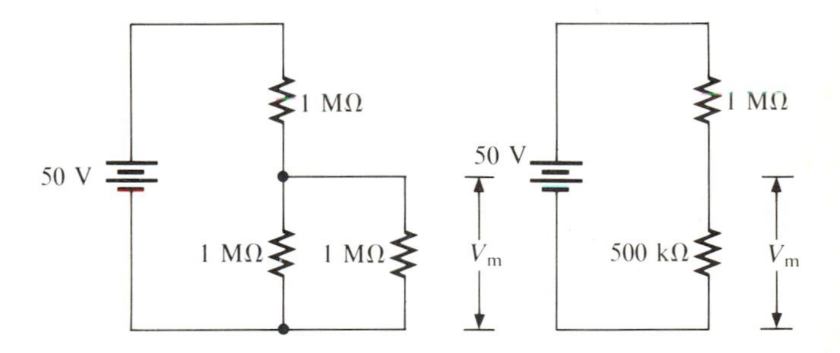

FIGURE 4–20
Meter set on the 50-V range

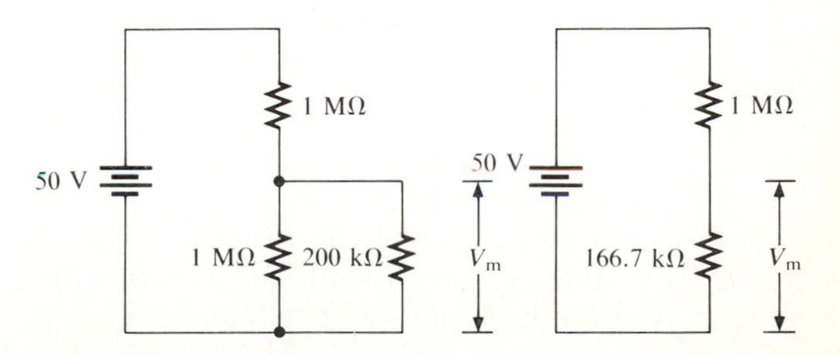

FIGURE 4–21
Meter set on the 10-V range

This corresponds to an error of (25 − 7.144)/25 = 71.4%!

On the 2.5-V scale,

$$R_m = 2.5\text{ V} \times 20\text{ k}\Omega/\text{V} = 50\text{ k}\Omega \quad (4\text{–}59)$$

The 1-MΩ resistance in parallel with the 50-kΩ resistance equals 47.6 kΩ, as shown in Figure 4–22. Therefore,

$$V_m = 50\text{ V}\left(\frac{47.6\text{k}\Omega}{1047.6\text{ k}\Omega}\right) = 2.27\text{ V} \quad (4\text{–}60)$$

This corresponds to an error of (25 − 2.27)/25 = 91%!

If 2.27 V is dropped across the combination of the meter and the resistance that is being measured, according to Kirchhoff's voltage law, 50 V − 2.27 V = 47.3 V is being dropped across the other 1-MΩ resistance. This voltage may be high enough to damage that resistance.

The problem here is that the meter, although an excellent one, is used in a situation where it affects the circuit so much that the error is unacceptable. The following rules of thumb can be used to prevent this from happening (see Figure 4–23).

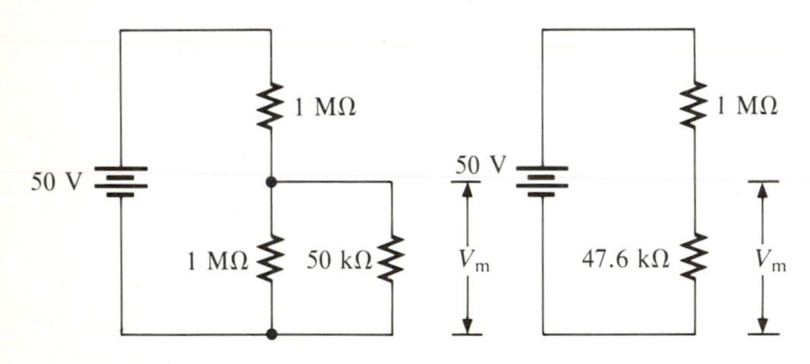

FIGURE 4–22
Meter set on the 2.5-V range

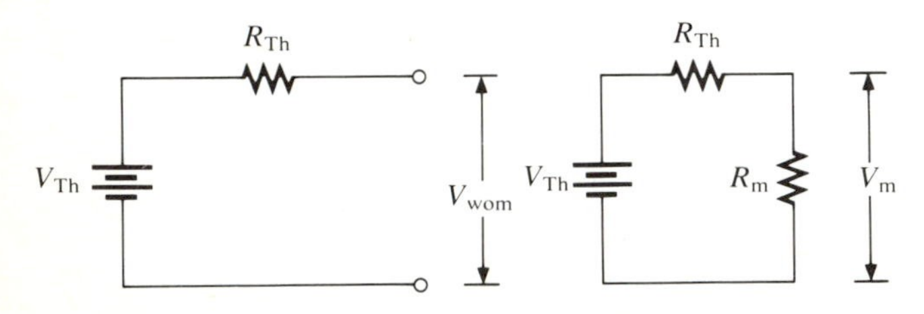

FIGURE 4–23
Circuit for minimizing loading effects

The voltage without a meter is $V_{wom} = V_{Th}$ (the Thevenin equivalent of the resistance across which the reading is being taken). The voltage with a meter (V_m) can be found using voltage division:

$$V_m = V_{Th}\left(\frac{R_m}{R_m + R_{Th}}\right) \quad (4\text{–}61)$$

where R_{Th} is the Thevenin equivalent of the resistance in the circuit under consideration. The accuracy can be expressed as

$$\frac{V_m}{V_{wom}} = \text{accuracy} \quad (4\text{–}62)$$

From equation (4–60),

$$\frac{V_m}{V_{wom}} = \frac{R_m}{R_m + R_{Th}} = \text{accuracy} \quad (4\text{–}63)$$

If we want to ensure 99% accuracy = 0.99, we calculate

$$\frac{R_m}{R_m + R_{Th}} = 0.99$$

$$R_m = 0.99R_m + 0.99R_{Th} \quad (4\text{–}64)$$

$$R_m - 0.99R_m = 0.99R_{Th}$$

$$0.01R_m = 0.99R_{Th}$$

and so

$$R_m \cong 100\ R_{Th} \quad (4\text{–}65)$$

or, to be safe, we use

$$R_m > 100\ R_{Th} \quad (4\text{–}66)$$

If we want to ensure 95% accuracy = 0.95, it can be shown that

$$R_m \cong 20\ R_{Th} \quad (4\text{–}67)$$

or, to be safe, we use

$$R_m > 20\ R_{Th} \quad (4\text{–}68)$$

These accepted rules of thumb can be used as guides for taking measurements while minimizing loading error.

In a similar manner we can derive guidelines for taking current measurements. For 99% accuracy,

$$R_m < \frac{1}{100}R_{Th} \quad (4\text{–}69)$$

and for 95% accuracy,

$$R_m < \frac{1}{20}R_{Th} \quad (4\text{–}70)$$

In these cases, R_m is the resistance of the ammeter used in the circuit.

SERIES-TYPE OHMMETER

In Figure 4–24, the zeroing resistor R_z is a variable series dropping resistor, R_m is the meter resistance,

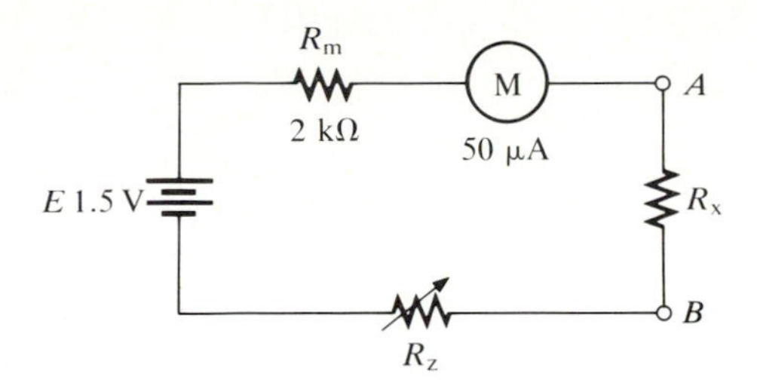

FIGURE 4–24
Series-type ohmmeter circuit

E is a battery, and R_x is the resistance to be measured. For this example, a 50-μA, 2-kΩ meter movement and a 1.5-V battery will be used.

The meter will read ∞ or infinity when terminals A and B are open since no current flows through it. This reading will be considered an indication of infinite resistance (∞), or open circuit. If A and B are shorted (zero ohms) and R_z is set for full-scale deflection,

$$R_z + R_m = \frac{1.5\ \text{V}}{50\ \mu\text{A}} \tag{4–71}$$

Since $R_m = 2$ kΩ,

$$R_z = 30\ \text{k}\Omega - 2\ \text{k}\Omega = 28\ \text{k}\Omega \tag{4–72}$$

The full-scale deflection on the meter face will be zero ohms, giving the points indicated in Figure 4–25. If a resistance R_x is used to give half-scale deflection, 25 μA will be flowing through the circuit. Therefore,

$$\begin{aligned} R_x + R_m + R_z &= \frac{1.5\ \text{V}}{25\ \mu\text{A}} \\ R_x &= 60\ \text{k}\Omega - 2\ \text{k}\Omega - 28\ \text{k}\Omega \\ &= 30\ \text{k}\Omega \end{aligned} \tag{4–73}$$

Continuing, R_x to give quarter-scale deflection is

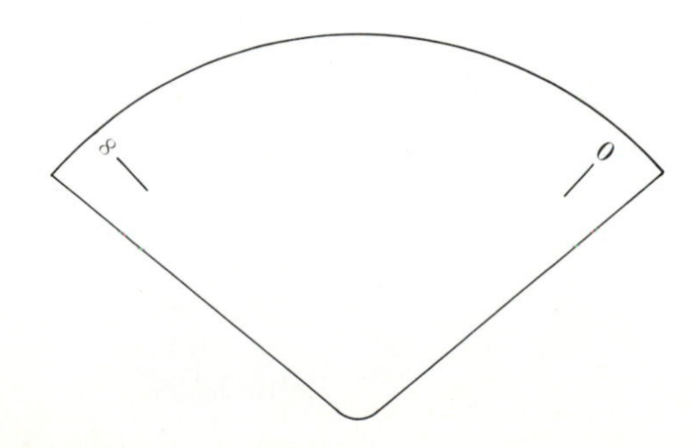

FIGURE 4–25
Ohmmeter scale

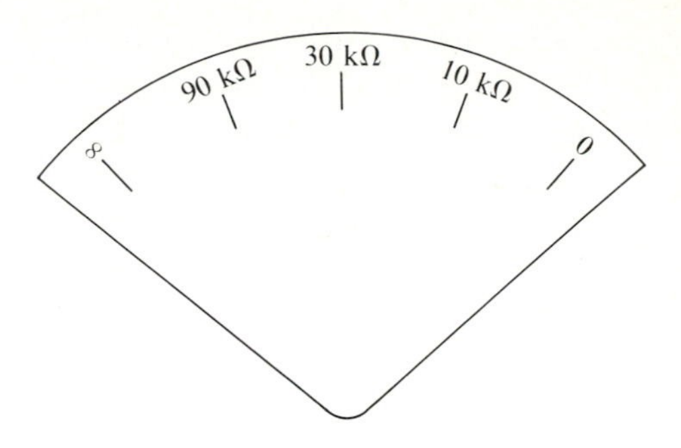

FIGURE 4–26
More detailed ohmmeter scale

$$\begin{aligned} R_x + R_m + R_z &= \frac{1.5\ \text{V}}{12.5\ \mu\text{A}} \\ R_x &= 90\ \text{k}\Omega \end{aligned} \tag{4–74}$$

and R_x to give three-quarter-scale deflection is

$$\begin{aligned} R_x + R_m + R_z &= \frac{1.5\ \text{V}}{37.5\ \mu\text{A}} \\ R_x &= 10\ \text{k}\Omega \end{aligned} \tag{4–75}$$

The meter face looks like Figure 4–26. This is not only a nonlinear scale but also a reversed scale. Many other points can be labeled to produce a useful ohmmeter.

What would happen if E were to lower in value due to age, say to 1.35 V? When the terminals are open, the meter still indicates ∞. But with the terminals shorted,

$$I = \frac{1.35\text{V}}{R_m + R_z} = \frac{1.35\text{V}}{30\ \text{k}\Omega} = 45\ \mu\text{A} \tag{4–76}$$

This is no longer full-scale deflection. Therefore, it does not indicate zero ohms. This is why a variable resistance was used for R_z: Resistance R_z can be reduced to restore full-scale current deflection:

$$\begin{aligned} R_z + R_m &= \frac{1.35\ \text{V}}{50\ \mu\text{A}} \\ R_z &= 27\ \text{k}\Omega \end{aligned} \tag{4–77}$$

The R_z is the zero adjust on this ohmmeter. It compensates for small changes in battery voltage but should be considered only for small changes, as we shall see.

Assume that R_z is now 27 kΩ. Resistance R_x for half-scale deflection was 30 kΩ. Using that resistance now, we get

$$I_m = \frac{1.35\text{V}}{R_x + 30\ \text{k}\Omega} = 23.7\ \mu\text{A} \tag{4–78}$$

This value is less than half-scale current and indicates not 30 kΩ but a larger resistance value (remember the reverse scale!). If zeroing an ohmmeter proves difficult, the best course is to replace the battery or batteries.

SHUNT-TYPE OHMMETER

The previous circuit has serious limitations. If a meter that would measure 300 Ω at half-scale were designed, the control of current needed would be far too complicated to be practical. The circuit in Figure 4–27 would be more useful.

When $R_x = \infty$ (infinity, or open circuit), zero deflection will be obtained, which will be labeled. When $R_x = 0\ \Omega$ (short circuit), 1.5 V must be dropped across the meter movement and R_z. In addition, 1.5 V must be dropped across the shunt 300-Ω resistance. Using Ohm's law,

$$R_z + 2\text{ k}\Omega = \frac{1.5\text{ V}}{50\ \mu\text{A}} = 30\text{ k}\Omega \qquad R_z = 28\text{ k}\Omega \tag{4–79}$$

A resistance of 300 Ω in parallel with 30 kΩ equals 297 Ω. If $R_x = 300\ \Omega$, the meter will have

$$1.5\text{ V}\left(\frac{297\ \Omega}{597\ \Omega}\right) = 0.746\text{ V} \tag{4–80}$$

across it. The current through the meter will be 24.87 μA, or almost half-scale current. If the battery ages to 1.35 V as in the previous section, the current through the meter will drop to 45 μA when $R_x = 0\ \Omega$. This gives an error similar to that in the series-type ohmmeter. If R_z is adjusted to compensate for this,

$$R_z + 2\text{ k}\Omega = \frac{1.35\text{ V}}{50\ \mu\text{A}} = 25\text{ k}\Omega \tag{4–81}$$

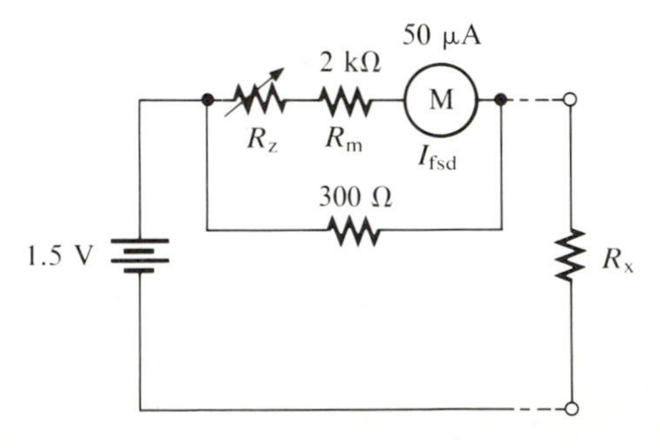

FIGURE 4–27
Shunt-type ohmmeter

A resistance of 300 Ω in parallel with 25 kΩ equals 296.4 Ω. If $R_x = 300\ \Omega$, the voltage across the meter branch is 0.67 V, and the current through the movement equals 24.8 μA. This represents a smaller percentage of error than that produced in the series-type ohmmeter circuit. The shunt design reduces error caused by aging batteries and allows the range to be changed by "plugging in" a resistance in parallel with the meter equal to the half-scale reading desired.

Most commercial ohmmeter circuits are more sophisticated than the examples given. These circuits are presented to give a general idea of how ohmmeter circuits work and how they respond to naturally occurring internal changes.

SUMMARY

The dc analog meter movement is a flexible device that can be adapted to measure dc current and voltage or resistance. Proper shunting enables the movement to measure current ranges far larger than that for which it was designed. Multipliers or series dropping resistors can be used to adapt the meter movement to measuring dc voltages. If a battery is added to the circuit, the meter can be used to measure resistance (usually with a reversed

FIGURE 4–28
Simpson 260 VOM

scale). Series- and shunt-type ohmmeters can be configured. The shunt-type circuit is superior because it is easily adapted to many ranges and because battery decay affects the precision of reference readings less.

The Simpson 260, shown in Figure 4–28, is an industrial standard volt-ohmmeter that incorporates the best of the ideas discussed in this chapter.

EXERCISES

1. If the meter in the circuit of Figure 4–29 indicates full-scale deflection, what value of shunt resistance is required to reduce the meter deflection to half-scale?

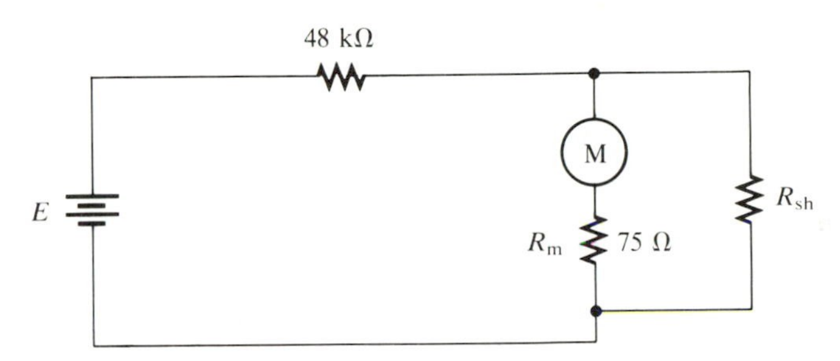

FIGURE 4–29
Circuit for Exercise 1

2. Explain why a meter should be used to indicate as close to full-scale deflection as possible.
3. Given a meter movement that indicates 1 mA full-scale and has an internal resistance of 65 Ω, design a meter that will indicate 50 mA full-scale. Draw the circuit diagram and give the component values. What is the resistance of the new meter?
4. Using a 100-μA meter movement with 1250 Ω internal resistance, design a meter that will indicate 2 mA full-scale. What is the resistance of this new meter?
5. In Exercise 3, we have to come up with a very small resistance. How can this be done?
6. If a piece of AWG #38 wire is used in Exercise 5, how long will the wire have to be to provide the necessary resistance for the shunt in Exercise 3? (AWG #38 copper wire resistivity = 660 Ω/1000 ft.)
7. If the wire in Exercise 6 is coiled for convenience, what effect will the inductance of the coil have on the performance of the meter?
8. Design a meter that would use a 25-mA movement (5 Ω) and have ranges of 100 mA and 500 mA full-scale.
9. What does *make-before-break* mean?
10. Using a 100-μA meter movement with internal resistance of 825 Ω, design a meter using an Ayrton shunt having the following ranges: 200 μA, 500 μA, 1 mA full-scale deflection.
11. Give an advantage of the Ayrton shunt circuit over the simpler circuit developed earlier.
12. How does shunting a meter movement affect the loading effect of the meter?
13. Design a voltmeter using a 1-mA, 50-Ω meter movement that will read 50 V full-scale.
14. Design a voltmeter that will read 1, 5, 10, and 50 V using a meter movement that reads 50 μA full-scale and has 2 kΩ internal resistance.

15. Repeat Exercise 14 using an alternative circuit arrangement.
16. What is the sensitivity of a meter that uses a meter movement with a full-scale deflection of 100 μA?
17. What is the resistance of a meter with a sensitivity of 5 kΩ/V if the meter is used on the 10-V range?
18. What resistance should a meter have if it is used to measure the voltage across the circuit in Figure 4–30 with 95% accuracy? with 99% accuracy?

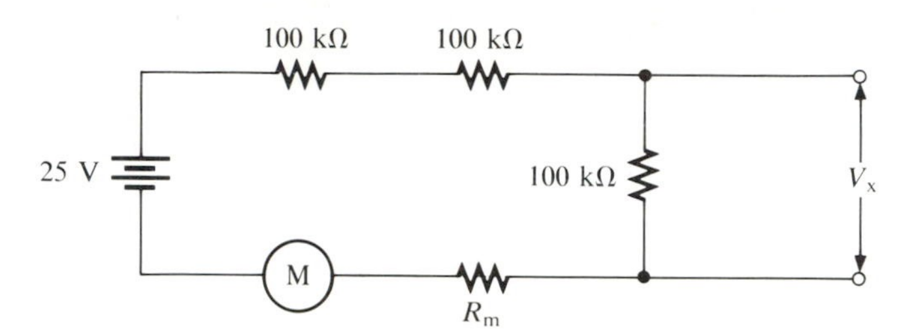

FIGURE 4–30
Circuit for Exercise 18

19. What accuracy would be expected from the voltage measurement in Figure 4–31?

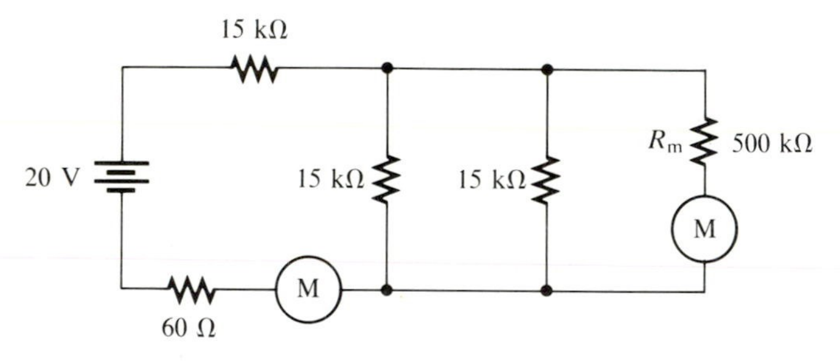

FIGURE 4–31
Circuit for Exercise 19

20. What resistance should an ammeter have to measure the current in Exercise 18 with 95% accuracy? with 99% accuracy?
21. What would be the expected accuracy of the current reading in the circuit in Exercise 19?
22. Using a 1-mA, 50-Ω meter movement and a 9-V battery in a series-type ohmmeter circuit, what will be the half-scale resistance reading?
23. For Exercise 22, calculate the R_x that will give quarter-scale and three-quarter-scale readings.
24. In Exercise 23, approximately what reading will you get if $R_x = 9$ kΩ and the battery ages down to 8.5 V?
25. Using a 1-mA, 50-Ω meter movement, design a shunt-type ohmmeter that will indicate 500 Ω half-scale. $V = 1.5$ V.
26. Why does a shunt-type ohmmeter exhibit less error when the battery ages than does a series-type ohmmeter?

LAB PROBLEM 4–1
Measuring Internal Resistance

Equipment

One meter movement, two variable resistors; two dry-cell batteries; one DVM; one low-voltage power supply.

Object

To design simple circuits that can be used to indirectly measure the internal resistance of batteries and meter movements. This problem is designed to have the student begin to develop problem-solving techniques. The student is presented with basic problems to be solved using fundamental measuring techniques and Ohm's and Kirchhoff's laws.

Procedure

Using the *indirect method*, that is, using current and voltage in an Ohm's law calculation, design a circuit that will measure the internal resistance of a battery. Use this circuit to measure the resistance of at least two batteries.

Design a circuit with which you can measure the internal resistance of an analog meter movement. As above, use the indirect method. Measure R_{in} for two meters with different ranges. Can you check your results with a high-resistance DVM without damaging the meter movements? If so, do so. If not, explain.

Design a circuit and use it to measure the internal resistance of a DVM on a dc range. How do your results compare with the figures in Table 4–1 (p. 47)?

LAB PROBLEM 4–2
Ayrton Shunt

Equipment

One meter movement; three variable resistors; one low-voltage power supply.

Object

To design and breadboard a multirange ammeter using an Ayrton shunt.

Procedure

Select a sensitive meter movement (50–100 μA full-scale deflection). Measure its internal resistance indirectly. Design a three-range ammeter using an Ayrton shunt. The design should allow for readings that are 2, 5, and 10 times the original full-scale reading of the meter. For example, if a 100-μA meter movement were used, it would be designed to read 200, 500, and 1000 μA full-scale. Breadboard your meter, test it, and have your lab instructor check it. It is understandable that exact results may not be obtained. Prepare a test circuit consisting of a 5-V source, your meter (R_m), and a decade box. By Ohm's law,

$$I = \frac{5\text{ V}}{(R_m + R_{decade})} = 200\ \mu\text{A},\ 500\ \mu\text{A},\ 1000\ \mu\text{A} \tag{4–82}$$

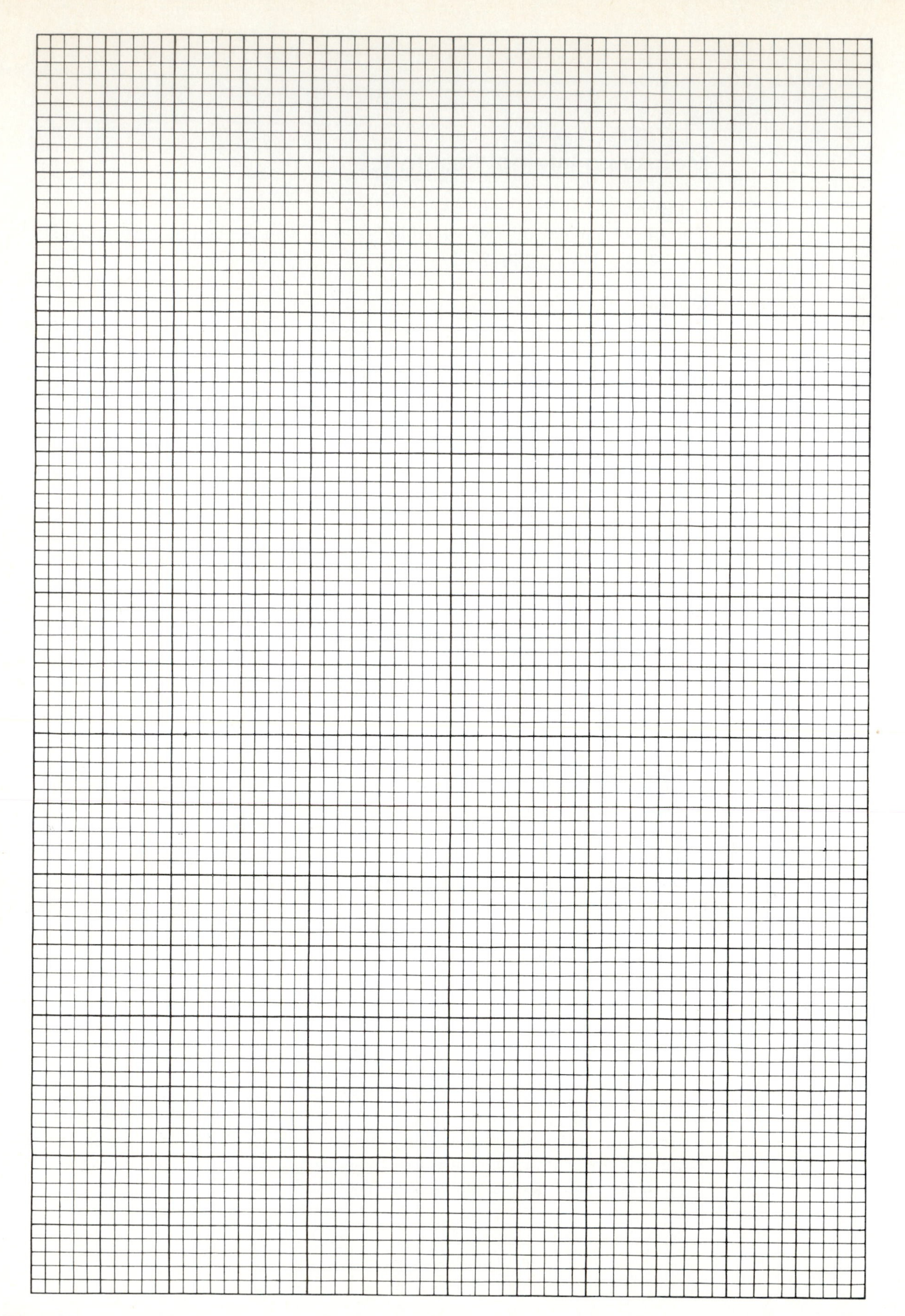

FIGURE 4–32
Complete the calibration graph.

to test the full-scale deflection of each setting of your circuit design. After checking your meter with an independent power supply circuit, graph the results on the calibration graph in Figure 4–32. Use the scale reading from your new meter for the horizontal axis and the current measured using a DMM for the vertical axis. Graph all three calibration curves together. If this is not possible, use one graph for each meter calibration.

PROGRAMMING PROBLEM 4–1 Two-Range Ayrton Shunt

Object

To write a BASIC program that will have as its input the internal resistance of the meter movement, the full-scale current rating of the movement, and the *two* current ranges you wish to have the meter read, and that will have as its output the two resistance values required to complete the meter design.

Procedure

Begin with a 100-μA meter movement with an internal resistance of 825 Ω. Design the Ayrton shunt that will extend the meter's range to read 300 μA and 1000 μA.

PROGRAMMING PROBLEM 4–2 Three-Range Ayrton Shunt

Object

To write a BASIC program that will have as its input the internal resistance of the meter movement, the full-scale current rating of the movement, and the *three* current ranges you wish to have the meter read, and that will have as its output the three resistance values required to complete the design.

Procedure

Begin with a 100-μA meter movement with an internal resistance of 825 Ω. Design the Ayrton shunt that will extend the meter's range to read 200, 500, and 1000 μA full-scale.

5 ac Analog Meters

OUTLINE

OBJECTIVES
REVIEW OF ac TERMINOLOGY
☐ Peak
☐ Peak-Peak
☐ rms
☐ Average (Effective or dc Equivalent)
RECTIFIERS
LINEARITY
LEAKAGE
PEAK DETECTOR
PEAK-PEAK DETECTOR
TRUE rms DETECTOR
SENSITIVITY
FREQUENCY CONSIDERATIONS
BALANCED-BRIDGE ELECTRONIC VOLTMETER
SUMMARY
EXERCISES
LAB PROBLEM 5–1—ac Detectors and Bandwidth Measurement

OBJECTIVES

After completing this chapter, you will be able to:

- ☐ Convert from peak to peak-peak to rms to average voltage and current values.
- ☐ Describe and use rectifiers.
- ☐ Explain the operation of ac voltmeters and ammeters.
- ☐ Discuss the circuitry used for peak and peak-peak detectors.
- ☐ Relate the sensitivity of an ac meter to loading problems.
- ☐ Describe the effect of frequency on an ac meter.
- ☐ Describe a true rms detector and use it in circuit applications.

Analog ac meters are basically dc meter movements with a rectified ac signal as an input. This chapter illustrates the conversion of a dc meter movement to an ac detector using discrete components. Most modern ac meters include a true rms converter and specialized large-scale integration (LSI) components to convert the analog input to a digital output. Since most field service technicians are involved with low current measurements, the material focuses on low-current ac coverage. Typical weaknesses of ac meters are also presented.

REVIEW OF ac TERMINOLOGY

Peak

Two measurements that relate to each other and appear in mathematical formulas are frequency and amplitude. *Frequency* is the rate of repetition of a signal and is the inverse of the period (time) of one complete cycle of the signal. *Amplitude* is the *peak* value of voltage (V_{pk}) or current (I_{pk}). It is the measurement from the zero or reference line to the highest point on the waveform (Figure 5–1). Amplitude is a difficult measurement to take with an oscilloscope but one that can be taken directly us-

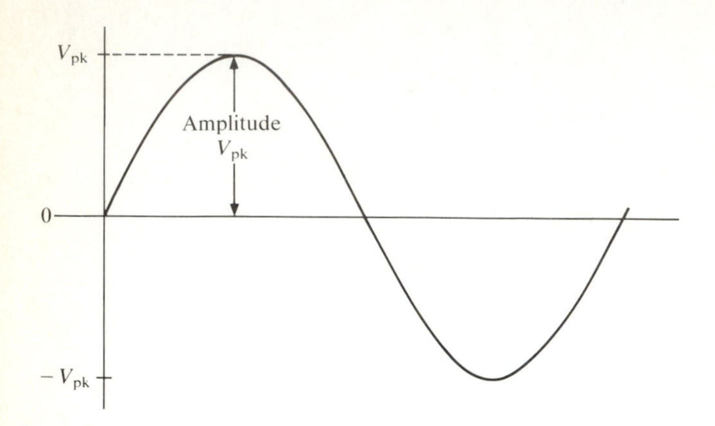

FIGURE 5–1
Peak measurement

ing an ac meter designed for that purpose. In mathematics, amplitude is an electrical parameter used in the analysis of many complex waveforms.

Peak-Peak

A quantity that is easier to measure with an oscilloscope is the *peak-peak* voltage or current (Figure 5–2). The peak-peak voltage will be abbreviated V_{pk-pk}, and the peak-peak current will be written I_{pk-pk}. The relationships between the amplitudes, or peak values, and the peak-peak values are

$$V_{pk-pk} = 2V_{pk} \tag{5–1}$$

and

$$I_{pk-pk} = 2I_{pk} \tag{5–2}$$

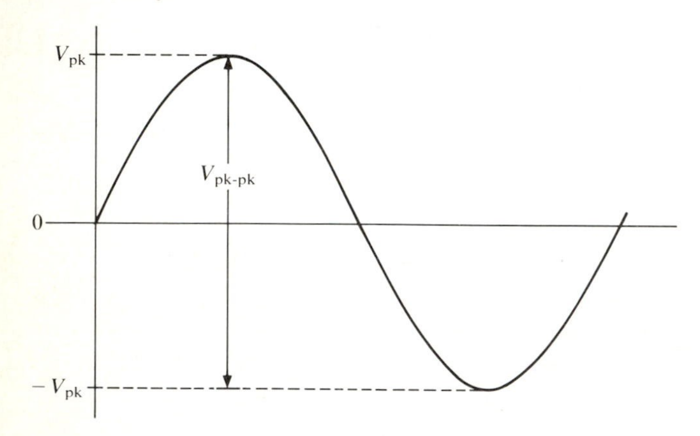

FIGURE 5–2
Peak-peak measurement

rms

A useful quantity is the *root mean square* (rms) value. This value is important because it is the most likely ac value used (or given). It represents the value of ac voltage or current that will dissipate the same amount of heat through a resistance as a numerically equal dc voltage or current. For example, 6 V rms will cause the same heat dissipation in a resistance as 6 V dc. $V_{rms} = 0.707V_{pk}$, as shown in Figure 5–3.

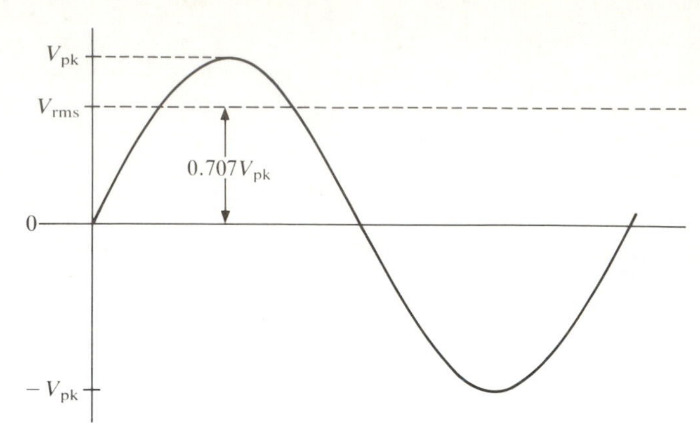

FIGURE 5–3
rms measurement (of a pure sinewave)

Average (Effective or dc Equivalent)

A dc meter movement responds to a rectified ac sinewave. It will indicate the *average,* or dc equivalent, of the signal if the frequency is high enough ($f > 30$ Hz). At lower frequencies, the needle will try to follow the changes in voltage or current level (Figure 5–4); therefore, the results are not very useful.

With half-wave rectification, the relationship between the peak value and the average value for a pure sinewave is

$$V_{avg} = 0.318V_{pk} \tag{5–3}$$

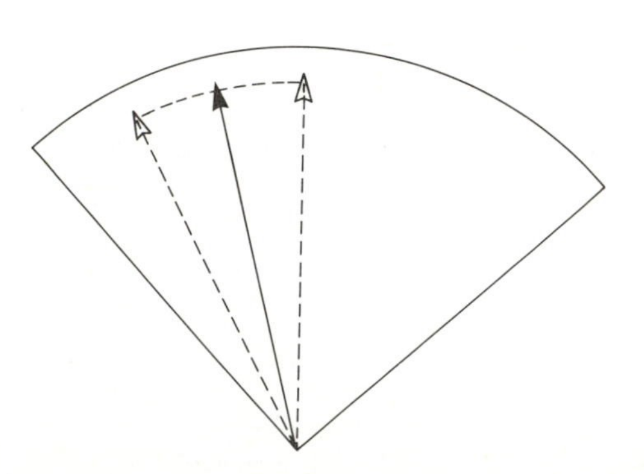

FIGURE 5–4
Reaction of a dc meter to a low-frequency half-wave rectified sinewave

If a full-wave rectifier is used to produce a dc from an ac input consisting of a pure sinewave, the average or effective voltage can be calculated as follows:

$$V_{avg} = 0.636V_{pk} \quad (5\text{–}4)$$

and the current calculation is

$$I_{avg} = 0.636I_{pk} \quad (5\text{–}5)$$

A useful relationship exists between rms voltage and current (usually the given input) and the dc equivalent that will drive the meter movement (V_{avg} or I_{avg}). Since

$$V_{rms} = 0.707V_{pk} \quad (5\text{–}6)$$

and

$$V_{avg} = 0.636V_{pk} \quad (5\text{–}7)$$

dividing equation (5–7) by equation (5–6) yields

$$\frac{V_{avg}}{V_{rms}} = \frac{0.636\ V_{pk}}{0.707\ V_{pk}} \quad (5\text{–}8)$$

Therefore, for full-wave rectification,

$$V_{avg} = 0.9V_{rms} \quad (5\text{–}9)$$

and for half-wave rectification,

$$V_{avg} = 0.45V_{rms} \quad (5\text{–}10)$$

Thus, for a given input voltage or current, the voltage or current measured with a dc meter movement will be a value given by these relationships. For example, a 10-V rms input to a full-wave rectifier will give a 9-V average output. The same 10-V rms input to a half-wave rectifier will give a 4.5-V average output. In a similar manner, a 50-mA rms input to a meter using a full-wave rectifier will yield a 45-mA average current output reading. The same 50-mA rms input current to a meter using a half-wave rectifier will yield a 22.5-mA average current output reading.

RECTIFIERS

The easiest way to produce a dc voltage or current from an ac input is to use a half-wave rectifier as shown in Figure 5–5. The forward-biased diode will usually have a dynamic resistance of less than 100 Ω. The meter must be protected whether it is to be used as a voltmeter (multiplier resistor) or as an ammeter (shunt).

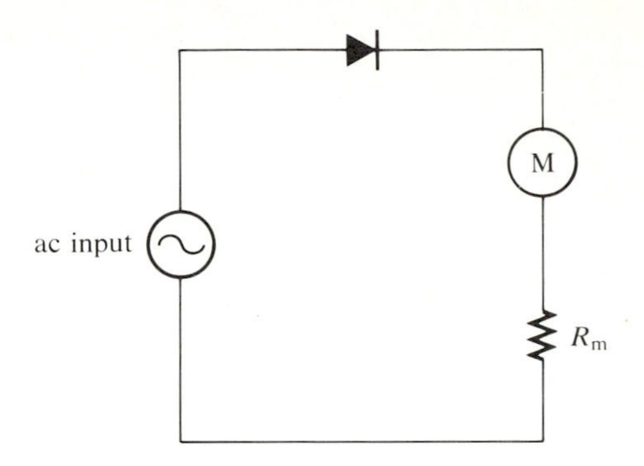

FIGURE 5–5
ac ammeter using a half-wave rectifier and a dc meter movement

EXAMPLE 5–1

A 1-mA meter movement with an internal resistance of 50 Ω is to be used with a 50-V ac peak voltage input (see Figure 5–6). The meter movement with a half-wave rectifier will respond to the average current—that is, a 1-mA full-scale deflection meter movement will read 1 mA = I_{avg}.

$$V_{avg,in} = 0.318V_{pk} = 0.318(50\text{ V}) = 15.9\text{ V (avg)} \quad (5\text{–}11)$$

Assume that the voltage drop across the diode is 0.7 V. The voltage across the meter resistance is

$$V_m = (1\text{ mA})(50\ \Omega) = 0.05\text{ V} \quad (5\text{–}12)$$

The voltage across the multiplier resistance is

$$V_{R_s} = V_{in} - V_{diode} - V_{meter} = 15.9\text{ V} - 0.05\text{ V} - 0.7\text{ V} = 15.15\text{ V} \quad (5\text{–}13)$$

Therefore

$$R_s = \frac{V_{R_s}}{I} = \frac{15.15\text{ V}}{1\text{ mA}} = 15.15\text{ k}\Omega \quad (5\text{–}14)$$

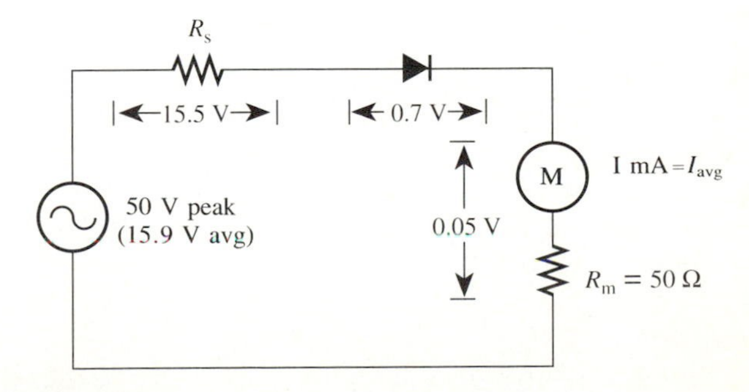

FIGURE 5–6
ac voltmeter using a half-wave rectifier and multiplier resistor

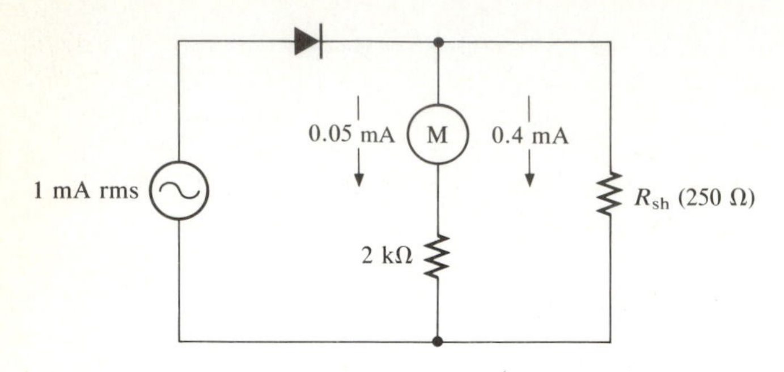

FIGURE 5–7
ac milliammeter using a current shunt

Because there might not be a 15.15-kΩ resistor available, a slightly larger resistor value would be acceptable. This value would not allow for full-scale deflection and would thus protect the meter movement.

EXAMPLE 5–2
If a 50-μA meter movement with 2kΩ internal resistance were used to measure 1 mA rms input current, the current that would reach the meter would be 0.45 mA. The meter could handle only 0.05 mA; the other 0.4 mA of current would have to be shunted around the meter. In this case,

$$V_m = V_{sh} \quad \textbf{(5–15)}$$

and

$$I_m R_m = I_{sh} R_{sh} \quad \textbf{(5–16)}$$

so that

$$R_{sh} = \frac{R_m I_m}{I_{sh}} \quad \textbf{(5–17)}$$

giving

$$R_{sh} = 2\ \text{k}\Omega\left(\frac{0.05\ \text{mA}}{0.4\ \text{mA}}\right) = 250\ \Omega \quad \textbf{(5–18)}$$

(see Figure 5–7).

EXAMPLE 5–3
Compare the sensitivity of a 50-μA, 2-kΩ meter movement when used (a) in a dc meter, (b) in a half-wave rectified ac meter, and (c) in a full-wave rectified ac meter.

Solution:
a. The sensitivity of a dc meter is

$$S = \frac{1}{I_{fsd}} = \frac{1}{50\ \mu\text{A}} = 20\ \text{k}\Omega/\text{V} \quad \textbf{(5–19)}$$

b. For a half-wave rectified meter,

$$S = \frac{0.45}{I_{fsd}} = \frac{0.45}{50\ \mu\text{A}} = 9\ \text{k}\Omega/\text{V} \quad \textbf{(5–20)}$$

c. For a full-wave rectified meter,

$$S = \frac{0.9}{I_{fsd}} = \frac{0.9}{50\ \mu\text{A}} = 18\ \text{k}\Omega/\text{V} \quad \textbf{(5–21)}$$

LINEARITY

Two problems remain to be solved. The first is caused by the nonlinear characteristic of the diodes at low voltage values (Figure 5–8). Nonlinear characteristics could result in an inaccurate reading at low scale readings. The easiest way to minimize nonlinearity is to place a *bleeder* resistor in parallel with the meter movement. The value of this resistance is usually equal to the meter resistance (Figure 5–9). The resistor causes diode D_1 to conduct

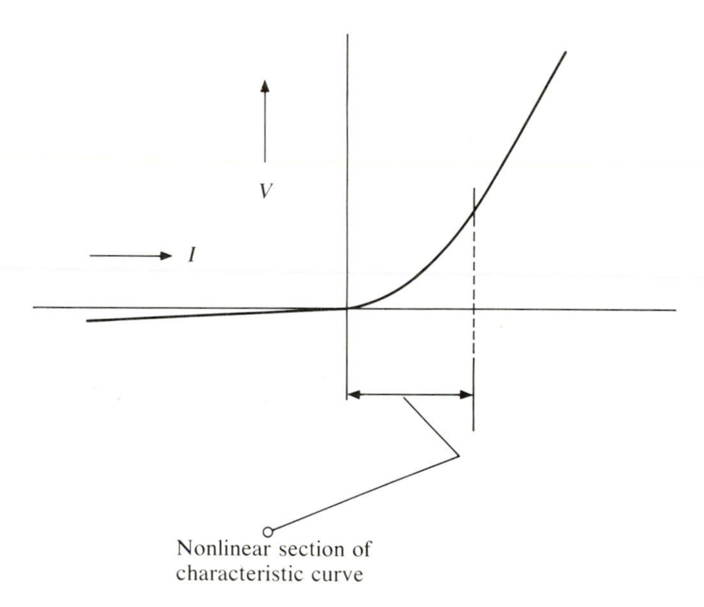

FIGURE 5–8
Nonlinear section of the characteristic curve

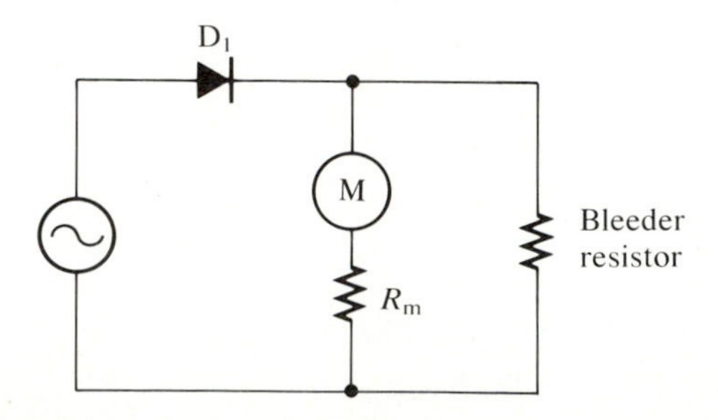

FIGURE 5–9
Application of a bleeder resistor

more quickly, moving it out of the nonlinear region rapidly.

LEAKAGE

The second problem may arise if a diode has significant leakage resistance in the reverse-biased condition. Reverse leakage could lead to an error in the meter indication (Figure 5–10). The simplest way to prevent this error is to use an extra diode as shown in Figure 5–11.

During the first half of a sinewave signal, diode D_1 will be forward biased and diode D_2 will be reverse biased. During the second half of the input signal, D_1 will be reverse biased and D_2 will be forward biased. During the second part of the cycle, when the rectifier D_1 is reverse biased, the protector diode (D_2) will shunt any leakage current away from the rectifier (and meter movement), eliminating the problem.

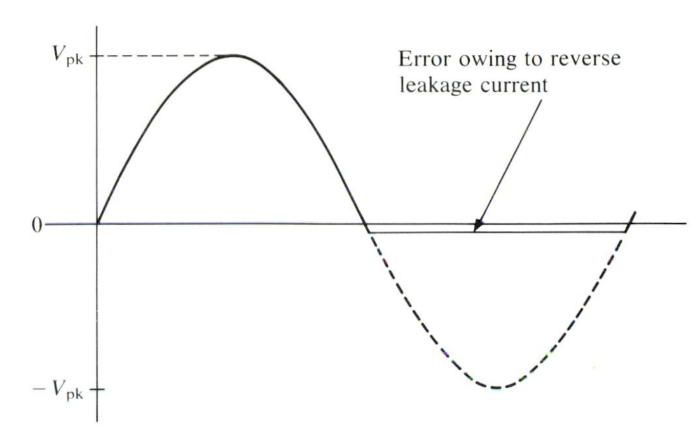

FIGURE 5–10
Results of reverse leakage current

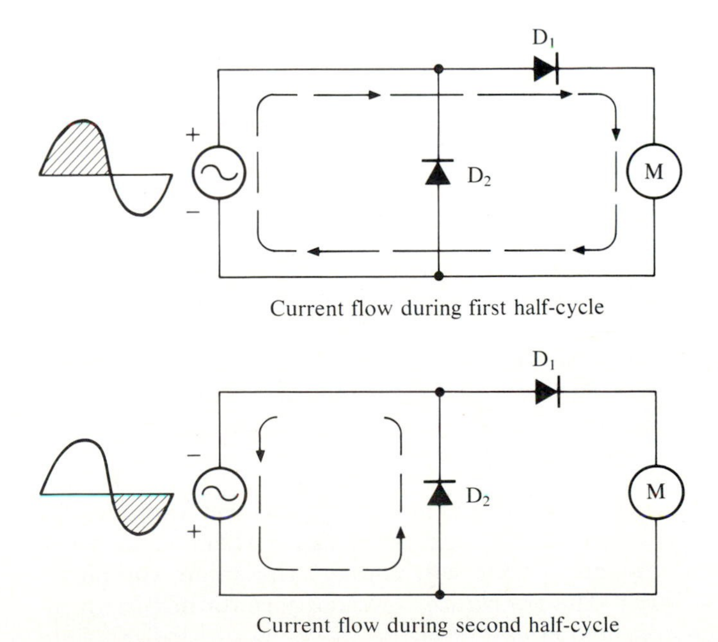

FIGURE 5–11
Correction of reverse leakage problem

PEAK DETECTOR

A detector indicates the value being measured in a circuit. A peak detector can be produced by placing a capacitor in parallel with the meter movement (Figure 5–12A). The capacitor will charge through the low resistance of the forward-biased diode(s) and discharge through the parallel combination of the resistance of the reverse-biased diode(s) and the load. This combination usually produces a longer time constant and gives results similar to those shown in Figure 5–12B.

If we consider an approximation that the voltage discharges from $V_{initial}$ to V_{final} in one cycle, and

$$f = \frac{1}{T} \tag{5–22}$$

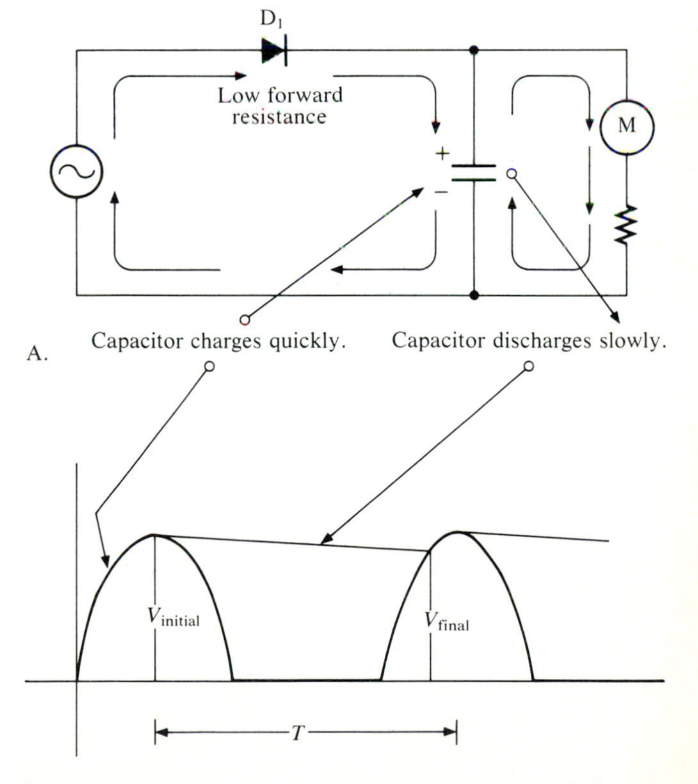

FIGURE 5–12
A. A peak detector circuit. B. Waveform resulting from a peak detector circuit.

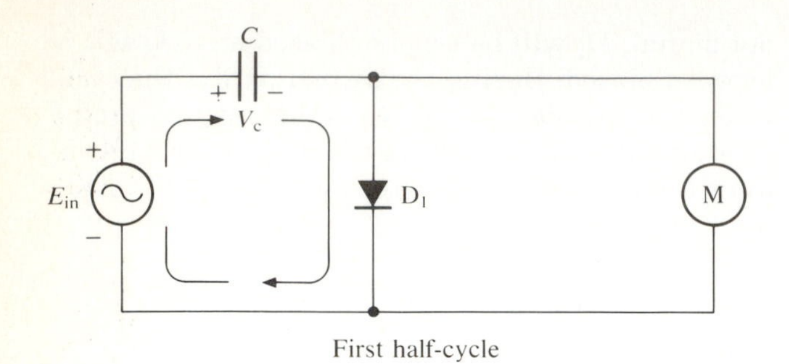

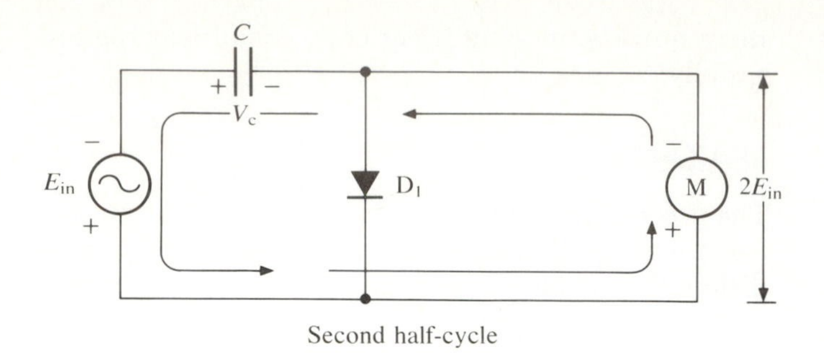

FIGURE 5–13
Voltage doubler as a peak-peak detector

and

$$V_{\text{final}} = V_{\text{initial}} e^{-t/RC} \quad (5\text{–}23)$$

we can use the following approximation:

$$e^x = 1 + \frac{x}{1!} + \frac{x^2}{2!} + \frac{x^3}{3!} + \ldots \quad (5\text{–}24)$$

which is called the Maclaurin expansion. With $x = 0.1$, we have

$$\begin{aligned} e^{0.1} &= 1 + \frac{0.1}{1} + \frac{0.1^2}{2 \cdot 1} + \frac{0.1^3}{3 \cdot 2 \cdot 1} + \ldots \\ &= 1 + 0.1 + 0.005 + 0.000167 \quad (5\text{–}25) \\ &\quad + \ldots = 1.105167 \end{aligned}$$

The calculated or tabular value of $e^{0.1}$ is 1.1051709 (enter 0.1 on a calculator and press the button "e^x" to check this value).

The approximation works well if $x < 1$. It works even better if $x \ll 1$, in which case

$$e^x = 1 + x \quad (5\text{–}26)$$

This leads to the conclusion that

$$V_{\text{final}} = V_{\text{initial}}\left(1 - \frac{T}{RC}\right) \quad (5\text{–}27)$$

if $-T/RC$ is the exponent of e.

If the percentage of error in the decay of voltage from peak is considered to be $V_{\text{final}}/V_{\text{initial}}$, then

$$\frac{V_{\text{final}}}{V_{\text{initial}}} = 1 - \frac{T}{RC} = \text{accuracy} \quad (5\text{–}28)$$

If 99% accuracy is desired,

$$\begin{aligned} 0.99 &= 1 - \frac{T}{RC} \quad (5\text{–}29) \\ \frac{T}{RC} &= 0.01 \end{aligned}$$

and

$$T = \frac{1}{f} \quad (5\text{–}30)$$

which gives

$$\frac{1}{fRC} = 0.01 \quad (5\text{–}31)$$

or

$$\frac{1}{RC} = 0.01f \quad (5\text{–}32)$$

If the peak detector to be designed is to yield 99% accuracy, the relationship between the meter resistance and the shunt capacitor is

$$RC = \frac{100}{f} \quad (5\text{–}33)$$

PEAK-PEAK DETECTOR

A peak-peak detector is simply a voltage doubler (Figure 5–13). During the first half of the cycle, diode D_1 is forward biased, which allows capacitor C to be charged quickly through the small resistance of the forward-biased diode. During the second half of the cycle, the diode is reverse biased and the total voltage ($E_{in} + V_c$) is placed across the meter movement. This arrangement yields a peak-peak voltage as shown in Figure 5–14. After any changes that affect circuit operation, the face of the meter movement must be calibrated to reflect the input voltage or current values.

TRUE rms DETECTOR

Since rms values relate to the power content of a waveform, one way to measure them is to use a transducer that will convert the input voltage to heat. The transducer is usually contained in an integrated circuit.

The input is an ac voltage, and the output is a true rms voltage that is an equivalent dc voltage.

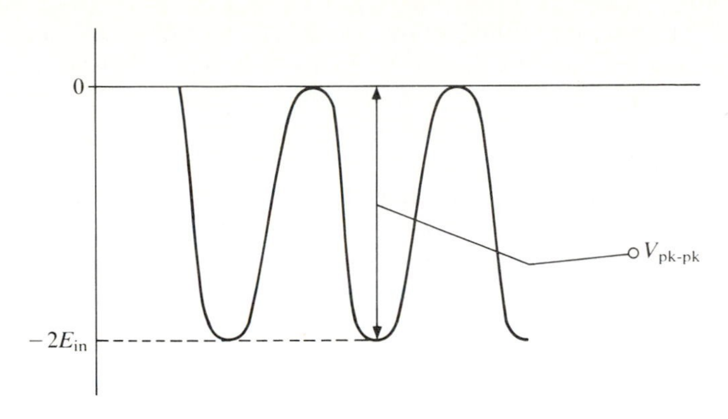

FIGURE 5–14
Waveform resulting from a peak-peak detector

The transducer converts the heat energy developed by the ac voltage to a dc voltage which can then be detected and measured by the dc meter movement.

In summary, V_{pk} is the quantity referred to as the amplitude (A) in the mathematical expression

$$v = A \sin(2\pi ft) \quad \textbf{(5–34)}$$

where v is the instantaneous voltage amplitude, f is the frequency, and t is the time. Its value is used in mathematical calculations and derivations. Quantity $V_{pk\text{-}pk}$ is most easily measured using an oscilloscope. Quanity V_{avg} is the result of the effect of a rectifier on a signal, and V_{rms} is the equivalency between the power dissipation of an ac and a dc signal level. The V_{rms} is the value most often indicated by digital multimeters.

The equations for the these quantities are as follows:

$$V_{pk} = \text{amplitude} \quad \textbf{(5–35)}$$

$$V_{pk\text{-}pk} = 2V_{pk} \quad \textbf{(5–36)}$$

$$V_{avg} = 0.318V_{pk} \text{ for a half-wave rectified sinewave} \quad \textbf{(5–37)}$$

$$V_{avg} = 0.636V_{pk} \text{ for a full-wave rectified sinewave} \quad \textbf{(5–38)}$$

$$V_{rms} = 0.707V_{pk} \text{ for a pure sinewave} \quad \textbf{(5–39)}$$

Since many signals used in electronics are nonsinusoidal in nature, it is important to remember that most common laboratory test equipment is calibrated for pure sinewaves unless specifically designed for other uses.

SENSITIVITY

The sensitivity of ac meters is considerably less than that of dc meter movements. It can be calculated by dividing the total input resistance of the meter by the full-scale voltage. But a quicker method can be used. For a half-wave rectifier meter,

$$S = \frac{0.45}{I_{fs}} \quad \textbf{(5–40)}$$

and for a full-wave rectifier,

$$S = \frac{0.9}{I_{fs}} \quad \textbf{(5–41)}$$

Remember that the sensitivity for a dc meter is

$$S = \frac{1}{I_{fs}} \quad \textbf{(5–42)}$$

FREQUENCY CONSIDERATIONS

A meter movement contains coils of wire that cause problems when ac measurements are taken. The coils represent an inductive reactance to the ac; in addition, a *stray capacitance* occurs between the windings (Figure 5–15).

If an ac component is allowed to reach the meter movement, the inductive reactance of the meter coil will increase linearly with the frequency. The capacitance between adjacent windings decreases in proportion to frequency, providing a low-impedance path that bypasses the meter coil.

These problems cause a limited frequency response in ac measurements (Figure 5–16). The bandwidth for a good ac meter may be very limited. Frequency response is typically between 50 Hz and 4500 Hz. This range is surprisingly small. The technician should be aware of the limitations of

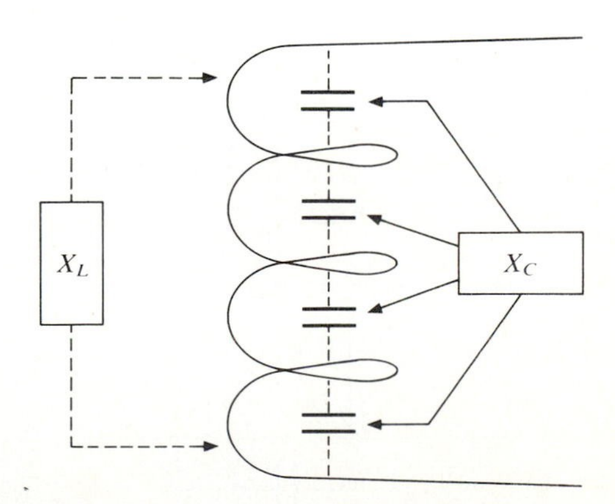

FIGURE 5–15
Effects of inductive reactance and interwinding capacitance on the frequency response

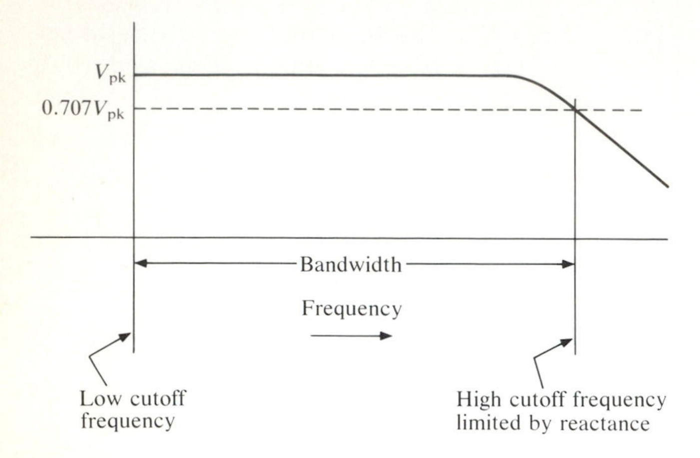

FIGURE 5–16
Frequency response curve showing bandwidth limits

standard meters. When higher frequencies are to be encountered, meters designed to take these measurements should be used.

Another possible reason for the comparatively low frequency response of an ac meter can be found by examining the diodes used to rectify the input. These diodes allow current to flow in the forward-biased direction but not in the opposite direction. When the diode is reverse biased, a depletion zone exists between the P-type material and the N-type material (Figure 5–17). This is a nonconducting zone between two conductors, which is the structure of a capacitor. The capacitance is very small (probably 5 pF). At 1 kHz, the reactance will be

$$X_C = \frac{1}{2\pi fC} = \frac{1}{2\pi(10^3)(5 \times 10^{-12})} = 31.8\ \text{M}\Omega \quad \textbf{(5–43)}$$

At 1 MHz, this drops to

$$X_C = \frac{1}{(2\pi)(10^6)(5 \times 10^{-12})} = 31.8\ \text{k}\Omega \quad \textbf{(5–44)}$$

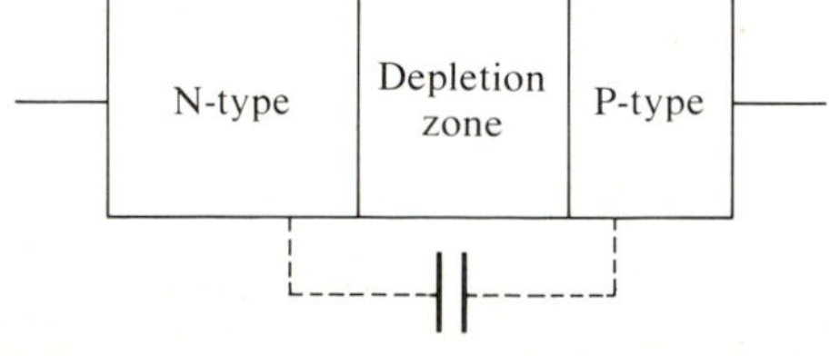

FIGURE 5–17
Leakage capacitance at a reverse-biased P-N junction

Stray capacitance allows leakage of the ac in the reverse direction, lowering the average current or voltage indicated on the meter movement.

In summary, ac meters can be designed to measure all common ac parameters. Average, peak, peak-peak, and rms currents and voltages can be measured using a dc meter movement, rectifiers, and simple electrical components. The greatest weaknesses of the meters designed using the techniques presented in this chapter are the nonlinearity of the low-scale readings and the relatively poor frequency response. Meters especially designed to read ac currents and voltages with higher frequencies are available. These designs use more sophisticated circuitry than that discussed here. The technician choosing an ac meter must know the range of application of the meter. A proper choice can be made by studying the specifications published by the manufacturer.

BALANCED-BRIDGE ELECTRONIC VOLTMETER

Figure 5–18 illustrates one type of electronic voltmeter. In this circuit, FET 1 and FET 2 form the upper half of a bridge circuit. Resistances R_2 and R_3, plus a fraction of R_1, form the lower half of the bridge circuit. The meter is an analog instrument activated by an unbalanced bridge condition. Since devices are never "exactly the same" when power is applied, either FET 1 or FET 2 will conduct a little more than the other device. To bring the bridge into balance when there is no voltage applied to the input, R_1 may be adjusted. This adjustment will cause the meter to read zero (balanced bridge condition).

When an unknown dc voltage causes the dynamic resistance of FET 1 to lower, the bridge becomes unbalanced and the meter deflection is an indication of the unknown voltage. If the unknown dc voltage causes the dynamic resistance of FET 1 to increase, the unbalanced bridge condition will tend to cause the meter movement to deflect in the other direction. A changeover switch is usually provided to reverse this condition so that the meter movement deflects only in the forward direction, allowing the voltmeter to measure unknown voltages of either polarity.

Resistance R_4 and capacitor C form a filter to remove any stray fluctuations picked up by the leads. If the instrument is to be used for ac voltage readings, the filter is removed and the necessary rectifiers and filters are applied to the meter move-

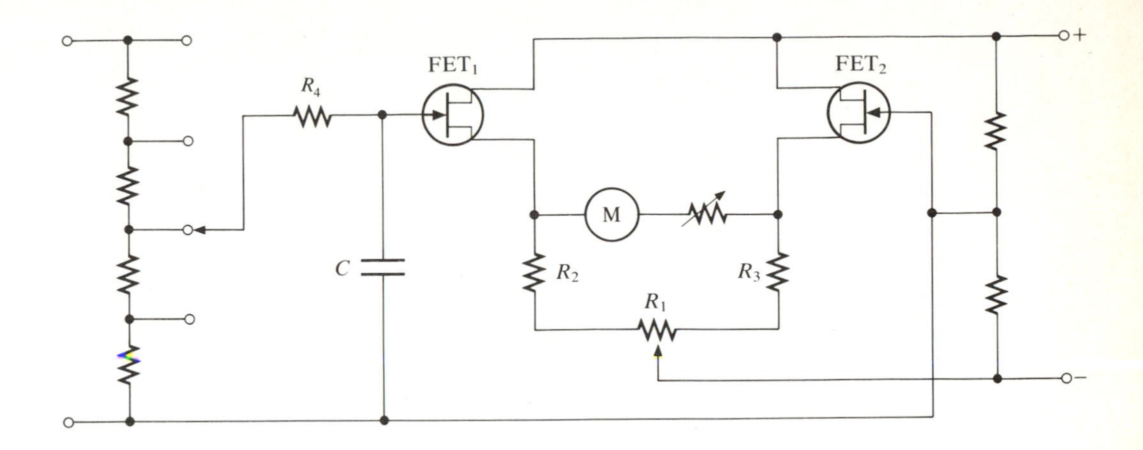

FIGURE 5–18
Balanced-bridge voltmeter

ment. A selection switch would be used in the meter portion of the circuit.

Older versions of the balanced-bridge electronic voltmeter circuit employed vacuum tubes. Advantages of solid-state circuitry include lower power, which can be provided by batteries, and the more rugged nature of solid-state devices. Solid-state instruments can be used in field service situations where line power may not be available and where handling of the instruments may be less gentle than on the technician's bench.

SUMMARY

Ac analog meters are dc meter movements with rectifiers, bleeders, and some filtering if desired. All shunting and multiplier resistance circuits used with dc meter movements are the same as if the dc meter were to be used to measure dc. The principles of conversion of a dc meter movement to read ac may have useful applications in other measurement situations.

EXERCISES

1. An ac detector is to be designed using a half-wave rectifier with a diode whose forward resistance is 100 Ω. If the meter movement is 50 μA with internal resistance of 2 kΩ, what value of multiplier resistance must be used for a 10-V rms input voltage?
2. Repeat Exercise 1 using a full-wave bridge rectifier. The forward resistance of the diodes is 100 Ω, and the reverse resistance is infinite.
3. If a meter movement has 10 kΩ internal resistance and the voltage of a 20-kHz signal is to be measured, what value of capacitance must be used to make the meter movement a peak detector yielding 95% accuracy? 99% accuracy?
4. Why is a bleeder resistor used in designing an ac meter?
5. Why is a second diode used in designing a half-wave rectifier ac detector?
6. If a 1-mA, 50-Ω meter movement is to be used in the design of an ac detector using a half-wave rectifier, what will be the sensitivity of the instrument?
7. If a 50-μA, 2-kΩ meter movement is to be used in the design of an ac detector using a full-wave rectifier, what will be the sensitivity of the instrument in an ac application? If this meter movement were to be used in the design of a dc meter, what would the sensitivity of the instrument be?
8. Why is the frequency response of an ac meter usually quite low?
9. Convert 15 V rms to volts peak, volts peak-peak, and volts average.

10. What is one advantage of taking a voltage reading instead of a current reading if you have a choice?

LAB PROBLEM 5–1
ac Detectors and Bandwidth Measurement

Equipment

One meter movement; four diodes; one function generator; one filter capacitor; one lab standard oscilloscope

Object

To design, build, and calibrate an ac detector, and to measure its bandwidth or frequency response.

Procedure

Select a dc meter movement, preferably one rated less than or equal to 1 mA full-scale deflection. Measure the internal resistance of the meter movement (see Chapter 4, Lab Problem 4–1). Use a full-wave bridge rectifier (either discrete components or a chip will do).

Calculate the multiplier resistance needed to protect the meter if a peak voltage of 3 V is to be used. Use this resistance (or a bigger one to get started) and wire an ac detector. Use a variable-frequency sinewave generator as the input ac voltage source. Since your meter will yield the average or effective dc value, vary the input voltage from zero to 3 V peak and graph the meter indication versus input voltage.

Insert a capacitor to make your meter a peak detector. Use the calculations in the text to ensure 99% accuracy.

Vary the input voltage from zero to 3 V peak and graph the meter indication versus input voltage. Compare these results with those obtained using a larger value of capacitance and a lower value of capacitance.

With the input voltage set at 3 V peak, vary the frequency of the generator from a very low frequency (about 30 Hz) to such a frequency that the indication shown by your meter "falls off" to lower than 0.707×3 V. It is important that you check the output of the frequency source each time *before* you wire your meter to the source. Not only does the meter load down the generator, but the generator loads down the meter. Use a lab standard oscilloscope to check the output of the ac generator.

Design a 6-V peak-peak detector using the voltage doubler technique illustrated in the text. Vary the input voltage from zero to 3 V peak and graph the meter indication versus input voltage.

If you are using a function generator with square-wave and triangular-wave options, compare the reaction of your meter, which is calibrated for a sinewave input, to a square-wave or triangular-wave input. Evaluate the results.

6 Digital Multimeters

OUTLINE

OBJECTIVES

HISTORY OF DIGITAL MULTIMETERS
- ☐ Tube-Driven Models
- ☐ Transistorized Models
- ☐ Integrated Circuit Models
- ☐ Large-Scale Integration Models

CRITERIA FOR SELECTING A DMM

TYPES OF A/D CONVERSION
- ☐ Ramp or Counter Type
- ☐ Successive Approximation Type
- ☐ Dual-Slope/Integrator Method

A TYPICAL DMM (THE FLUKE 8060A)
- ☐ Functional Description
- ☐ The Microcomputer
- ☐ The Measurement Acquisition Chip
- ☐ A/D Conversion Cycle
- ☐ Voltage Measurement
- ☐ Current Measurement
- ☐ Resistance Measurement
- ☐ Conductance Measurement
- ☐ Continuity Indication
- ☐ Frequency Measurement

MULTIMETER SAFETY

FREQUENCY CONSIDERATIONS

SUMMARY

EXERCISES

LAB PROBLEM 6–1—
Digital Multimeters

OBJECTIVES

After completing this chapter, you will be able to:

- ☐ Distinguish between analog and digital multimeters (DMMs).
- ☐ Discuss the historical development of DMMs.
- ☐ List the criteria for choosing a DMM.
- ☐ Describe the different types of A/D converters.
- ☐ Discuss the major characteristics of DMM operation.
- ☐ List the safety procedures to follow when using a DMM.
- ☐ Discuss the strengths and weaknesses of DMMs.
- ☐ Apply the field maintenance procedures for analyzing problems and repairing DMMs.

Digital multimeters (DMMs) have evolved from the development of logic techniques and the technical need for and subsequent development of fast, accurate, and inexpensive components and circuits. Virtually all DMMs operate the same way. An analog voltage or current input is produced which is input to an analog-to-digital (A/D) converter. The A/D converter digitizes the input, and the results are decoded and displayed on a digital readout device (LED or LCD). With the introduction of the microcomputer to the meter, internal control and monitoring of processes became easy. Most DMMs are automatically zeroed many times each second to ensure accuracy. The introduction of CMOS hardware has led to input resistance capabilities that virtually eliminate meter loading. Input resistances of greater than 10,000 MΩ with a nominal value of 80,000 MΩ on 200-mV to 20-V scales are available at reasonable cost. The ohmmeter function remains the same, with the meter providing a current through the unknown resistance and the resulting voltage compared with a known voltage across a known resistance. The results are then fed through the same process, and the results are displayed on the output. An additional capability of the DMM of assistance to the technician is the built-in software diagnostic self-tests that aid in troubleshooting the device.

HISTORY OF DIGITAL MULTIMETERS

Tube-Driven Models

Early models of voltmeters with digital readouts featured nixie tubes for the readout elements and tubes for the A/D conversion and the logic electronics involved. These instruments were very heavy, very large, and very expensive. They had metal cases and were somewhat limited in their range of functions. These devices required a 120-V line voltage, which limited their portability. Their reset time between readings was slow, but if enough money was spent on their construction, reasonable accuracy was attainable.

Transistorized Models

In the 1960s, transistorized models became popular. With the introduction of the LED and newer sampling techniques, DMM use became widespread in industry. The size, weight, cost, and power requirements were reduced so that handheld battery-operated models could be obtained at reasonable prices, making DMMs available to the independent technician.

Integrated Circuit Models

Integrated circuitry made it possible to develop meters with very high input impedances and greater reliability, at a lower cost with a wider range of functions. The introduction of the LCD further reduced power requirements. It became possible to purchase a DMM for under $150.

Large-Scale Integration Models

Today's DMMs are controlled by custom LSI chips and may consist of one specialized LSI chip to process the input and a microprocessor to control the functions and the output along with some external components. The input impedance can be greater than 10,000 MΩ for dc ranges up to 20 V, and some DMMs can be purchased for under $50.

CRITERIA FOR SELECTING A DMM

When selecting a DMM, the technician should consider the following items.

1. *Precision.* Is a 3½, 4½, or greater digit readout needed? The ½ in the 3½ and 4½ refers to the high-order digit, a 0 or a 1 (one binary bit), which doubles the precision.
2. *Packaging.* Would a bench or a handheld model be more appropriate?
3. *Construction.* Will the instrument be used in a hostile environment? If so, look for a hard plastic case (e.g., NORYL, which is used in making football helmets).
4. *Protection.* Can the instrument be overloaded? Can an incorrect function be used without damaging the instrument?
5. *Warranty.* Warranties are usually available for periods from 90 days to three years. Of course the longer the warranty, the more expensive the instrument will be.

A.

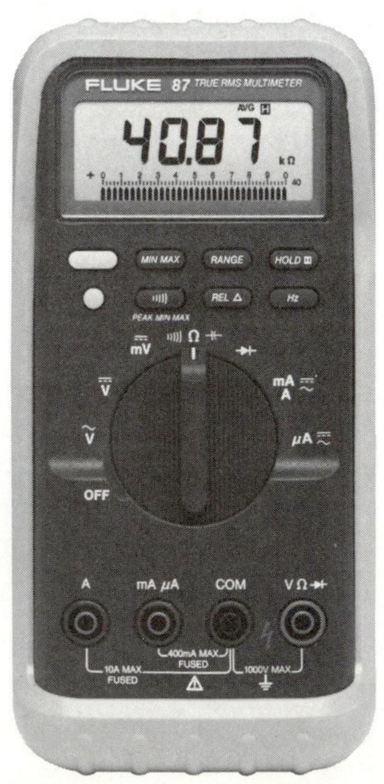

B.

FIGURE 6–1
A. Simpson 474 DMM. B. Fluke 87 DMM.

6. *Calibration guarantee.* Guarantees also correspond to instrument prices.
7. *Special functions.* Are frequency measurement or temperature measurement capabilities needed? Is a continuity test or an analog output important?
8. *Digital versus analog.* Many DMMs now provide an analog output in the form of a rapidly responding bar graph function. It is sometimes easier and more useful to see the differences between readings as shown by an analog meter rather than using a DMM. Figure 6–1A depicts a Model 474 Simpson DMM, and Figure 6–1B shows a Fluke Model 87 DMM, which provides both analog and digital outputs.

TYPES OF A/D CONVERSION

Ramp or Counter Type

Figure 6–2 shows the block diagram for a ramp- or counter-type A/D converter. An analog input is applied to one input terminal of the comparator. A start signal clears the binary counter and causes the ramp generator to begin its excursion from a reference point. At the same time, the R-S flip-flop is set. With an enabling signal appearing at one input to the AND gate, the clock pulses are gated through to the binary counter. When the voltage level of the ramp generator equals the analog voltage input, the output of the comparator changes and resets the R-S flip-flop, causing the output voltage level to change and thus turn off the AND gate. The binary counter has counted a number of clock pulses proportional to the analog input voltage level. This number is latched and then displayed on the digital output devices.

Successive Approximation Type

Figure 6–3 shows the block diagram for a successive approximation–type A/D converter. The converter works on the *comparison* principle, also called the *successive approximation* method. In successive approximation, an input analog voltage

FIGURE 6–2
Ramp or counter type A/D converter

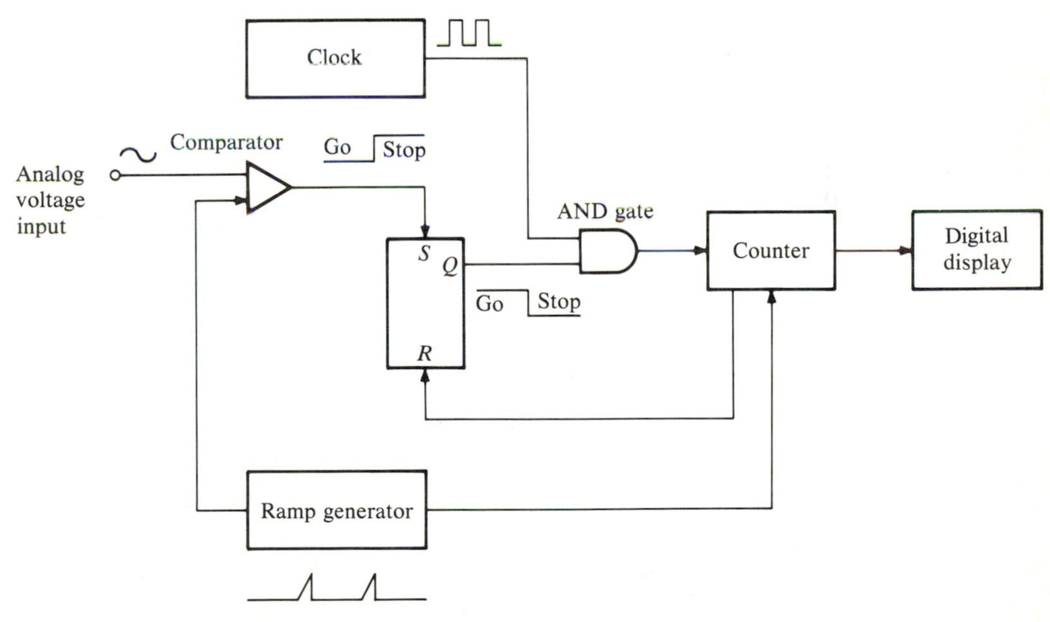

FIGURE 6–3
Block diagram of a successive approximation-type A/D converter

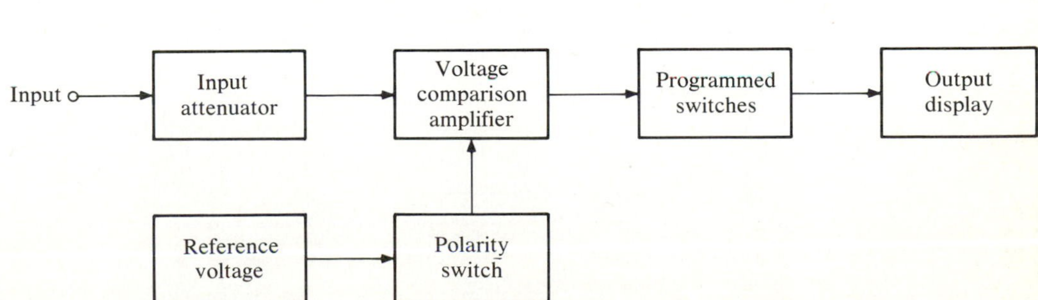

(V_{in}) is attenuated and compared to a reference voltage (V_{ref}). If V_{in} is smaller than V_{ref}, it is compared to $\frac{1}{2}V_{ref}$. If V_{in} is larger than this new reference (now V_{ref}), it is incremented by one-half its own value. If this new value ($V_{ref} + \frac{1}{2}V_{ref}$) is greater than V_{in}, the increment is discarded and one-half the discarded increment is added to V_{ref}. Voltage V_{ref} is thus incremented and compared to V_{in} until reasonable accuracy is achieved. The result is then displayed.

EXAMPLE 6–1

Assume that the input analog voltage is 6.75 V and that the reference voltage is 8 V. Table 6–1 shows the successive approximation method used to reach a result, which is displayed.

Dual-Slope/Integrator Method

In Figure 6–4, the switches are analog gates operated by the digital section of the A/D converter. Timing is usually a function of either an external crystal-controlled oscillator or a clock internal to one of the chips. The first of three steps is the autozero (AZ) period (Figure 6–5).

During this period, a ground reference voltage is applied to the input of the A/D converter. Ideally the output of the comparator would go to zero, but input offset voltage errors accumulate in the amplifier loop and appear at the comparator output as an error voltage. The error voltage is used to charge the autozero capacitor, which will be used as an offset correction voltage during the integrate and read periods.

At the end of the autozero period, the autozero switches are opened and the integrate switch is closed. The voltage is buffered and passed on to the integrator to determine the rate at which the integrator capacitor is charged. The capacitor charges to a level proportional to the input voltage.

During the read period, the integrate switch is opened and the read switch is closed. A known reference voltage is applied to the input. This volt-

TABLE 6–1
Example of the successive approximation method

Comparison	V_{in}	V_{ref}	Comparison	Result	Sum
1	6.75 V	8 V	6.75 V < 8 V	Discard	0 V
2	6.75 V	0 + ½(8 V)	6.75 V ≥ 4 V	Keep	4 V
3	6.75 V	4 V + ½(4 V)	6.75 V ≥ 6 V	Keep	6 V
4	6.75 V	6 V + ½(2 V)	6.75 V < 7 V	Discard	6 V
5	6.75 V	6 V + ½(1 V)	6.75 V ≥ 6.5 V	Keep	6.5 V
6	6.75 V	6.5 V + ½(0.5 V)	6.75 V = 6.75 V	Keep	6.75 V (display)

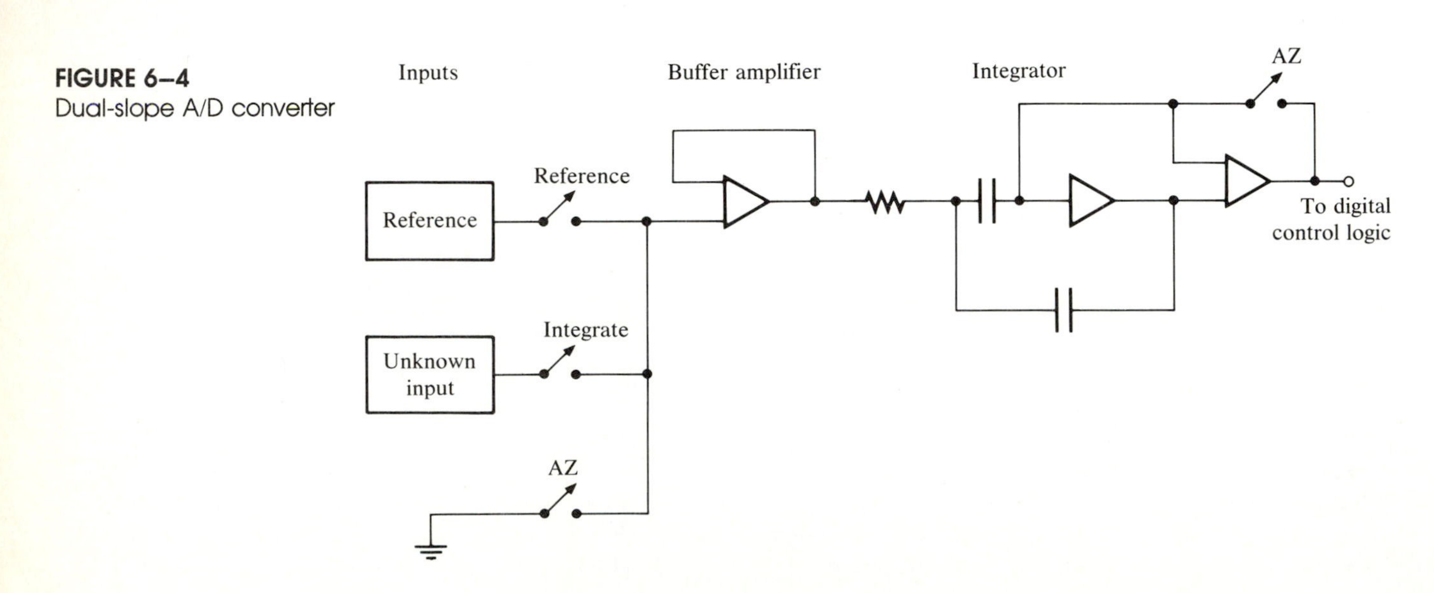

FIGURE 6–4
Dual-slope A/D converter

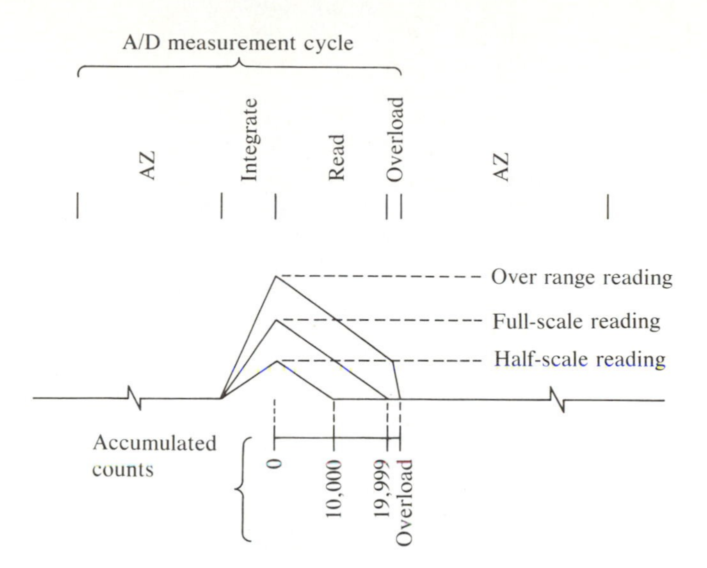

FIGURE 6–5
Timing cycle for a dual-slope A/D converter

age is switched so as to be opposite in polarity to the charge on the integrator capacitor. The autozero capacitor discharges the integrator capacitor at a fixed rate to the autozero level (which included the offset correction voltage). A series of clock pulses is counted, starting at the beginning of the read period and finishing at the end of the read period. The number of pulses is proportional to the level of the input voltage. This information is decoded and used to drive an LED display.

A TYPICAL DMM (THE FLUKE 8060A)

Functional Description

The following is a description of the theory of operation of the Fluke 8060A DMM (Figure 6–6A). This instrument was selected because it is an industrial standard and its operation is representative of many common DMMs.

The major components of the measurement system are two CMOS integrated circuits (Figure 6–6B). One is a 4-bit microcomputer, and the other is a CMOS IC known as the *measurement acquisition chip* (MAC). The microcomputer selects the appropriate measurement function in the MAC according to the switches or buttons pushed by the operator. It also controls the measurement cycles, performs calculations on measured data, and drives the display. The MAC measures the conditioned input signals with the A/D converter or frequency counter and controls the power supply and the continuity tone generator. Communication between the MAC and the microcomputer is through a 4-bit bidirectional bus and four control lines.

The input signals are routed by the range and function switches through the appropriate signal conditioners for input filtering and scaling (see Figure 6–7). Input signals for all measurement functions except frequency are converted to a proportional dc analog voltage that is applied to the A/D converter. The dual-slope A/D converter converts the dc analog voltage to a digital number that is sent to the microcomputer. Input signals for frequency measurement are ac voltages buffered by the ac converter and applied to the frequency counter in the MAC. The frequency counter supplies the digital number to the microcomputer.

The Microcomputer

The 4-bit CMOS microcomputer senses switch positions by reading status registers in the MAC, and senses buttons (input switches) that have been pushed through input lines connected directly to the microcomputer. The microcomputer processes the information and then selects the appropriate digital and analog configurations in the MAC by writing into an array of MAC control registers.

The operation of the instrument is controlled by software routines stored in the microcomputer memory. The routines include the normal turn-on routine as well as self-test routines. When the instrument is first turned on, the microcomputer performs the self-test routine that checks the LCD segments and the interface to the MAC. While the LCD segments are on, the microcomputer exercises the bus and checks the internal registers in the MAC to make sure it has control over them. If the microcomputer detects a problem with the MAC interface, it stays in the self-test routine with the LCD segments on until the error is corrected or the instrument is turned off.

If this self-test routine shows no errors and no other self-test options have been selected by the operator, the microcomputer begins the normal operating routine, which consists of four steps:

1. The microcomputer reads the function and range selections and checks the four pushbuttons to determine the mode the operator has selected. It then selects either the A/D converter (for measurement of voltage, current, resistance, conductance, continuity) and diode test or the frequency counter.

FIGURE 6–6
A. Fluke 8060A multimeter. B. Block diagram.

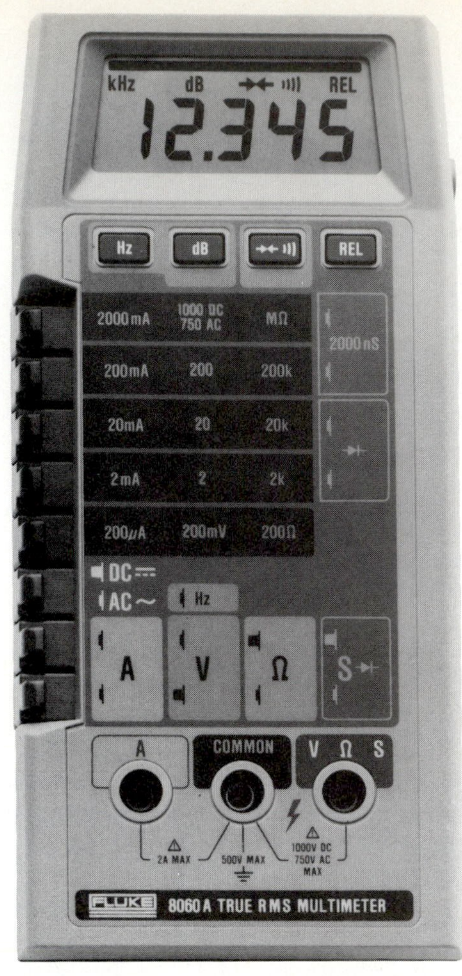

A.

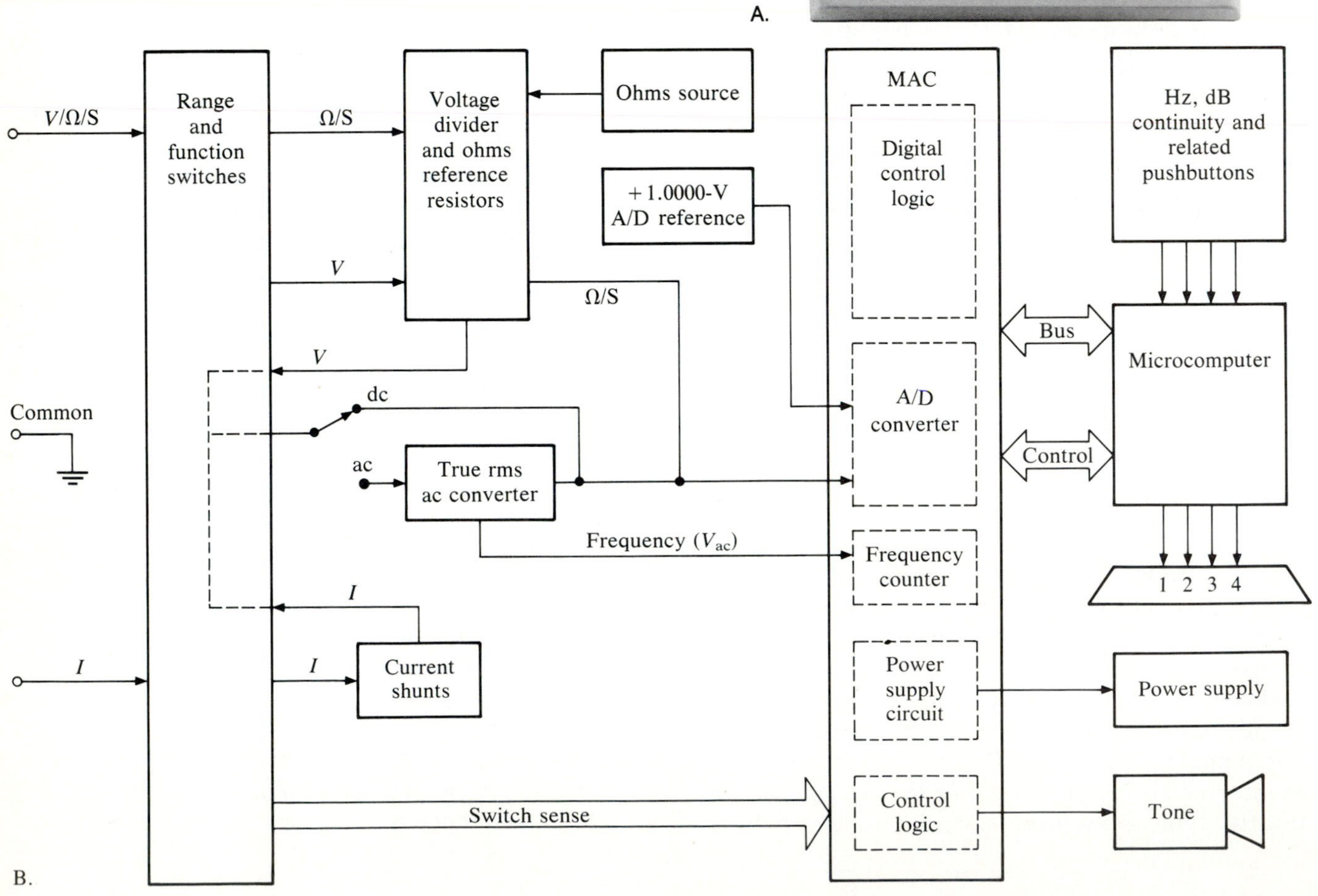

B.

2. The microcomputer initiates either the A/D measurement cycle or the frequency measurement cycle.
3. The microcomputer processes the data obtained in the measurement cycle, including calculations for the dB, relative offset, and megohm range or frequency autoranging.
4. The microcomputer displays the results, which remain on the display until it is updated.

After the results are displayed, the routine returns to the first step.

The Measurement Acquisition Chip

Figure 6–7 shows a block diagram of the MAC. The digital control logic includes a buffer and decoder, read and write logic, status and control registers, and logic controls for the continuity function. The power supply control uses the calibrated 1-V A/D reference voltage obtained from a bandgap reference diode to regulate the 5.2-V main power supply for the instrument. When the continuity function is selected and continuity is detected, the MAC generates the tone by supplying a square wave to the external piezoelectric transducer.

A/D Conversion Cycle

The heart of the MAC is the dual-slope A/D converter. A block diagram of the analog portion of the A/D converter is shown in Figure 6–8. The internal buffer, integrator, and comparators work in conjunction with external resistors and capacitors to convert the dc analog voltage to a digital number. The internal switches are field-effect transistor (FET) switches controlled by the microcomputer and MAC digital control logic. The switchable integrator gain depends on the function and range selected. The three-stage measurement cycle was discussed previously (refer to Figure 6–5). If, during the read period, the counter reaches the maximum number of counts for a full-scale reading (19,999 counts) and the integrator capacitor charge has not reached the initial autozero voltage, the microcomputer "knows" an overrange reading has been taken. The microcomputer places "OL" on the display and commands the A/D converter to go into the overload (OL) period, which rapidly returns the integrator voltage to the initial autozero voltage.

Voltage Measurement

Both the ac and dc voltage ranges use an overvoltage-protected 10-MΩ input divider as shown in

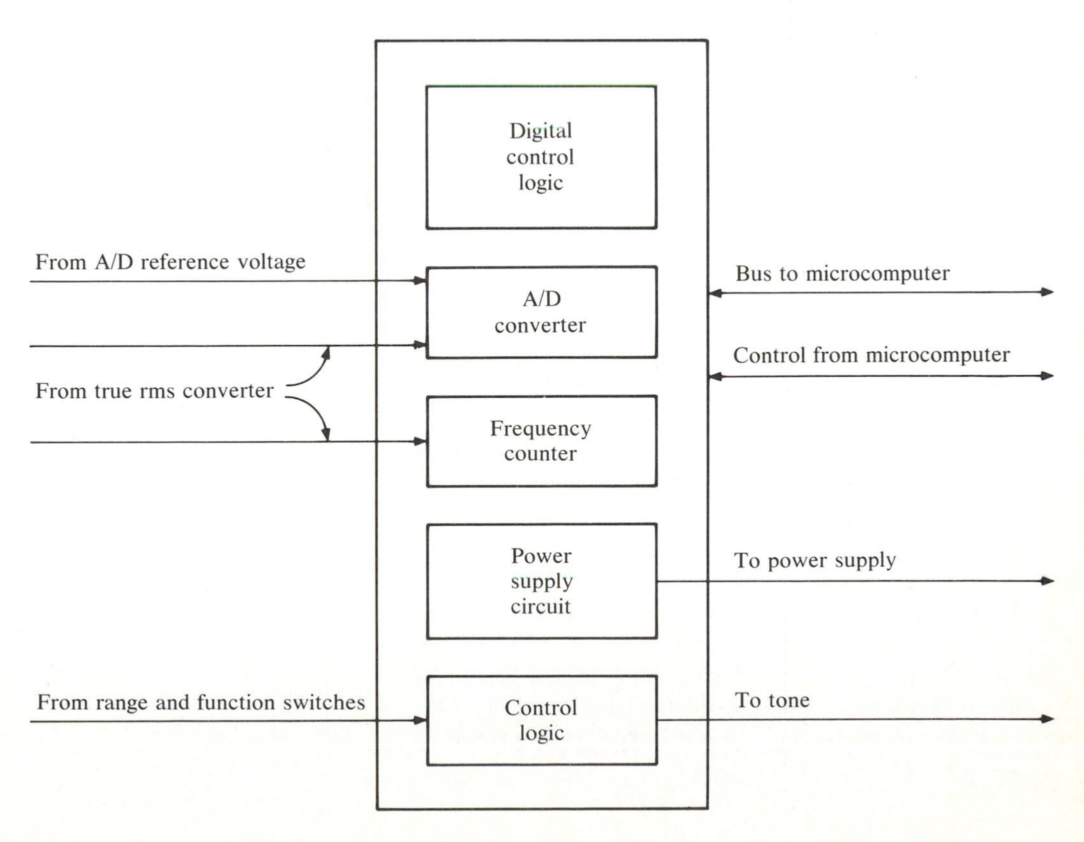

FIGURE 6–7
Block diagram of the measurement acquisition chip (MAC)

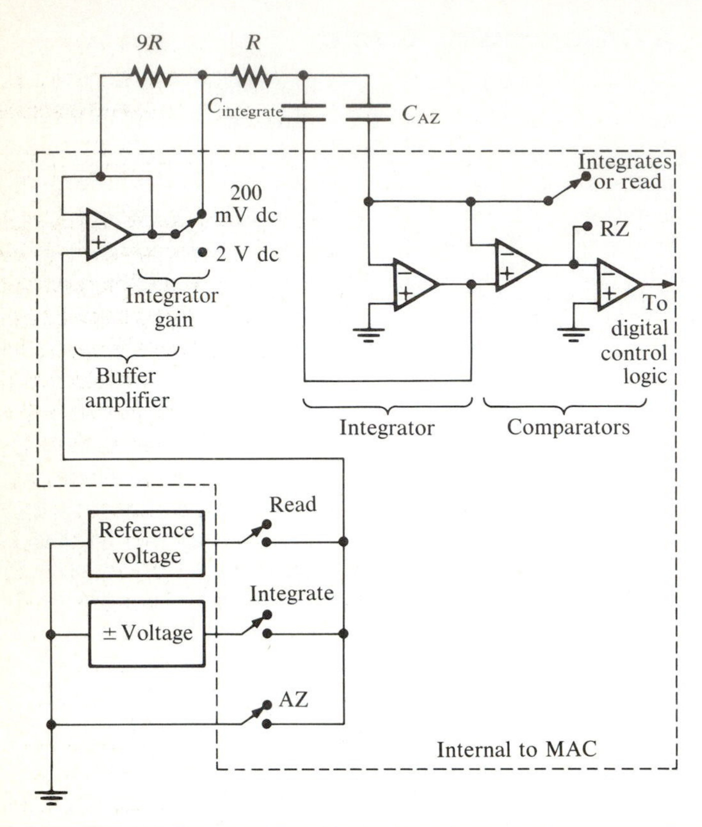

FIGURE 6–8
Analog portion of the A/D converter

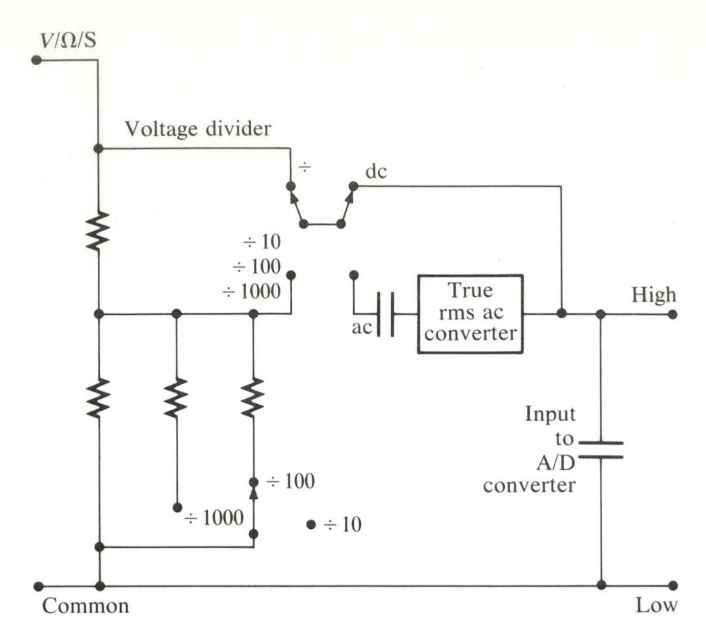

FIGURE 6–9
Voltage measurement circuitry

Figure 6–9. The overvoltage protection includes two 2-W fusible resistors and four metal-oxide varistors for high-voltage clamping. Depending on the range selected, lower leg resistors of the divider are connected to ground to perform the input signal division.

The dc input voltages for all ranges are divided by the appropriate factor of 10 to produce a proportional dc signal which is then filtered and applied to the input to the A/D converter. The microcomputer compensates by decreasing the integrator gain in the A/D converter by a factor of 10. The integrator gain is also reduced by a factor of 10 in the 1000-V dc voltage range, which uses the same divider as the 200-V dc range. The ac input voltages are divided with the same divider arrangement as the dc input voltages, except that the 2-V ac voltage range is divided by 10. The divider output signals for ac voltages are ac-coupled to the input of a true rms ac converter, which produces a current output. This negative dc representation is applied through a calibrated scaling resistor. The resultant negative voltage is filtered and applied to the input of the A/D converter.

Current Measurement

Current measurements are made using a double fuse–protected, switchable, five-terminal current shunt to perform the current-to-voltage conversion required by the A/D converter. Figure 6–10 shows a diagram of current measurements.

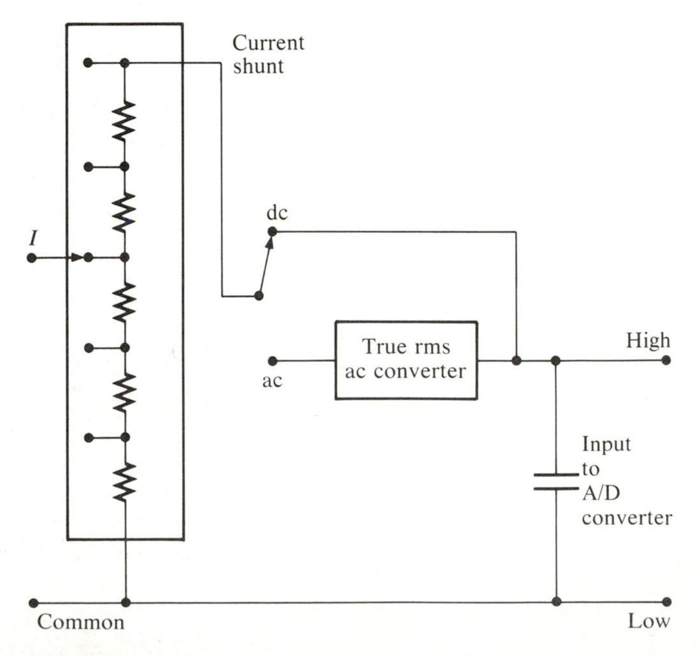

FIGURE 6–10
Current measurement

When the dc current function is selected, the dc voltage drop across the shunt is filtered and applied to the input of the A/D converter. When the ac current function is selected, the ac voltage drop across the shunt is ac-coupled to the input of the true rms ac converter. The dc representation of the ac voltage is filtered and applied to the input of the A/D converter.

Resistance Measurement

Resistance measurements are made using a ratio technique as shown in Figure 6–11. When the resistance function is selected, a series circuit is formed by the ohms source, a reference resistor from the voltage divider (selected by the range switches), and the external, unknown resistor. The ratio of the two resistors is equal to the ratio of the voltage drop across each of them. Since the voltage drop across the reference resistor and the value of the reference resistor are known, the value of the second resistor can be determined. Input protection during resistance measurements consists of a thermistor and a double-transistor clamp. The operation of the A/D converter during the resistance measurement is basically as described in the previous sections, with a few differences. During the integrate period, the voltage drop across the unknown resistor charges the integrate capacitor. During the read period, the voltage across the known resistor (stored on the flying capacitor) discharges the integrate capacitor. The length of the read period is a direct indication of the value of the unknown resistance.

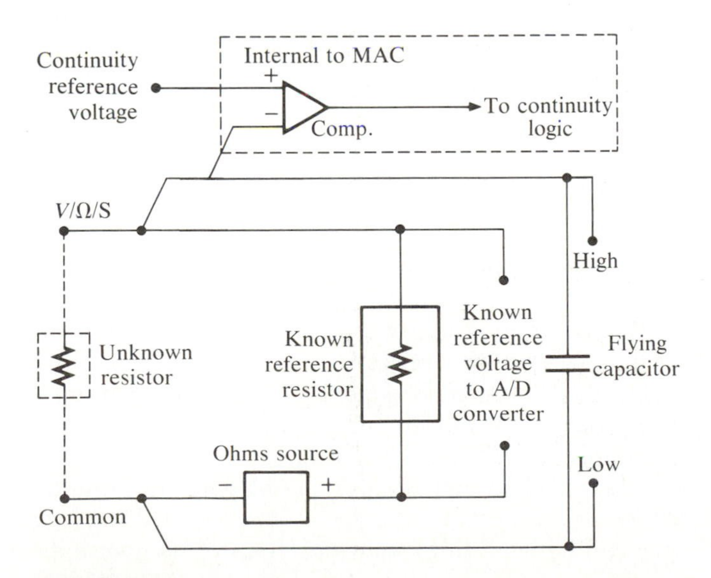

FIGURE 6–11
Resistance measurement, conductance measurement, and continuity indication

Conductance Measurement

Conductance measurements are made using basically the same ratio technique as in resistance measurements. The main difference is that the function of the range and unknown resistors in the A/D measurement cycle is reversed so that the smaller voltage is applied during the integrate period. This setup minimizes error caused by noise. During the integrate period, the voltage drop across the known resistor charges the integrate capacitor. During the read period, the voltage drop across the unknown resistor discharges the capacitor. The display presents a reading of the conductance, which is the reciprocal of resistance.

Continuity Indication

The 8060A determines whether continuity exists in the circuit under test by comparing the voltage drop across the external circuit with a continuity reference voltage. If the voltage drop across the external circuit is less than the reference voltage, the comparator sends the appropriate signal to the continuity logic (see Figure 6–11). The continuity logic notifies the microcomputer, which turns on the visible indicator or produces an audible tone if that function is selected by the operator.

Frequency Measurement

Frequency measurement is shown in Figure 6–12. The ac input signal is divided by the voltage divider and buffered by the true ac rms converter. The signal is then applied to a comparator in the MAC for counting. The counter gate is controlled

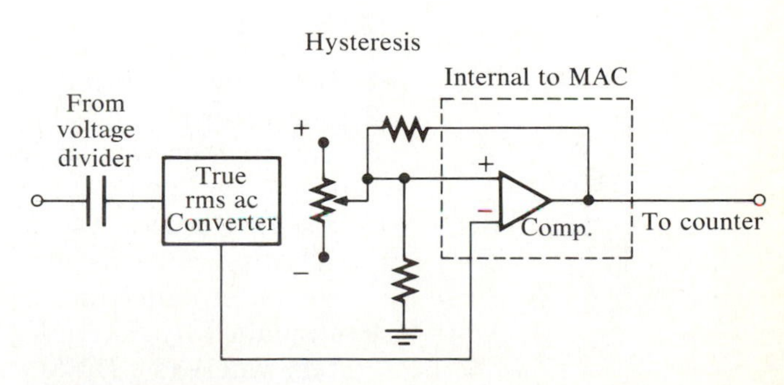

FIGURE 6–12
Frequency measurement

by the microcomputer, and the range is automatically selected by the microcomputer's software. For very low frequency input signals, the counter actually measures the period of the input signal. The microcomputer then inverts the measurement to derive the frequency.

MULTIMETER SAFETY

The following tips are suggested for the protection of both you and your DMM:

1. Replace leads that have damaged insulation or exposed metal.
2. Use meters with recessed jacks, shrouded connectors, and finger guards.
3. Check the continuity of the test leads.
4. Make sure the meter is working correctly.
5. Select the proper function and range for your measurement.
6. Electrically disconnect the "hot" line first.
7. Disconnect unnecessary power supplies and discharge high-voltage capacitors before testing.
8. Unsheltered test conditions may allow for wet conditions or condensation, directly lowering the insulation values and creating current paths. Shield your meter from moisture.
9. Only specially designed and approved (and clearly labeled as such) meters should be used in hazardous locations (i.e., potentially explosive atmospheres). Simply disconnecting test leads from live circuits in an explosive environment might create a dangerous spark.

FREQUENCY CONSIDERATIONS

Whenever a transistor amplifier is encountered in a DMM circuit, the reverse-biased junction between the base and collector provides a capacitive path for the incoming signal. The signal "sees" a low reactance path from the base to the collector and a relatively high impedance path from the base to the emitter (amplifier). The effective amplification is reduced at higher frequencies, limiting the frequency response of the DMM (Figure 6–13). Most common DMMs are limited in their use to low frequencies, unless specifically designed for higher-frequency work.

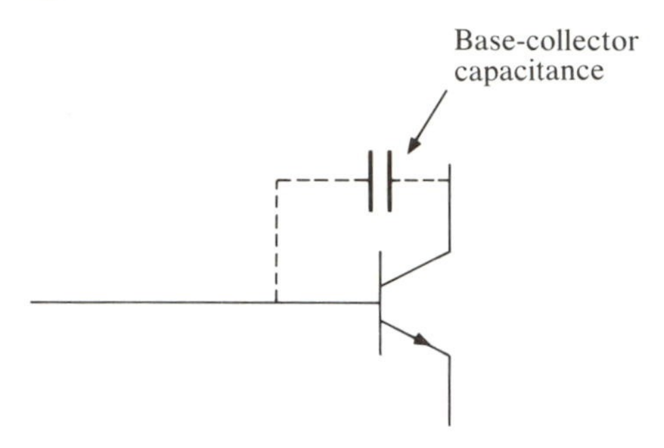

FIGURE 6–13
Base-to-collector capacitance becomes important at high frequencies.

SUMMARY

The digital multimeter is a product of technological advances during the last twenty years. DMMs are inexpensive, accurate, and mobile. They allow limited service repairs and have some functional limitations. DMMs are a well-accepted innovation in industry.

EXERCISES

1. List three advantages of DMMs over analog meters.
2. List three advantages of analog meters over DMMs.
3. You turn on your DMM and the display does not go on. What should you do?
4. If you get 000 on the display of your DMM but do not get a response when you measure a dc voltage, what should you do?
5. What precautions should you take if you must remove the MAC or the microcomputer chip from the DMM?
6. Explain why an analog meter can be used to measure the emf of its own battery whereas a DMM cannot.
7. Why would you choose a meter with 10,000 MΩ input resistance rather than one with 10 MΩ input resistance?

8. What is the first thing that should be done before using a newly acquired DMM?
9. What does a continuous 888.8 digital output usually mean?
10. What advantage does an oscilloscope have over a DMM when measuring a 25-kHz signal voltage?

LAB PROBLEM 6–1
Digital Multimeters

Equipment

One DMM; one high-voltage probe; one electrolytic capacitor; one low-voltage power supply; two single-pole, single-throw switches; one each 0.1-, 0.25-, 0.47-, and 0.001-μF capacitors; one 200-kΩ resistor, ¼-W; one 2.2-MΩ resistor, ¼-W; one lab standard oscilloscope; old television set.

Object 1

To extend the range of a DMM to measure the leakage resistance of an electrolytic capacitor (large resistance measurements).

Procedure

Wire the circuit in Figure 6–14.

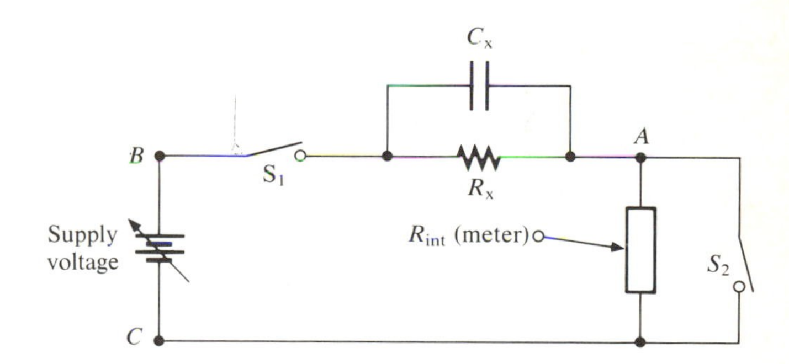

FIGURE 6–14
Circuit for measuring the leakage resistance of an electrolytic capacitor

R_{int} is the internal resistance of the DMM, and

$$R_x = R_{int}\left(\frac{V_{BC} - V_{AC}}{V_{AC}}\right)$$

Switches S_1 and S_2 are open. Use a 0.1-μF capacitor. Set the supply voltage at 20 V as measured on the DMM. Close S_1 to permit the capacitor to charge through the DMM toward 20 V. To reduce the charging time constant, close S_2 for about 2 s and then open it again. Change the meter scale to the lowest range on which a measurable voltage is indicated. After the capacitor action is stabilized, record this voltage in Table 6–2. Repeat the procedure for the other capacitor values listed. Compute the leakage resistance. If no measurable results are obtained, use a higher supply voltage. DO NOT exceed the WVDC of the capacitor.

TABLE 6–2
Complete the table.

Capacitance (μF)	V_{BC} (V)	V_{AC} (V)	R_x (MΩ)
0.1			
0.25			
0.47			
0.001			

Object 2

To observe the characteristics of a voltmeter in measuring sinusoidal and nonsinusoidal voltages, and to measure the bandwidth of the DMM frequency response.

Procedure

Wire the circuit in Figure 6–15.

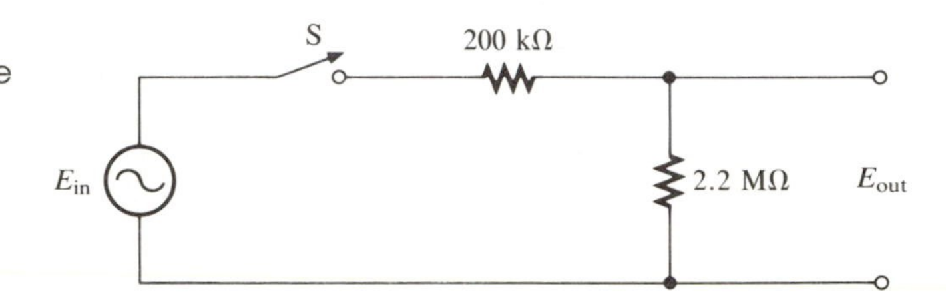

FIGURE 6–15
Circuit for measuring the bandwidth of a DMM

Take the indicated readings and enter them in Table 6–3. Set the peak-peak input voltage at 3 V for each trial. Check it with an oscilloscope at each setting. Make sure the bandwidth of the oscilloscope is wide enough to allow for an accurate reading (use a lab standard scope). Repeat the procedure using square and triangular waveforms at the input.

TABLE 6–3
Complete the table.

Frequency (Hz)	E_{in} Peak-Peak	E_{in} rms	E_{out} Peak-Peak	E_{out} rms
100				
500				
1,000				
5,000				
10,000				
100,000				

Object 3

To learn how to measure high voltages.

Procedure

Use a high-voltage probe to extend the measuring capacity of a DMM. Use this probe to measure the second anode voltage on a television receiver picture tube. FOLLOW THESE SAFETY PRECAUTIONS WITH GREAT CARE:

1. Know the equipment under test. Locate all high-voltage points first.

2. Do NOT work alone.

3. Work with ONE hand. Keep the other hand in your pocket.

4. Avoid contact between any part of your body and ANY object that can provide ground.

5. Keep hands, shoes, floor, probe, and test bench completely dry.

6. Use only an approved high-voltage probe when making high-voltage measurements.

7. Make sure the high-voltage probe has no dirt, grit, or grease on its surface or handle.

8. When using the probe, place it in such a manner that the guard ring is closer than your hand to any high-voltage point.

9. Do NOT try to adapt a high-voltage probe to any measuring instrument for which it was not specifically designed.

7 Oscilloscopes—Theory of Operation

OUTLINE

OBJECTIVES

WHAT IS A CATHODE RAY OSCILLOSCOPE?
- ☐ The Electron Beam
- ☐ The Beam Finder
- ☐ The Phosphor
- ☐ Graticules
- ☐ The Display

THE VERTICAL SECTION
- ☐ The Probe
- ☐ Picking a Probe
- ☐ Circuit Loading
- ☐ The Vertical Attenuator (Range Selector)
- ☐ Vertical Amplifiers/Phase Splitters
- ☐ The Delay Line

THE HORIZONTAL SECTION

THE TRIGGERING SECTION
- ☐ Internal Triggering
- ☐ Line Triggering
- ☐ External Triggering

LINEARITY OF SWEEP

THE HORIZONTAL SWEEP

DUAL TRACE

DELAYED SWEEP

SAMPLING OSCILLOSCOPES

DIGITAL STORAGE OSCILLOSCOPES

ANALOG STORAGE OSCILLOSCOPES

SUMMARY

EXERCISES

LAB PROBLEM 7–1

OBJECTIVES

After completing this chapter, you will be able to:

- ☐ Describe a cathode ray oscilloscope (CRO).
- ☐ Draw a block diagram of the functional parts of a CRO.
- ☐ Discuss the design and action of the cathode ray tube.
- ☐ Describe the path an incoming signal takes through the vertical part of the CRO.
- ☐ Discuss the functions of probes.
- ☐ Give the function of attenuators.
- ☐ Give the function of vertical amplifiers.
- ☐ Give the function of the delay line.
- ☐ Describe the methods of synchronizing the horizontal sweep with the vertical deflection.
- ☐ Explain the function of the trigger-level control.
- ☐ Describe external triggering.
- ☐ Explain the function of the ramp generator.
- ☐ Explain the function of the horizontal amplifier.

This chapter answers the question, How does an oscilloscope work? and does so on a level useful to the technician. In an age of microchips, troubleshooting an oscilloscope to the component level is not practical. In most cases, after preliminary tests have been made to determine whether the instrument is operating properly, either a relatively simple repair is performed or the equipment is returned to the manufacturer for more detailed analysis. It is essential that the technician know how the general parts of an oscilloscope operate so that a problem can be isolated, and possibly corrected locally.

Oscilloscope circuitry is broken down into vertical, horizontal, triggering, and display sections for analysis. The general operation of each section is presented. Chapter 8 contains a detailed discussion of the uses of the oscilloscope.

WHAT IS A CATHODE RAY OSCILLOSCOPE?

A *cathode ray oscilloscope* (CRO) is an *x-y* plotter. The *x*-axis corresponds to a signal placed on the horizontal control of the cathode ray tube (CRT), and the *y*-axis corresponds to a signal placed on the vertical control of the CRT (Figure 7–1). A signal placed on the horizontal plates causes the electron beam to deflect in the horizontal direction. A signal placed on the vertical plates causes the electron beam to deflect in the vertical direction.

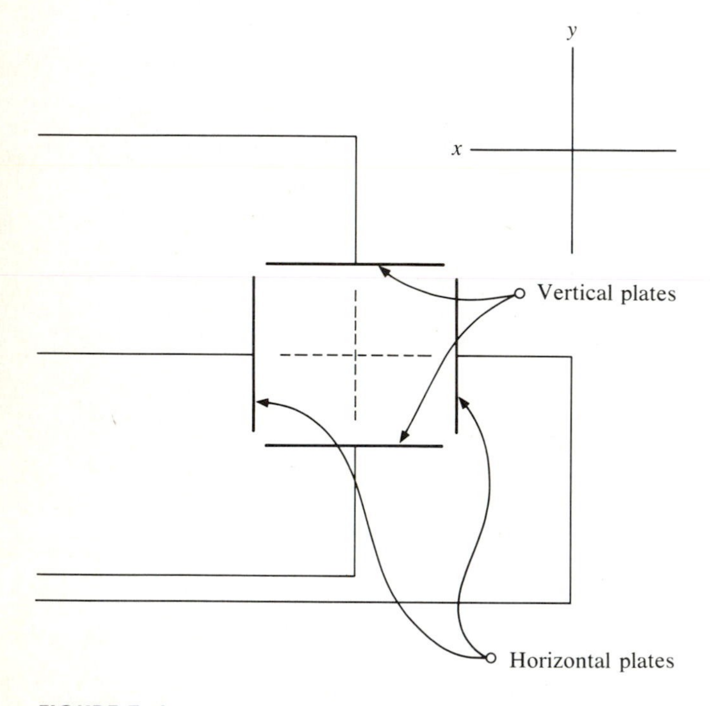

FIGURE 7–1
Vertical and horizontal plates provide position control of the electron beam.

The Electron Beam

The electron beam is generated in the *electron gun* section of the CRT. The electron gun contains a cathode designed to give off electrons, and it is here that the electron beam is shaped and focused. The electrons are accelerated to give them sufficient energy to produce a spot of light when they strike the face of the CRT (Figure 7–2).

In going from the electron gun to the CRT face, the electron beam passes through two sets of deflection plates. One pair of plates deflects horizontally, and the other pair deflects vertically. If an unbalanced potential is placed on the horizontal plates, the electron beam is made to deflect in the more positive direction. This happens because the electron beam is negative in character. A positive voltage on one of the plates will attract the electrons while a corresponding negative voltage on the opposite plate will repel the beam (Figure 7–3).

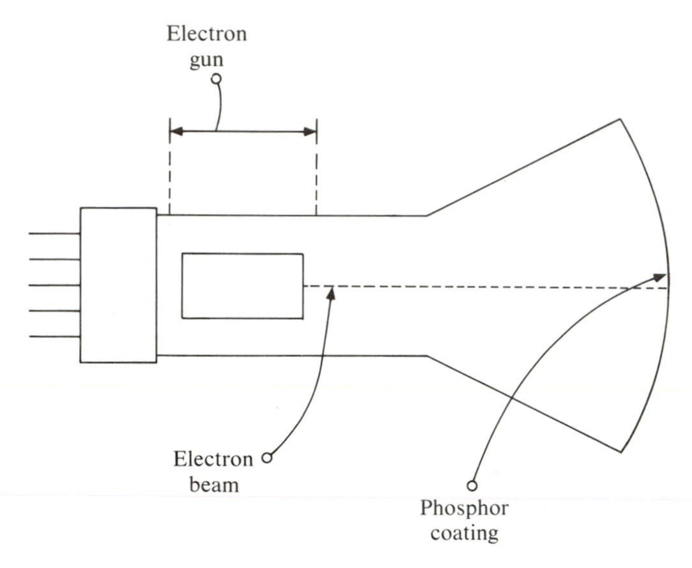

FIGURE 7–2
The electron gun produces the electron beam.

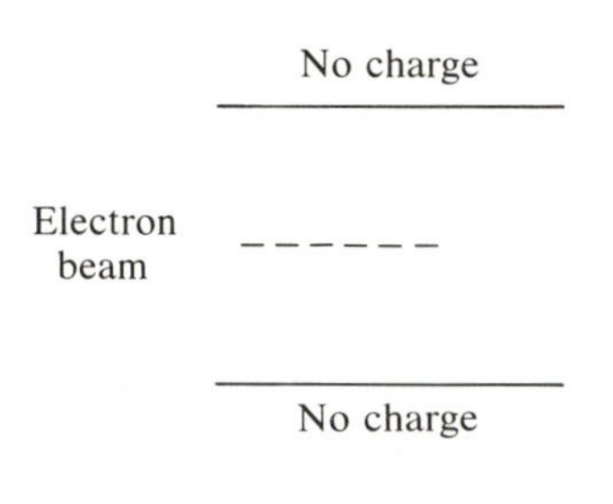

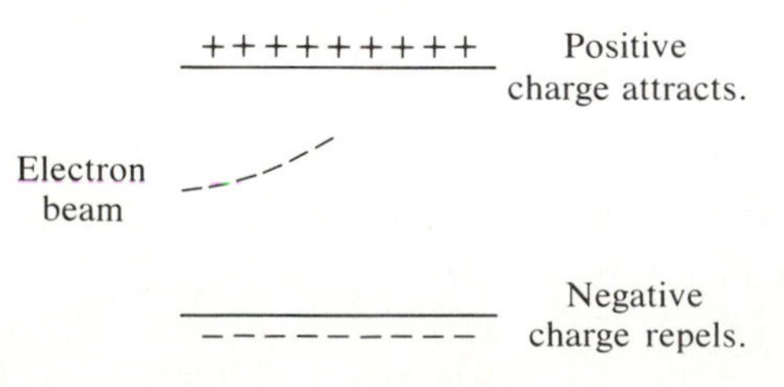

FIGURE 7–3
Electrostatic forces deflect the electron beam.

In practice, equal dc voltages (biases) are maintained on both plates, which should keep the electron beam centered (Figure 7–4). When the inputs produce an imbalance in the voltages across the horizontal plates, one plate becomes more positive than the other and the electron beam is caused to bend toward the more positive bias (Figure 7–5). These bias voltages are controlled at the front panel of the CRO by the horizontal centering control. A similar bias control is provided for the vertical deflection plates (vertical centering). With these two controls, the technician can move the light spot to any position on the face of the CRT.

Within the electron gun are high-voltage fields that focus and accelerate the electron beam. These fields are developed from the high-voltage power supply in the oscilloscope circuitry. The voltage is controlled by the focus control on the front panel, which concentrates the electron beam into a narrow stream of electrons, in turn producing a small, bright spot on the face of the CRT (Figure 7–6).

The control that affects the acceleration of the electrons in the beam is called the *brightness* control. The intensity of the spot of light produced when the electron beam impacts the face of the CRT depends on the speed of the electrons in the beam. If the electrons are traveling slowly upon impact, the light spot produced will be dim. If the electrons are traveling more rapidly, the light spot will be brighter (Figure 7–7).

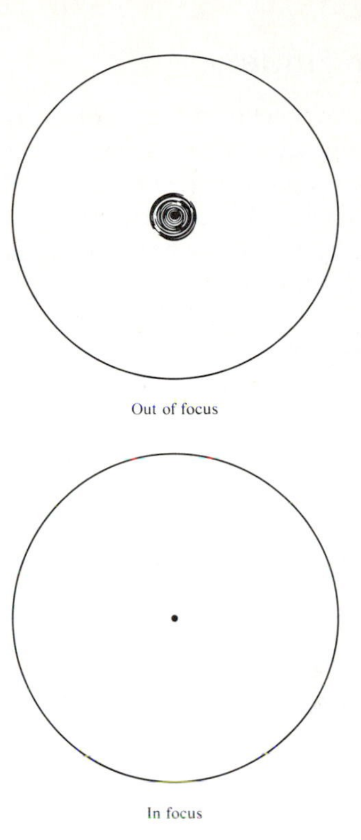

FIGURE 7–6
Focusing the electron beam

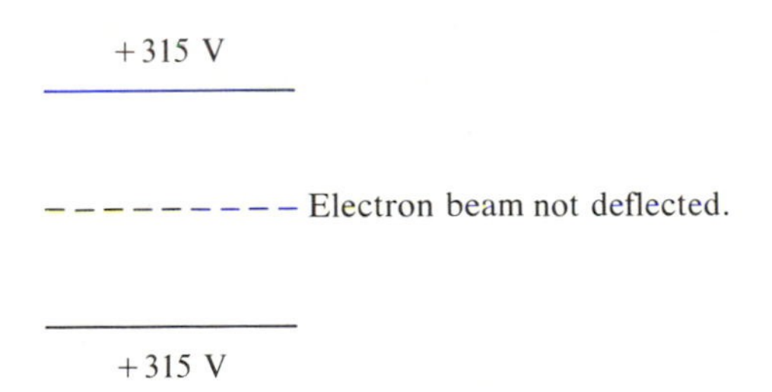

FIGURE 7–4
Biasing produces a centered beam.

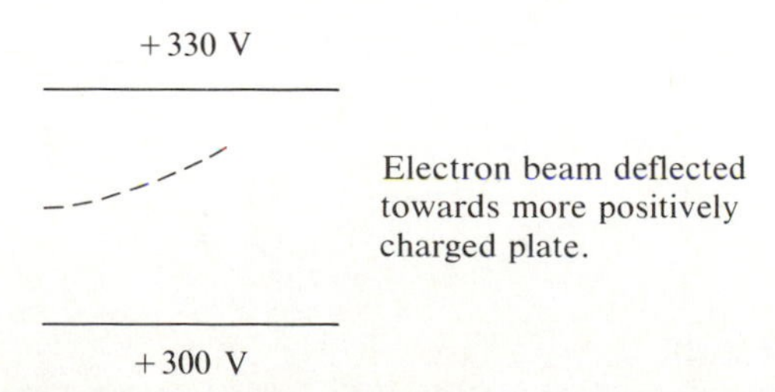

FIGURE 7–5
Unequal biasing produces a deflection.

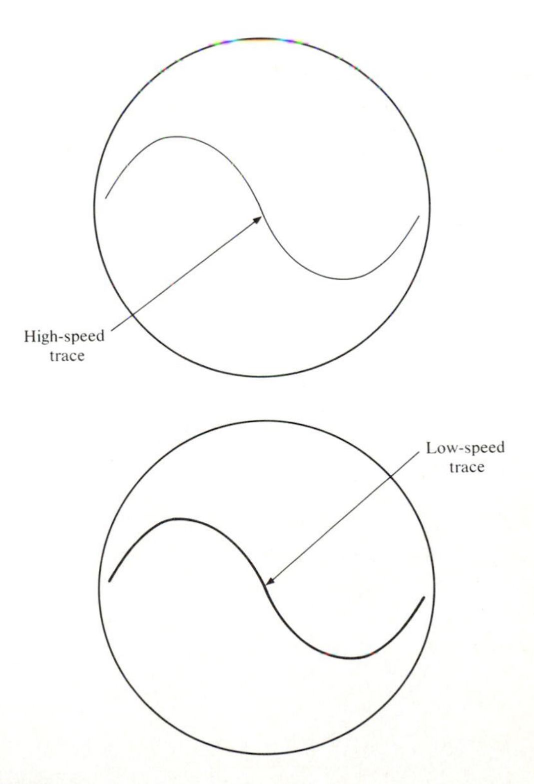

FIGURE 7–7
Trace speed affects brightness. The high-speed trace results in a dimmer display than that produced by the low-speed trace.

The Beam Finder

The *beam finder* contains circuitry that applies a very large voltage to the position controls. The voltage forces the electron beam into an area approximately the size of the graticule. If no trace appears on the face of the CRT when the oscilloscope is initially turned on, the beam finder can be used to determine if the trace is merely located off the screen.

The beam finder should be used only momentarily. Since the intensity of the beam is greatly increased, damage to the phosphor may result from prolonged use.

With the controls described so far, the technician can:

1. Focus the beam.
2. Control the beam's brightness.
3. Move the spot (beam) left and right.
4. Move the spot (beam) up and down.
5. Find the beam.

The Phosphor

The light spot on the face of the CRT is produced by the impact of the electron beam on the screen's phosphor coating. When phosphors are struck by electrons, they give off light. When an electron with considerable energy strikes an inner-shell electron of a phosphor atom, the inner-shell electron absorbs some of the energy from the incoming electron (Figure 7–8). This absorbed energy causes the inner-shell electron to rise to a higher energy level (Figure 7–9), producing an electron deficiency at the lower energy level, which will draw the electron back to its original (lower) energy level. As this occurs, the electron must give up some of its excess energy, which it does in the form of visible light. This return trip may be along direct or indirect paths. The wavelength (color) of the light depends on the amount of energy given off, and the

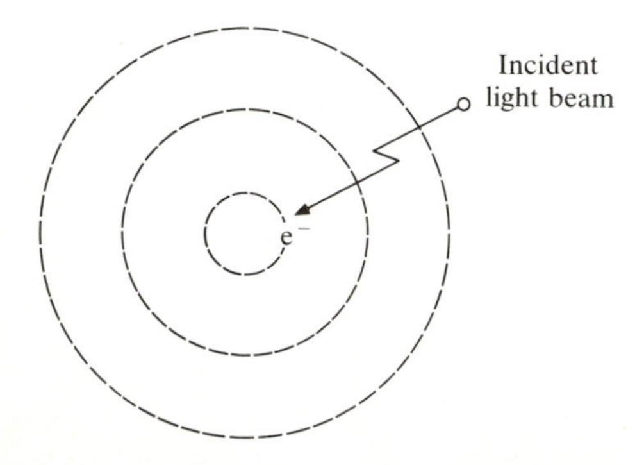

FIGURE 7–8
Light beam strikes an inner-shell electron.

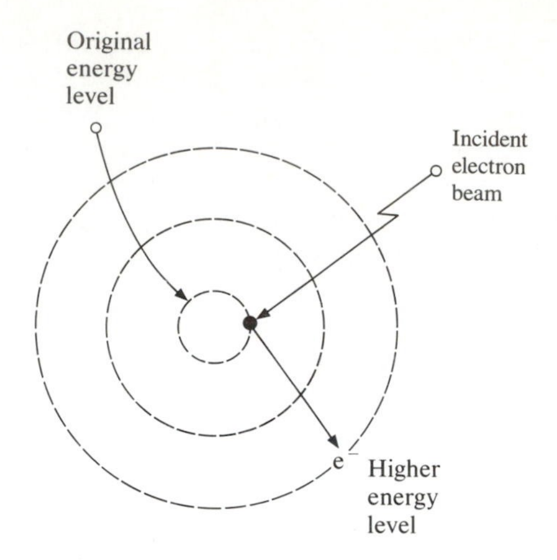

FIGURE 7–9
Inner-shell electron moves to a higher energy level.

duration of the light spot depends on the time required for the electron to reach ground state (Figure 7–10).

The time needed for this process may be short or long depending on the type of material and the amount of energy absorbed. These factors determine the *persistence* of the light beam on the face of the CRT. For high-speed phenomena, it is desirable that the light spot disappear quickly so that another can take its place. For low-speed phenomena, it is useful for the spot to remain until the entire trace is completed so that the composite signal can be seen. This situation requires a slower phosphor or one with longer persistence. Different phosphor materials provide different paths for the return of the electron to the lower energy levels. Differences in energy levels cause light energy of different wavelengths to be emitted. This is why oscilloscopes are available with different-colored

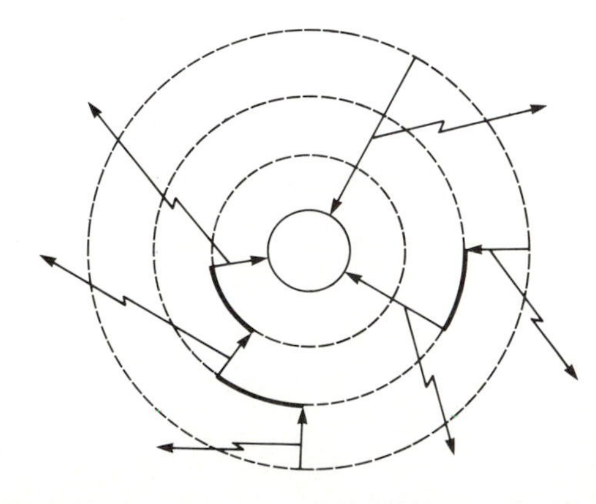

FIGURE 7–10
Electrons give off excess energy in the form of visible light, the wavelength dependent on the path. Note that different paths are possible.

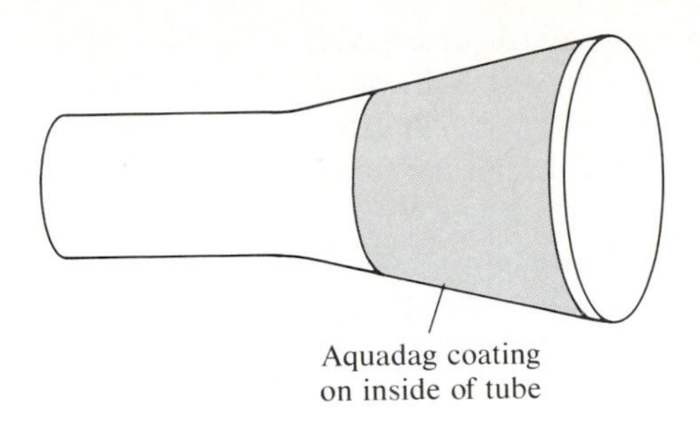

FIGURE 7–11

traces. A color can be selected that is most comfortable for the user.

The phosphor is applied in a very thin layer on the inside of the face of the CRT. After the electron beam has struck the phosphor, it must have a return path to complete the electric circuit. This path is provided by a gray conductive coating applied to the inside of the CRT. This coating, called the *aquadag,* is usually connected internally to a return path to the high-voltage power supply (Figure 7–11).

To protect the operator, the electron gun (source of electrons) is maintained at a very high negative voltage inside the CRO, and the face of the CRT is maintained at ground potential. Thus, although the face of the CRT is very positive relative to the electron gun, it is at ground potential relative to the user.

Graticules

If the CRT face is accidentally struck, the high-vacuum glass tube might implode, possibly injuring the user. To prevent this, the CRT has a very thick piece of glass as its face. This safety glass, however, can distort the image resulting in measuring inaccuracies.

The graph scale on the front of the CRT is called the *graticule.* This can be a plastic faceplate mounted on the outside of the face of the CRT, etched on the outside of the CRT, or etched on the inside of the CRT. If the graticule is etched on the outside of the CRT or is located on a plastic shield mounted on the outside of the face of the CRT, parallax can result because of the thickness of the glass. Etching the graticule on the inside of the face of the CRT minimizes this inaccuracy but raises the cost since the manufacturing process is more difficult (see Figure 7–12).

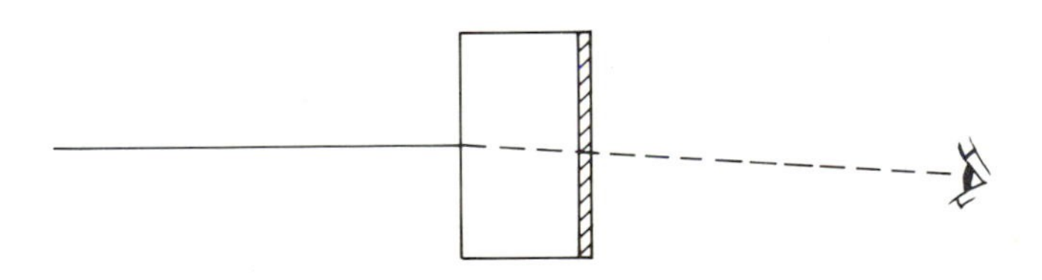

Graticule on outside of CRT face can cause parallax errors.

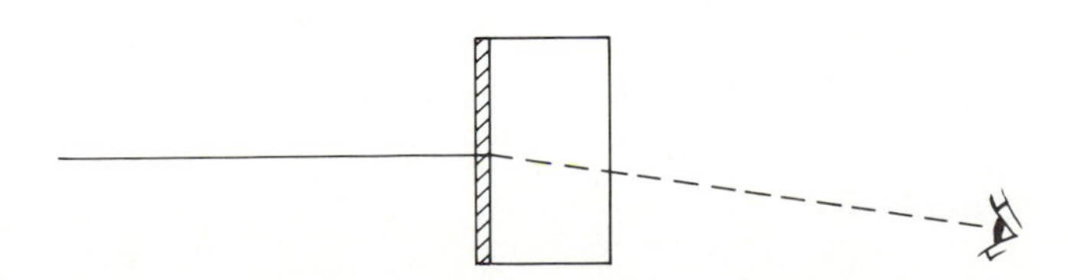

Graticule on inside of CRT face tends to eliminate parallax errors.

FIGURE 7–12
Parallax problem

The Display

An oscilloscope presents a two-dimensional image that is easy to see and understand. This display presents an accurate, detailed picture of time-related behavior of analog and digital phenomena.

Oscilloscopes consist of a vertical section, horizontal section, trigger section, and display. They can be categorized as conventional, sampling, or storage oscilloscopes. *Conventional* oscilloscopes are used to measure voltage, frequency, or phase. *Sampling* oscilloscopes have the capability of displaying very high speed phenomena in the gigahertz range.

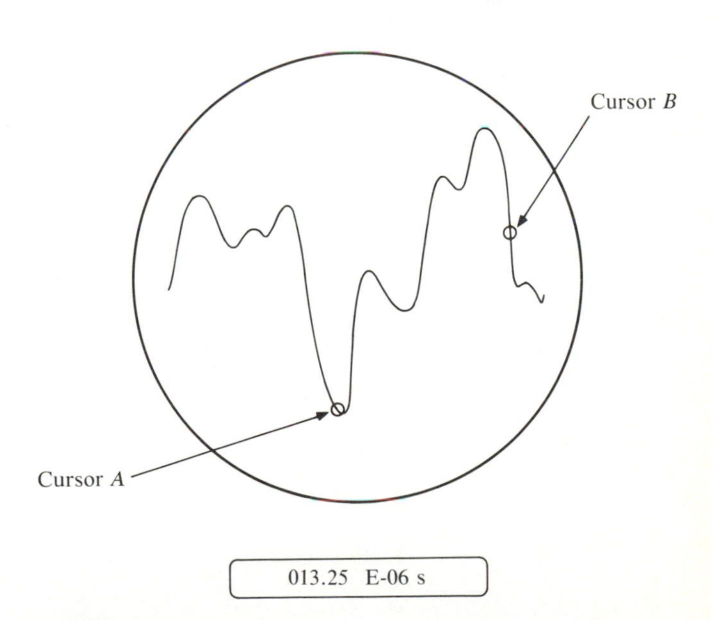

FIGURE 7–13
Oscilloscope with a two-cursor display. The readout is the time difference between point *A* and point *B* as indicated by the two cursors.

Storage oscilloscopes are capable of displaying a signal for seconds, hours, or indefinitely.

Oscilloscopes can be equipped with microprocessors that can perform intricate calculations using measurements indicated on the screen, releasing the operator to concentrate on the application of the oscilloscope and the interpretation of the results. Some oscilloscopes are available with two cursors (bright, movable light spots), which follow the waveform. The cursors can be placed at any location on the displayed waveform, enabling the operator to read from a digital display the time or frequency of a phenomenon occurring between the two cursors (Figure 7–13).

THE VERTICAL SECTION

The vertical section of the oscilloscope usually consists of a probe, an attenuator (range selector), amplifiers/phase splitters, and a delay line.

The Probe

The input signal is usually detected by the *probe,* which consists of the probe tip, barrel, and cable for connection to the oscilloscope input. Electronic components, usually found in the barrel of the probe, attenuate the input signal and compensate for bandwidth limitations. Figure 7–14 is an illustration of the components found in a 10:1 probe, also called a *times 10 (×10) multiplier*.

If a 10:1 probe is used, the amplitude of the signal shown on the CRT is 1/10 the original signal amplitude. The amplitude shown must be multiplied by 10 (hence the name times 10 multiplier). This also reduces capacitive loading of the oscilloscope by the source by a 10:1 ratio. The reduced capacitive loading factor will not be increased by subsequent amplification of the reduced signal. In some cases a *trimmer* capacitor is included to adjust for frequency limitations. The capacitor may be accessible through a small hole in the probe barrel and can be adjusted to improve high- or low-frequency response. With the probe attached to the square-wave calibration signal of the oscilloscope, proper adjustments can be made and the results observed. If the probe is limiting the low-frequency inputs of the incoming signal, the top of the calibration square wave will slope downward (Figure 7–15). If the probe is limiting the high-frequency inputs of the incoming signal, an inferior leading-edge (and trailing-edge) display of the calibration square wave will be produced (Figure 7–16).

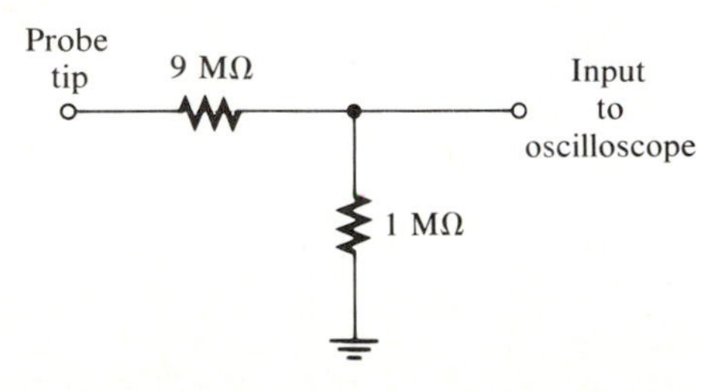

FIGURE 7–14
Voltage divider network in a 10:1 probe

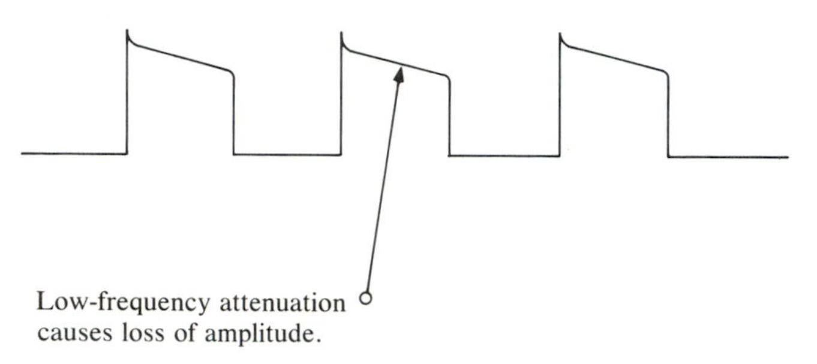

FIGURE 7–15

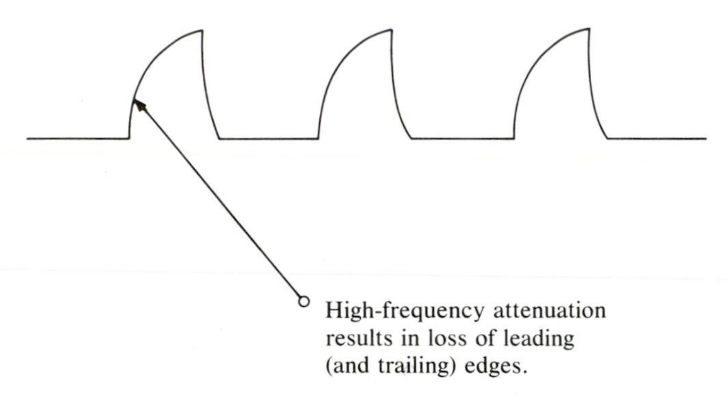

FIGURE 7–16

Adjustment of the trimmer capacitor must be made carefully since it usually involves a plastic screw that can be permanently damaged by rough handling with a screwdriver. Figure 7–17 illustrates the results of proper adjustment.

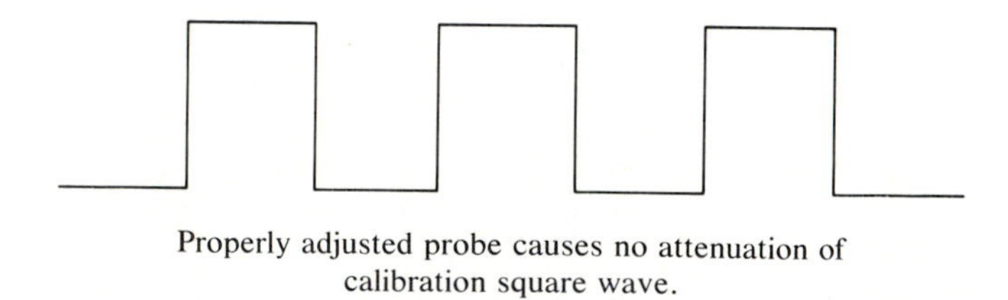

FIGURE 7–17

Generally, probes can be divided by function into *voltage-sensing* and *current-sensing* types. Voltage probes can be further divided into *passive* and *active* types. Table 7–1 illustrates the general

TABLE 7–1

Probe Types	Characteristics
×1 passive, voltage-sensing	No signal reduction, allowing maximum sensitivity at probe tip; limited bandwidths: 4–34 MHz; high capacitance: 32–112 pF; signal handling to 500 V
×10/×100/×1000 passive, voltage-sensing, attenuator	Attenuates signals; bandwidths to 300 MHz; adjustable capacitance; signal handling to 500 V for ×10, 1.5 kV for ×100, 20 kV for ×1000
Active, voltage-sensing, FET	Switchable attenuation; capacitance as low as 1.5 pF; more expensive, less rugged than other types; limited dynamic range; bandwidths to 900 MHz; minimum circuit loading
Current-sensing	Measures currents from 1 mA to 1000 A; dc to 50 MHz; very low loading
High-voltage	Signal handling to 40 kV

classification of probes and lists the characteristics of each probe type.

Picking a Probe

For most applications, the probes that were supplied with the oscilloscope are the ones that should be used. They will usually be attenuator probes. To make sure the probe can faithfully reproduce the signal for the oscilloscope, the compensation of the probe should be adjustable. Substitution probes should be avoided. However, if a substitution probe is necessary, it should be selected based on the voltage to be measured. For example, if a 50-V signal is to be observed and the largest vertical sensitivity of the oscilloscope is 5 V, the signal will take up 10 major divisions on the screen. This is a situation where attenuation is needed. A ×10 probe would reduce the amplitude of the signal to reasonable proportions.

Proper termination is important to avoid unwanted reflections of the signal to be measured within the cable. Probe/cable combinations designed to drive 1-MΩ inputs are engineered to suppress these reflections. But for 50-Ω scopes, 50-Ω probes should be used.

Circuit Loading

Circuit loading is resistive, capacitive, and inductive. For signal frequencies under 5 kHz, the most important loading component is resistance. To avoid significant circuit loading, a probe with a resistance of at least 100 times the circuit impedance should be used.

When measurements are made on a signal containing high-frequency components, inductance and capacitance become important. The addition of capacitance cannot be avoided when making connections, but it can be minimized.

One way to accomplish this is to use an attenuator probe since its design greatly reduces loading. Instead of loading the circuit with capacitance from the probe tip plus the cable plus the scope's own input, the ×10 attenuator probe introduces about 10 times less capacitance—as little as 10–14 pF. The penalty is the reduction in signal amplitude. Times 10 probes are adjustable to compensate for variations in oscilloscope input capacitance.

When high frequencies are measured, the probe's impedance changes with frequency. The specification sheet or manual for the probe will contain a chart showing this change. Another point to remember when making high-frequency measurements is to securely ground the probe with the shortest ground clip possible.

The Vertical Attenuator (Range Selector)

The input *attenuator,* or *range selector,* is a voltage divider network that allows many ranges of input

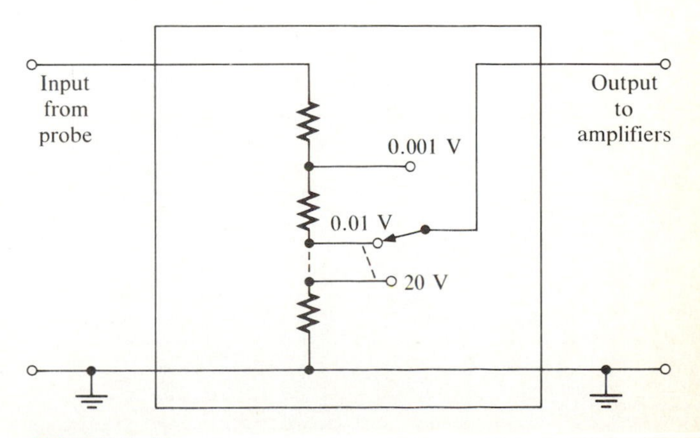

FIGURE 7–18
Attenuator (range switch)

and provides a constant range of output. The range selector is necessary since the oscilloscope can be used for measuring voltages in the millivolt or the kilovolt range. The vertical amplifiers operate at a relatively small voltage range. The attenuator protects the amplifiers by guaranteeing that they will always receive the same range of input voltages (Figure 7–18).

Vertical Amplifiers/Phase Splitters

Vertical amplifiers do several things. First, they must amplify the signals to provide enough voltage to cause proper vertical displacement. They must amplify these signals without distortion, and they must be separated from the input by a buffer amplifier. The buffer amplifier:

1. Acts as a class-A amplifier.
2. Buffers the vertical amplifier against any loading effects on the input end or the horizontal section.
3. Matches the impedance of the attenuator to the vertical voltage amplifier.

An emitter follower or source follower configuration is used as a buffer amplifier.

The next step may be a *phase splitter,* which usually is a push-pull amplifier or a paraphase amplifier that provides two antiphase output signals (180° out of phase). The phase splitter provides a push-pull effect on the vertical plates instead of a single-ended pull-against-ground. This action provides better control with less distortion since while there is a pull against one plate, there is a push against the other, resulting in a better balance in deflecting forces (Figure 7–19).

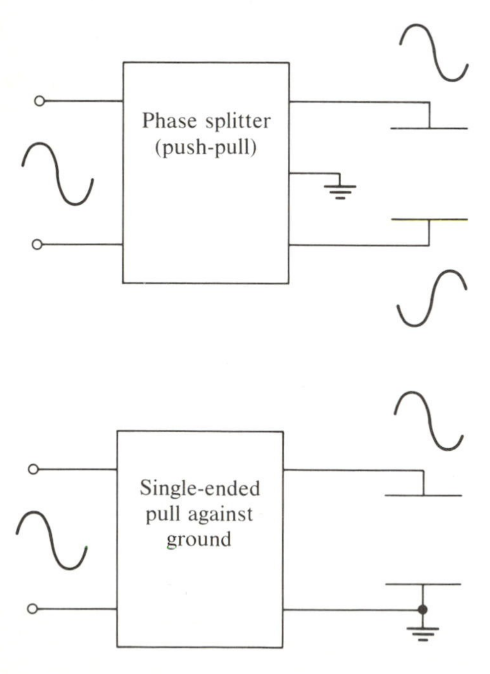

FIGURE 7–19
Two methods of vertical/horizontal control

The Delay Line

There is a *delay line* in the vertical section. The vertical section provides the pickup for the horizontal timing signal. The processing of the timing signal and the generation of the sweep ramp voltage take time (80–200 ns). The signal on the vertical line must be delayed by this amount of time. If this is not done, the signal will arrive at the vertical plates before the horizontal sweep begins, and the first part of the vertical trace will be lost.

Figure 7–20 presents the block diagram of the vertical section of an oscilloscope.

THE HORIZONTAL SECTION

The function of the horizontal section is to convert the time-base ramp to an appropriate driving voltage for the horizontal plates of the CRT. When a sawtooth wave is applied to the horizontal plates, the electron beam is moved from left to right across the CRT face as the ramp rises. The ramp falls to the original level and the trace returns to the left of the screen during the retrace portion of the horizontal signal (Figure 7–21). The trace is cut off, or *blanked,* during the retrace portion so that no trace is visible as the electron beam moves from right to left (Figure 7–22). Figure 7–23 illustrates the movement of the visible spot of the CRT as the horizontal ramp voltage goes through its cycle.

The *sweep generator* produces a sawtooth waveform that is processed by the horizontal amplifiers. It is then used to deflect the electron beam from left to right across the face of the CRT. The resulting sawtooth waveform should have the rate-of-rise amplitude and linearity that will result in a time-measuring reference, or *time base.*

THE TRIGGERING SECTION

Input signals of various shapes and amplitudes are applied to the trigger circuit. This circuit develops from these input signal pulses of uniform amplitude and shape and makes it possible to properly time the start of the sweep. The trigger circuit allows the technician to use either the positive or negative slope to start the waveform (Figure 7–24). The trigger also allows for the selection of any voltage level to trigger the sweep.

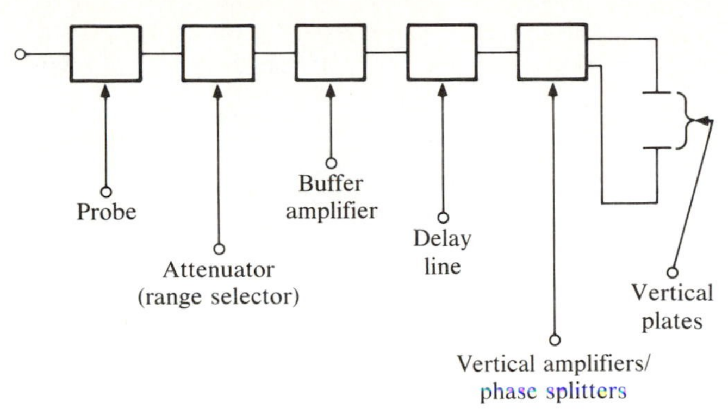

FIGURE 7–20
Block diagram of the vertical section of an oscilloscope

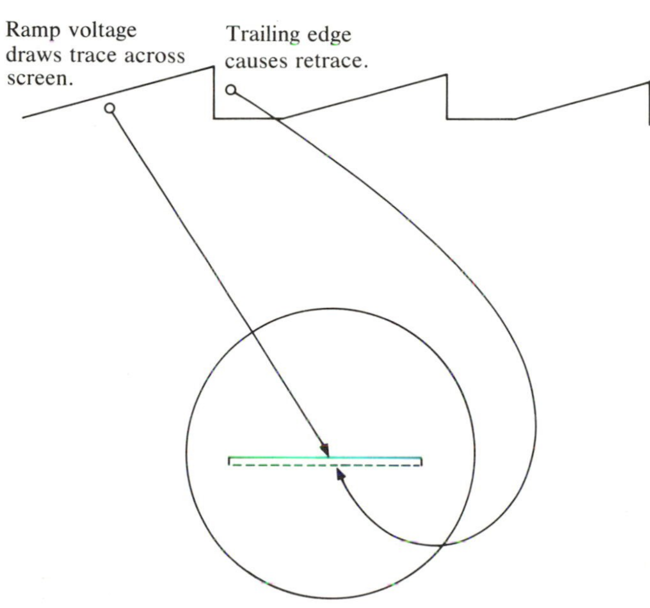

FIGURE 7–21
Action of trace and retrace

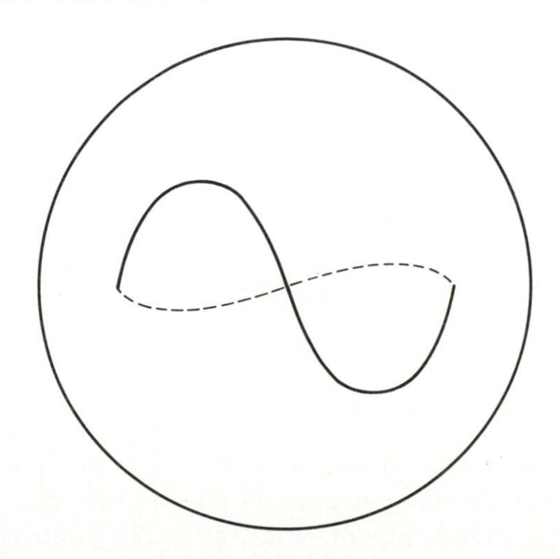

FIGURE 7–22

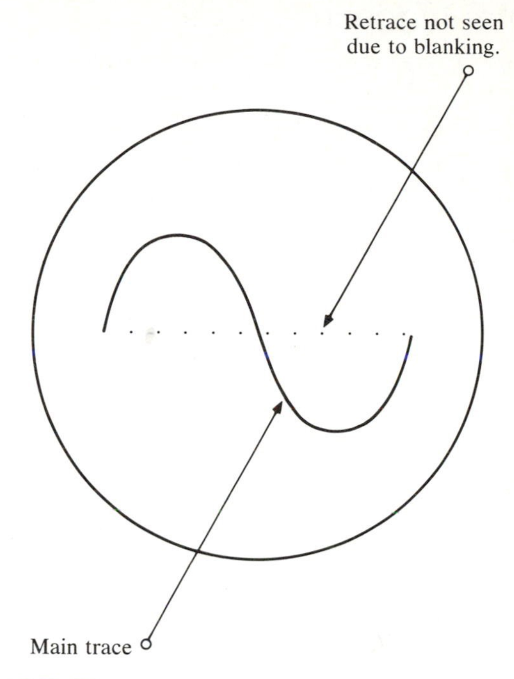

FIGURE 7–23
Effect of blanking the retrace

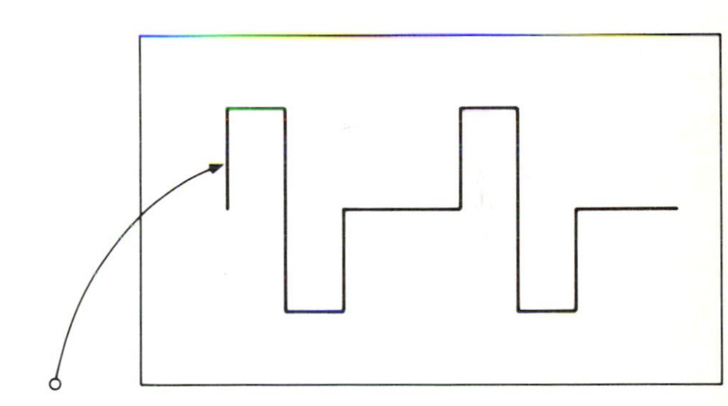

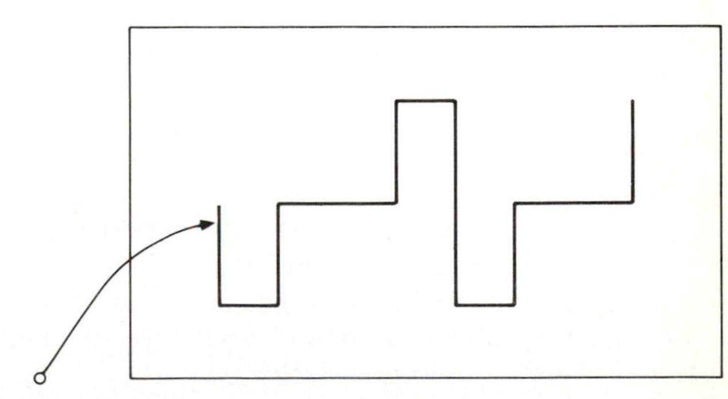

FIGURE 7–24
Triggering

Internal Triggering

In *internal* triggering, the sweep is triggered by the signal being viewed. This triggering is accomplished by using a sample of the signal from the vertical amplifier to begin the trace. The circuitry between the vertical amplifier and the trigger circuit must act as a buffer to keep the trigger circuitry from loading down the vertical amplifier. This circuitry may also have to process the signal by changing its voltage level to a level sufficient to cause the trigger circuit to operate. The trigger level is adjusted by a potentiometer that provides a threshold voltage for a comparator (Figure 7–25).

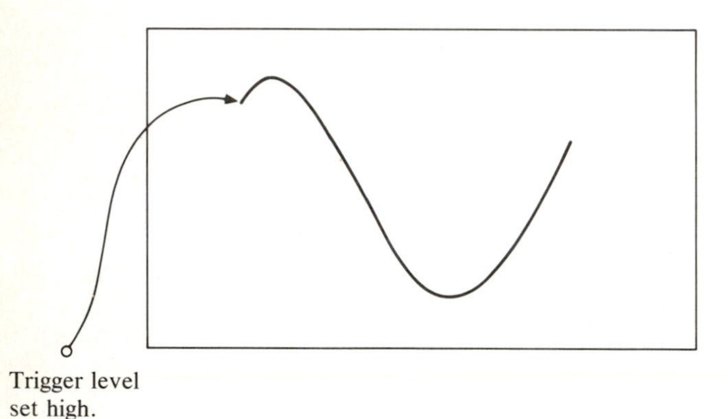

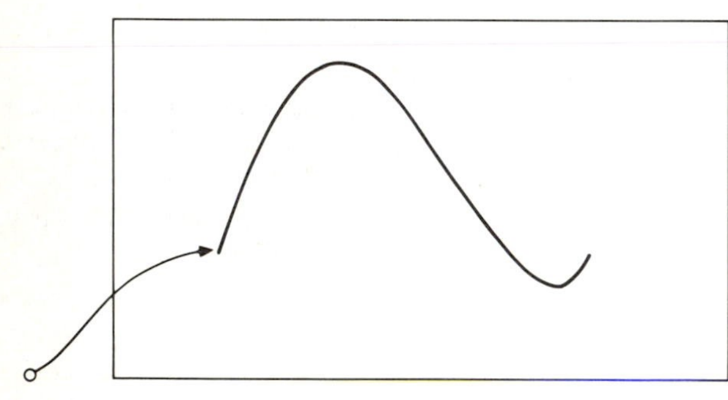

FIGURE 7–25
Adjusting the triggering level

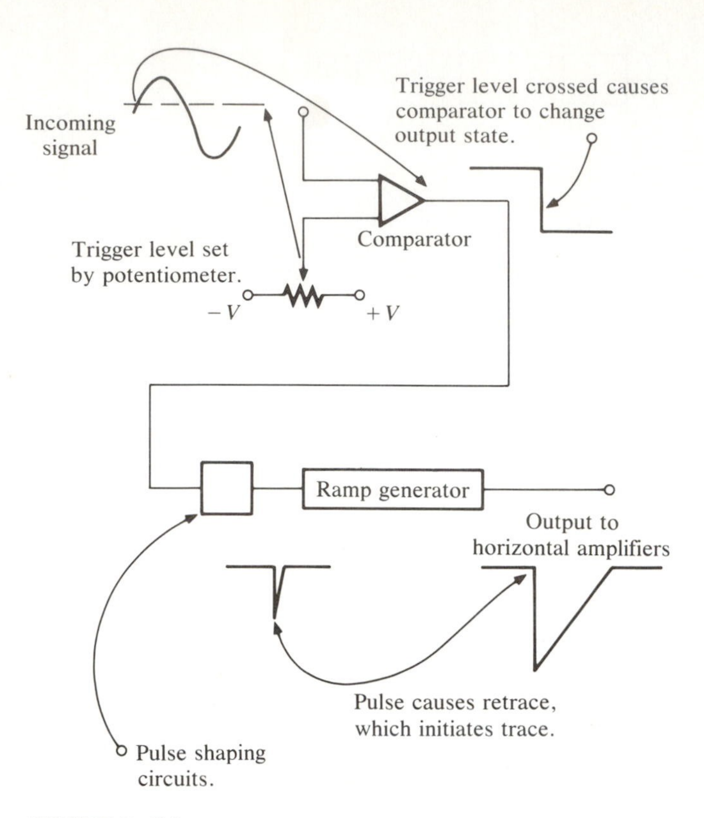

FIGURE 7–26
Block diagram of the horizontal section of an oscilloscope

When the signal from the vertical section reaches the voltage level established by the trigger-level control, the comparator changes state and the resulting pulse is processed and used to control the start of the sweep (Figure 7–26).

Line Triggering

Another triggering mode is *line* triggering. Many signals occur at powerline frequencies (60 Hz or 120 Hz). In line triggering, a sample of the line voltage is available through a transformer winding. The triggering automatically follows any variation in line frequency, making it easy to keep the display synchronized.

External Triggering

Occasions arise when the vertical signal must be observed in its relationship to other events, either different times or different frequencies. This comparison can be made using an *external* trigger. The net effect is to switch the trigger circuit to a terminal on the front panel of the CRO so that an independent trigger source can be used.

LINEARITY OF SWEEP

If the ramp voltage is nonlinear, the time/deflection relationship will not provide an accurate time (frequency) measurement (Figure 7–27). Linearity of sweep is usually ensured by properly designed circuitry. The sweep is controlled by a sawtooth waveform.

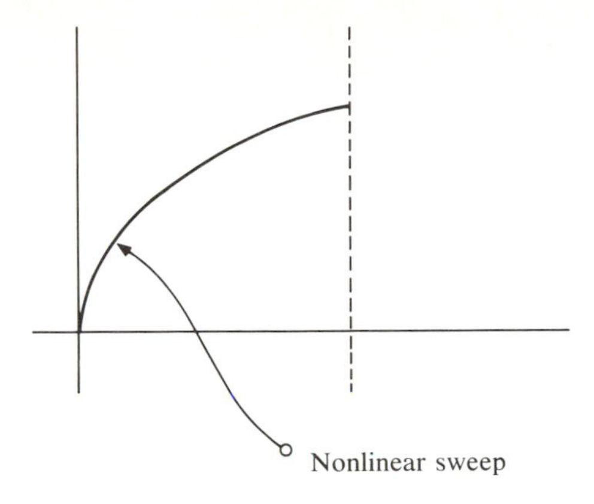

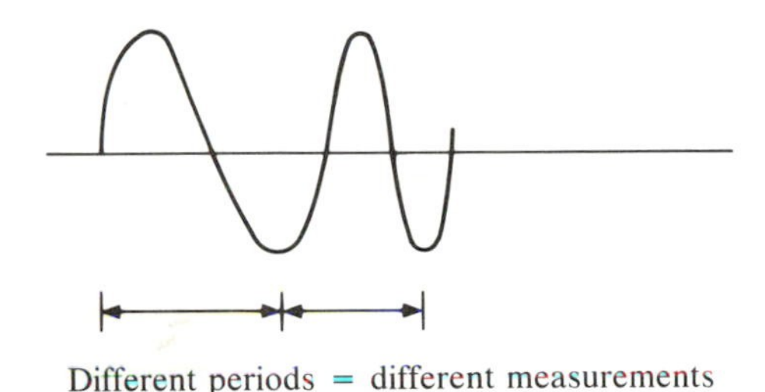

FIGURE 7–27
Nonlinear sweep

THE HORIZONTAL SWEEP

The waveform in Figure 7–28 is a voltage waveform that starts at a given level, rises at a linear rate to some maximum value (linear ramp), and then falls abruptly (retrace) to the original level (holdoff). During the ramp generation, an ac unblanking pulse is applied to allow sufficient beam intensity during sweep time. The unblanking pulse actually turns on the beam for the start of the trace. When the ramp reaches maximum voltage, a blanking pulse is applied to turn off the beam during retrace. The time during which the electron beam is cut off is called *holdoff.* The blanking pulse during holdoff allows the voltage of the sawtooth generator to return to the start level (and the beam to return to the left side of the CRT screen). It also allows time for any transients and retrace to disappear.

When displaying high-speed signals, the trace must be turned on after blanking. This is accomplished by the unblanking pulse. As a result, the start of any signal input to the vertical deflection section is not lost to the blanking but is visible.

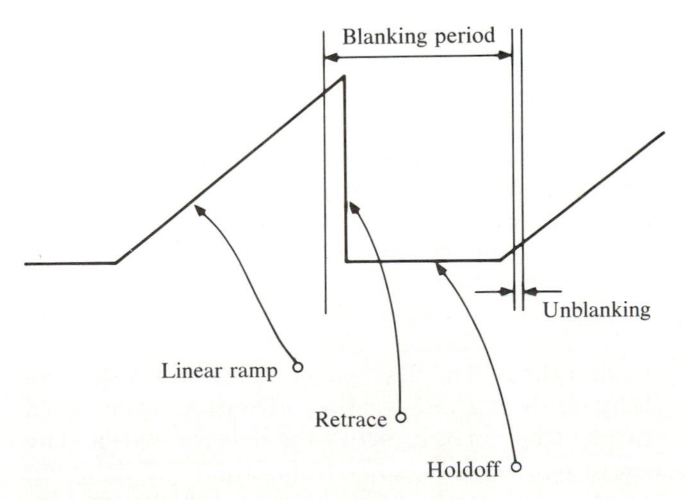

FIGURE 7–28
Linear sweep with blanking and unblanking

DUAL TRACE

Sweep switching is employed by many CRO models to allow two beams to share a single-trace CRT. The sweep-switching system allows the operator to use a single CRO as two systems. The individual sweeps may be labeled "A" and "B." Channel A and channel B can be displayed separately. "A alt. B" allows for channel A and channel B to be displayed alternately. This is accomplished by an electronic switch that processes the input to channel A and then the input to channel B. This can be observed by using a very slow sweep rate (Figure 7–29). This mode is useful for faster signals the technician wishes to observe simultaneously.

In the *chopped* mode, a sample of trace A is shown, and then a sample of trace B. This process is repeated until both traces have been completely shown. The chopping effect can be seen if a very fast sweep rate is used (Figure 7–30). Both the alternating and the chopped modes are dual-trace modes. In many cases only one of the traces may be synchronized at one time, but sometimes by depressing both the A and B channel selectors, both channels can be synchronized while in the alternating mode. Since channels A and B process the signals separately, they trigger independently and may be used for phase measurements. Chopped mode triggering cannot be used for this purpose because its triggering is at a single fixed voltage level.

Many CROs are equipped to accept an external horizontal deflection signal, producing an *x-y* or A-versus-B trace. This requires that a matching amplifier circuit be available for the *x-y* application so that the horizontal (*y*) and vertical (*x*) inputs are amplified equally and linearly. The results are *Lissajous figures.* When *x-y* applications are desired, the horizontal sweep generator is bypassed and the

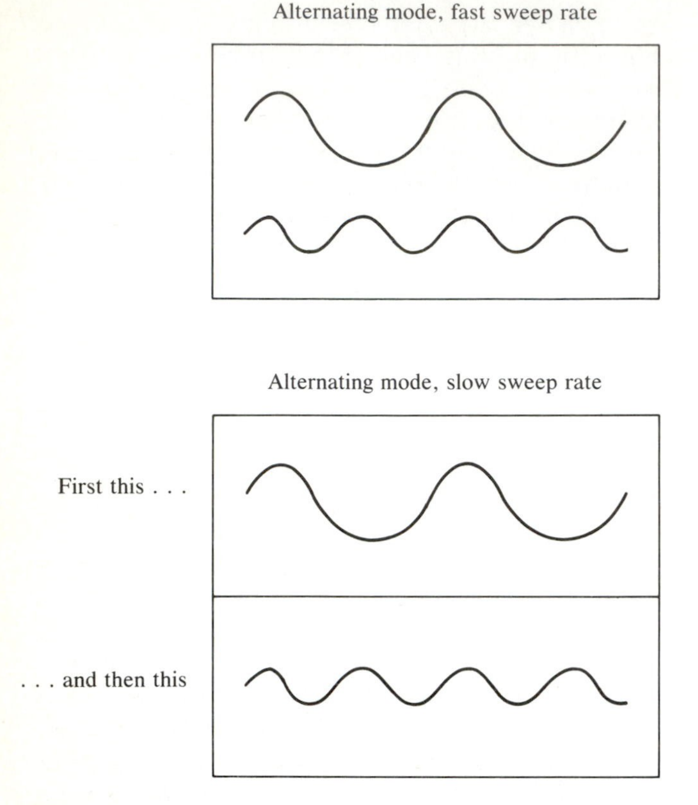

FIGURE 7–29
Alternating sweep mode

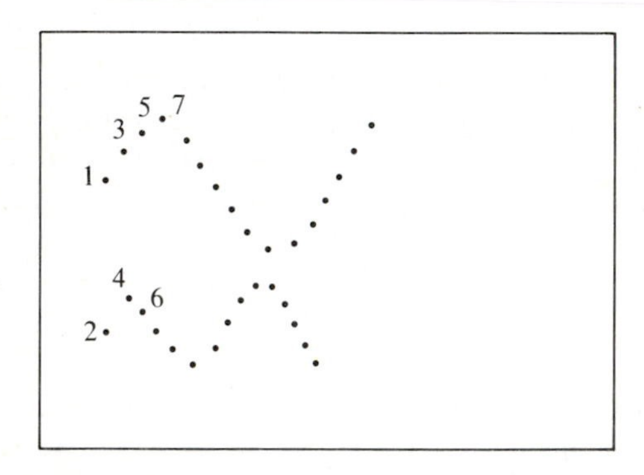

FIGURE 7–30
Chopped-mode plot. The chopped mode plots one point on channel A, one point on channel B, and so on.

signal on the horizontal (or channel B) input is amplified and applied directly to the horizontal plates. The production and use of Lissajous figures will be discussed in Chapter 8.

If a ×10 magnified mode is available and selected, the gain of the amplifier is such that the CRT beam would be driven off the screen. Since there is room for only 1/10 of the trace, the sweep time is reduced to 1/10 the normal sweep time to keep such a uselessly large signal from being displayed. The amplifier actually goes into saturation, and so with the sweep time limited, the signal (1/10 the original) appears wholly within the sweep window and thus this portion of the ramp causes the horizontal beam deflection (Figure 7–31).

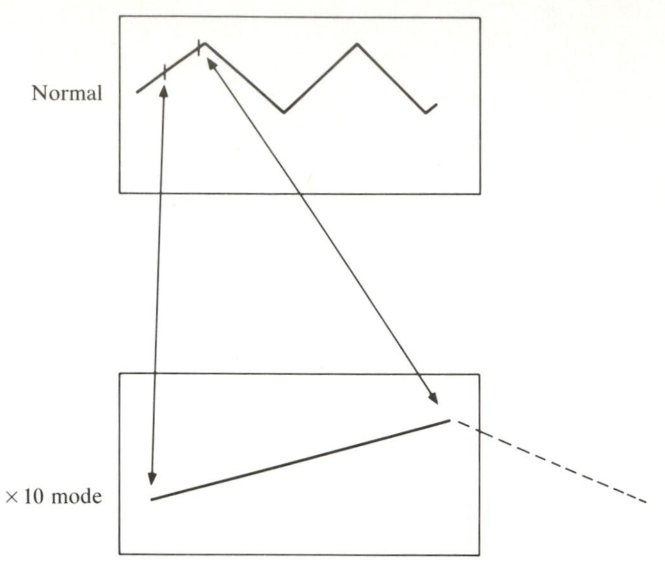

FIGURE 7–31
× 10 horizontal magnifier

The horizontal deflection factor is the voltage required to deflect the electron beam a particular number of graticule divisions. For example, a horizontal deflection factor of 10 V/cm means that a 100-V horizontal sweep voltage will be required to deflect the beam 10 cm across the CRT face in the horizontal direction.

DELAYED SWEEP

Oscilloscopes with delayed sweep capabilities allow the waveform to be observed at two different sweep rates. Many instruments permit a mixed-mode sweep and a delayed sweep (Figure 7–32).

Instruments with delayed sweep capabilities use a special trigger circuit called a *delay pickoff*. The purpose of this circuit is to start or enable a second (or delayed) sweep at a precisely controlled time. Delayed pickoff is measured from the start of the first or *delaying* sweep. Delaying-sweep measurements are based on the use of two calibrated linear sweeps. The first sweep, commonly called the *delaying* sweep, allows the technician to select a specific time for triggering the *delayed* sweep. The delayed sweep is usually faster than the delaying sweep and thus allows for better resolution and increased accuracy in time measurements.

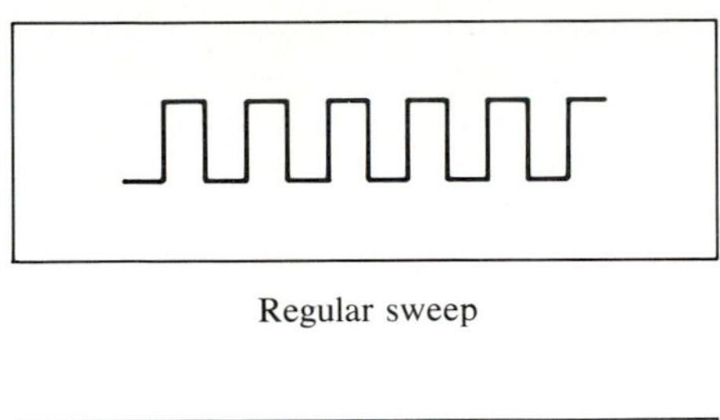

Regular sweep

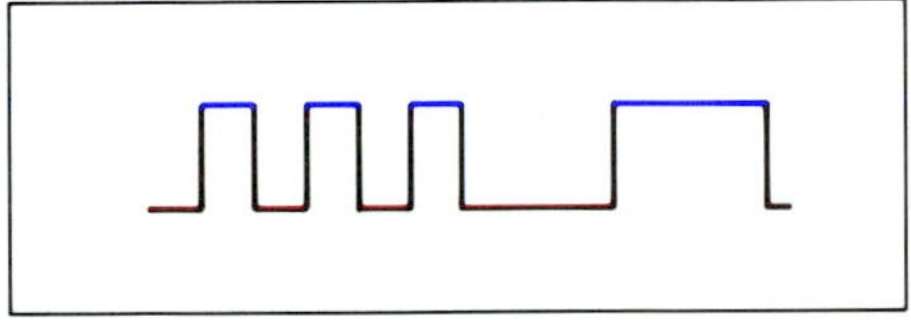

Mixed-mode sweep

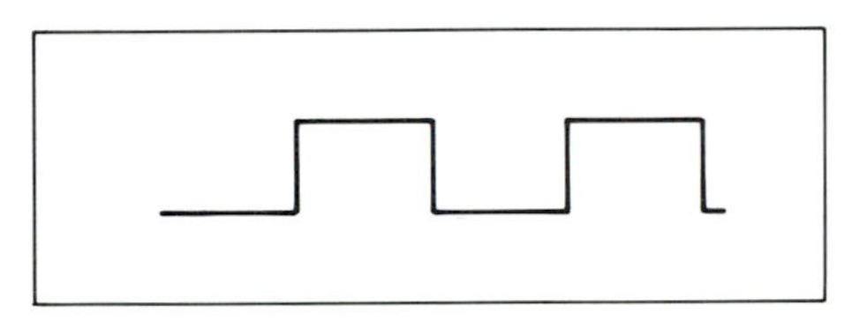

Delayed sweep

FIGURE 7–32
Main, × mixed, and delayed sweep modes

SAMPLING OSCILLOSCOPES

A technique known as *equivalent-time sampling* can be used to extend the frequency range of conventional oscilloscopes to very large bandwidths (e.g., 14 GHz). With equivalent-time sampling, the oscilloscope builds a picture of a waveform by selecting small bits of information during each signal repetition. Eventually enough information is compiled to reconstruct the entire waveform (Figure 7–33).

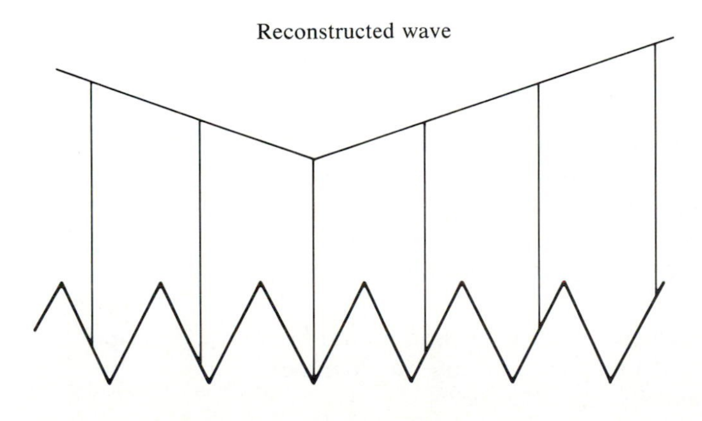

FIGURE 7–33
How a sampling oscilloscope samples

Since sampling oscilloscopes depend on a continuously repeating waveform, they are somewhat limited in their applications. They can attain useful high bandwidths, however.

Sampling oscilloscopes have been commercially available since about 1959. Their primary advantage is their ability to respond to small fast-changing signal voltages better than conventional oscilloscopes.

The most fundamental sampling principle is analogous to the principle of optical stroboscopes (strobes). A strobe generates very brief flashes of light which, when properly timed, make rapidly moving things appear to be at rest or in very slow motion. By generating very brief electrical pulses, sampling scopes graphically depict fast-changing voltages as if the changes were occurring slowly. The pulses allow the operator to "look" at a repetitive signal one point and one cycle at a time and picture the characteristics of the signal graphically over a period of many cycles.

While the faster response of sampling scopes is their principal advantage, they are also almost free from display aberations caused by overscanning the screen. In addition, the process of sampling lends itself to digitizing time measurements directly.

DIGITAL STORAGE OSCILLOSCOPES

Digital oscilloscopes store data representing waveforms in a digital memory. The waveform undergoes digitizing and then reconstruction. Digitizing consists of sampling and quantizing. Sampling is the process of building the waveform by selecting bits of information about the waveform at discrete points in time. Quantizing is conversion to a binary quantity by an analog-to-digital (A/D) converter. Digital storage oscilloscopes store the binary data. This data can be recalled, processed, or used to control other processes.

Although digital storage oscilloscopes show superior bandwidth—with the memory useful in many applications—they are expensive and are susceptible to errors when used with very narrow pulses. The reconstructed signal may not be a continuous presentation since binary data is incremental. The display may be a set of dots.

ANALOG STORAGE OSCILLOSCOPES

Figure 7–34 illustrates the operation of an analog storage oscilloscope. The flood gun "floods" the target with an electron beam at all times. Writing is accomplished by gating on the writing gun with the

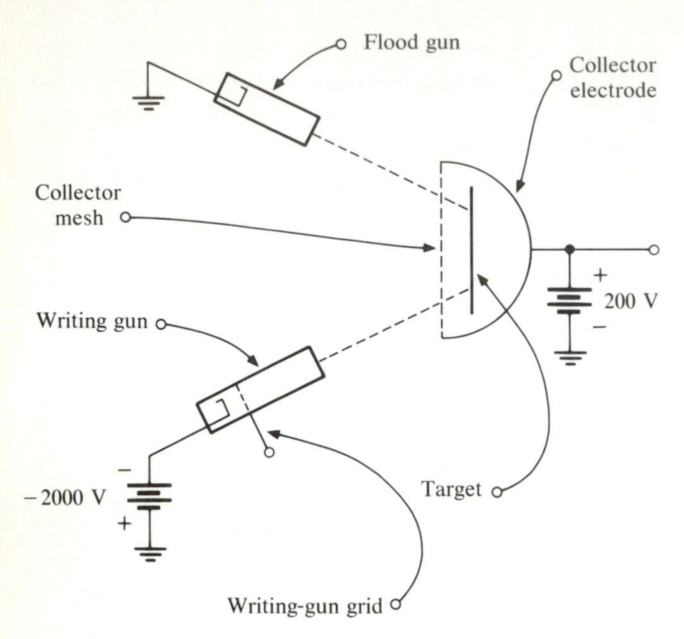

FIGURE 7–34
Diagram of analog storage oscilloscope

writing-gun grid. The combined electron beams are sufficient to cause the target to produce an image of the signal from the writing gun. The constant bombardment from the flood gun causes the target to retain the signal. Erasure is accomplished by pulsing the collector negative, which overcomes the stabilizing current due to the flood gun.

SUMMARY

Oscilloscopes accept electronic signals of many frequencies and amplitudes. The vertical section processes the signals and delivers them to the vertical deflection plates. The horizontal section picks off a sample of the vertical signal and uses it to trigger the sweep of the horizontal control voltage. This ramp voltage is delivered to the horizontal deflection plates. The result is an x-y plot on the face of the cathode ray tube. This plot is a representation of the behavior of the incoming signal that can be used for taking voltage, frequency, and phase measurements.

EXERCISES

1. How does an oscilloscope draw a graph?
2. What is a graticule?
3. Describe the operation of:
   a. Intensity control   b. Focus control
4. Describe how position (vertical and horizontal) controls work.
5. Why does the input coupling control usually contain a dc, an ac, and a ground setting?
6. Which section allows the operator to watch millivolt and kilovolt signals on an oscilloscope using the same set of amplifiers for both ranges?
7. Briefly describe the operation of the horizontal section of an oscilloscope.
8. Describe:
   a. Unblanking   b. Blanking   c. Retrace   d. Trigger level
9. What is the purpose of the delay line in the vertical section?
10. What is the difference between the alternating mode and the chopped mode in a dual-trace oscilloscope?
11. How does the trigger-level control work?
12. What is the difference between the slope (positive or negative) control and a control that inverts the signal?
13. Describe a use for each of the following:
    a. Internal triggering   b. External triggering   c. Line triggering
14. What is a 10:1 probe?
15. What is the difference between a dual-trace and a dual-beam oscilloscope?
16. What is the main reason for using a sampling oscilloscope?
17. What is unique about a storage oscilloscope?
18. What is a beam finder?

LAB PROBLEM 7–1
Oscilloscope Characteristics and Measurements

Equipment
One lab standard oscilloscope; one function generator; one 56-kΩ resistor, ¼-W; one 5-kΩ potentiometer; one single-pole, single-throw switch.

Object 1
To compare the bandwidths of different oscilloscopes.

Procedure
Use a laboratory standard CRO as a standard of reference in maintaining the output of the signal generator at a constant level for each of the frequency settings. Use a convenient ac voltage for all measurements. Increase the frequency of the signal source until significant falloff can be observed. The frequency at which falloff becomes significant represents the upper limit of the frequency response of the CRO. Use a different generator for higher frequencies if necessary. Record the results in Table 7–2. Repeat this procedure for three different types of oscilloscopes or as many as are available. Graph the results for all three scopes on Figure 7–35; label the curves carefully.

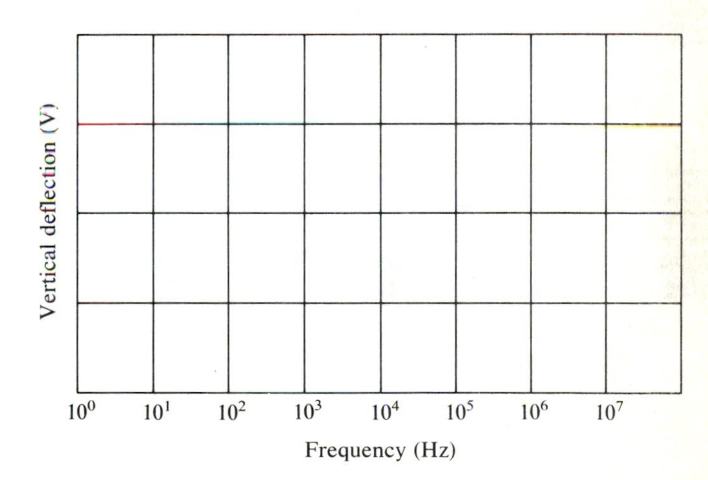

FIGURE 7–35
Complete the graph.

TABLE 7–2
Complete the table.

Oscilloscope 1		**Oscilloscope 2**		**Oscilloscope 3**	
Frequency (Hz)	*Deflection (V)*	*Frequency (Hz)*	*Deflection (V)*	*Frequency (Hz)*	*Deflection (V)*

Object 2

To measure the sensitivity of the vertical amplifier.

Procedure

Wire the circuit in Figure 7–36.

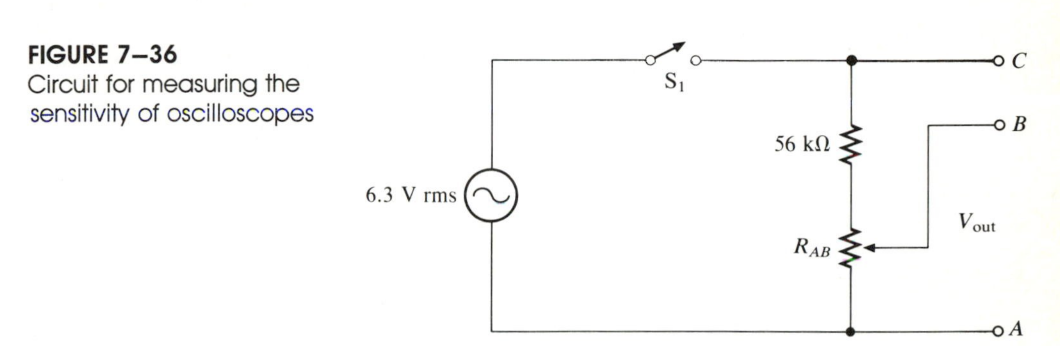

FIGURE 7–36
Circuit for measuring the sensitivity of oscilloscopes

Set the vertical attenuator for maximum scope sensitivity (gain). Connect the scope to measure V_{out} and adjust R_{AB} until a waveform that gives approximately 1-cm vertical deflection on the CRT. Open switch S_1 and measure R_{AB} and R_{AC}. Use the formula

$$\text{Sensitivity z } \frac{R_{AB}}{R_{AC}}(6.3)(2.828)(10) \text{ mV/cm}$$

Compare this to the rated sensitivity of the scope in the operation manual.

8 Oscilloscopes — Measurement

OUTLINE

OBJECTIVES
TURNING ON, AND CHECKING THE CALIBRATION
dc MEASUREMENT WITH AN OSCILLOSCOPE
ac MEASUREMENT WITH AN OSCILLOSCOPE
TIME AND FREQUENCY MEASUREMENTS
DUAL-TRACE METHOD OF PHASE MEASUREMENT
LISSAJOUS FIGURES
☐ Phase Measurements Using Lissajous Figures
☐ Frequency Measurements Using Lissajous Figures
A + B AND A − B
USING THE DELAYED SWEEP MODE
z-AXIS MODULATION
NOTES ON FIELD SERVICE
SUMMARY
EXERCISES
LAB PROBLEM 8–1—Oscilloscope Familiarization

OBJECTIVES

After completing this chapter, you will be able to:

☐ Turn on an oscilloscope and check to see if it is calibrated.
☐ Use an oscilloscope to measure dc and ac voltages.
☐ Use an oscilloscope to measure time and frequency.
☐ Use a two-channel oscilloscope to take phase measurements.
☐ Use Lissajous figures to take phase measurements.
☐ Use Lissajous figures to take frequency measurements.
☐ Measure the sum (A + B) or the difference (A − B) of the amplitudes of two signals.
☐ Use the delay sweep capability to isolate portions of a waveform.
☐ Take measurements with the oscilloscope in the delayed sweep mode.
☐ Use the *z*-axis modulation method for frequency measurements.
☐ Discuss the limitations on field service of an oscilloscope.

This chapter presents the applications of the oscilloscope as a measuring device starting with the initial setup. The oscilloscope is one of the most versatile instruments available. It can be used in many ways to measure voltage, frequency, and phase angle. Two-channel measurements are very useful and are presented. Also discussed are the use of Lissajous figures for various measurements, the use of the delayed sweep mode, and *z*-axis modulation. This chapter promotes an appreciation of the versatility and power of the oscilloscope as a measuring tool.

TURNING ON, AND CHECKING THE CALIBRATION

After turning on an oscilloscope, the technician should check the calibration of the instrument. A trace will appear if the oscilloscope is set for automatic triggering. A "no trace" condition may be caused by poor centering of the beam. Beam positioning can be checked by adjusting the vertical and horizontal position controls. A beam finder will locate the beam if it is out of the display area. This button should be depressed for only a short time since it causes a very bright spot to appear in the viewing area. The trace will also be absent if:

1. The intensity is set too low.
2. The focus is far out of adjustment.
3. The attenuator (range switch) setting is causing the signal to appear so big on the display that it cannot be found.
4. In a multichannel oscilloscope, the correct channel has not been selected.

Most oscilloscopes have a standard signal generator (usually a square-wave generator) which produces a calibration waveform. The technician should check the probe to see if it is a 10:1 probe and if it has other adjustments. Some oscilloscope probes have slide switches that can select ×1, ground, or ×10. The technician should make sure the input to the vertical section through the probe is not set on ground. Additionally, the vertical section usually contains an ac-GND-dc mode switch which should be checked to ensure that it is not in the ground position. Both the vertical and horizontal range switches may have a calibration adjustment, allowing selection of all values between those indicated on the dial (Figure 8–1).

In Figure 8–1, the 1-ms/cm and 5-ms/cm settings are automatic if the calibration adjustment is in the proper position. If the calibration adjustment is not in the CAL position, the multiplier will have a value between 1 ms/cm and 5 ms/cm. Although this capability may be useful for some applications, for measurement purposes it is important to know the exact multiplier value.

Once the probe is connected to the calibration signal terminal, the wave will indicate whether or not the oscilloscope is calibrated. The calibrate adjustment can be used to compensate for some small discrepancies. In the case of radical differences, the oscilloscope should be calibrated by trained technicians since other problems may be indicated.

Trace brightness should be kept at a comfortable minimum. If the trace is too bright, the electron beam producing it has too much energy and can physically damage the phosphor (called *burning*). In general, high-frequency traces require greater brightness than lower-frequency traces since the beam is sweeping so fast that it does not have time to impart much energy to any point of the phosphor. Slower traces have sufficient time to energize the phosphor particles, and therefore lower brightness levels should be used with them.

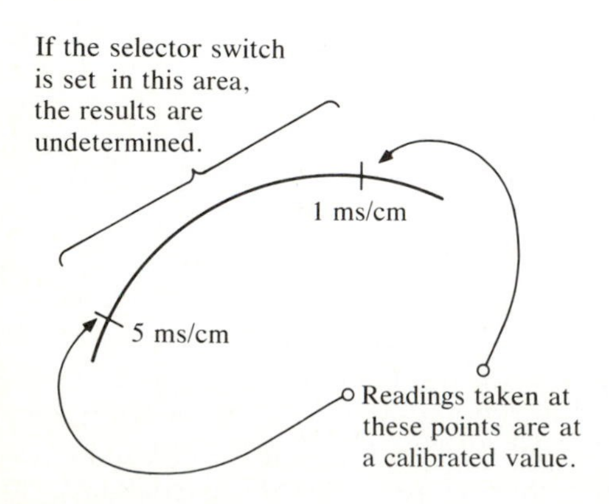

FIGURE 8–1
Calibration and settings of a CRO

dc MEASUREMENT WITH AN OSCILLOSCOPE

dc measurements on an oscilloscope can be made if there exists in the vertical input section a direct input line. ac input lines have a blocking capacitor to block out any dc level (Figure 8–2).

With zero dc on the input, the trace can be moved vertically until it lines up with one of the graticule lines, which will be used as a reference. One way to establish a zero dc reference is to switch the vertical input to ground. This can usually be done on the front panel or on the barrel of the probe (Figure 8–3).

This ground position is used not only as a zero reference input but also to discharge any capacitance that may be in the test circuit as the oscilloscope is switched from dc to ac. The dc to be measured is then applied to the vertical input. The trace will be deflected upward if the dc has a posi-

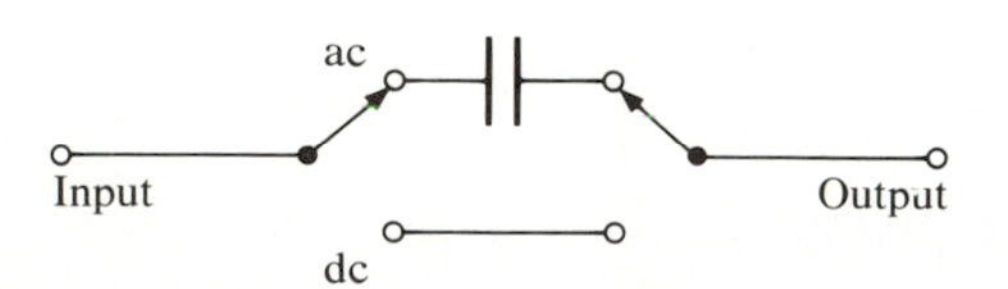

FIGURE 8–2
Circuit for the selector switch. The selector switch allows ac to pass while blocking dc.

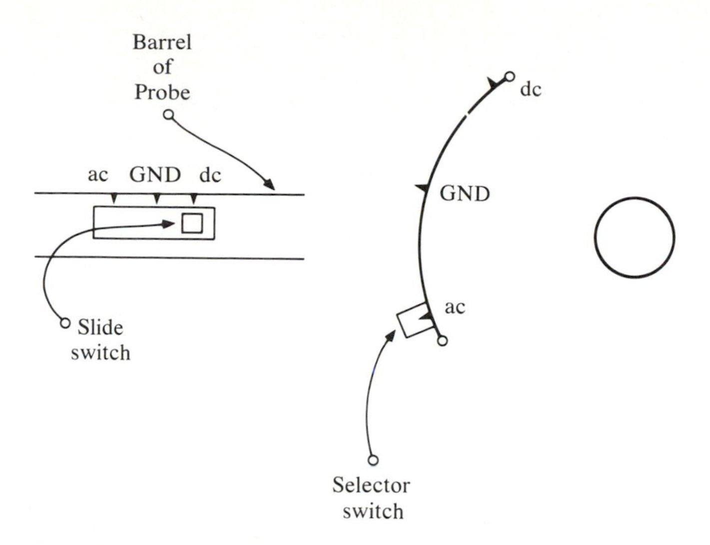

FIGURE 8–3
Selector switch. The ground connection between the dc and ac settings grounds out (discharges) capacitors in circuits under test, protecting the oscilloscope circuitry.

tive polarity, and downward if the input has a negative polarity (Figure 8–4).

The setting of the vertical attenuator (in volts per centimeter) can be used to determine the dc input level. The attenuator should be set to give maximum deflection, thus minimizing reading errors.

EXAMPLE 8–1
If the trace if deflected upward 2.5 cm when a dc level is applied to the vertical input of an oscilloscope and the vertical attenuator is set at 0.1 v/cm, what is the dc voltage?

Solution:

$$2.5 \text{ cm} \times 0.1 \text{ V/cm} = 0.25 \text{ V} \qquad \textbf{(8–1)}$$

This is a dc voltage with a positive polarity relative to ground.

If a $\times 10$ probe is used, the display is $\frac{1}{10}$ the magnitude of the actual input voltage. The results must be multiplied by 10 to get the correct value (hence the term $\times 10$ probe).

---

ac MEASUREMENT WITH AN OSCILLOSCOPE

The most accurate ac voltage measurement using an oscilloscope is the peak-peak voltage. Locating the ac zero level from which the peak value would be measured is difficult, and the rms and average values cannot be measured directly.

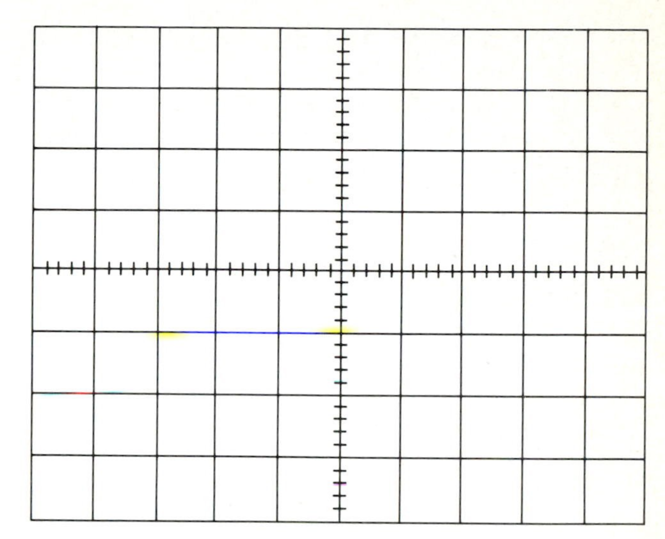

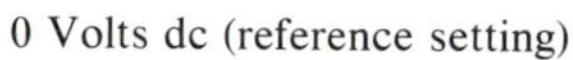
0 Volts dc (reference setting)

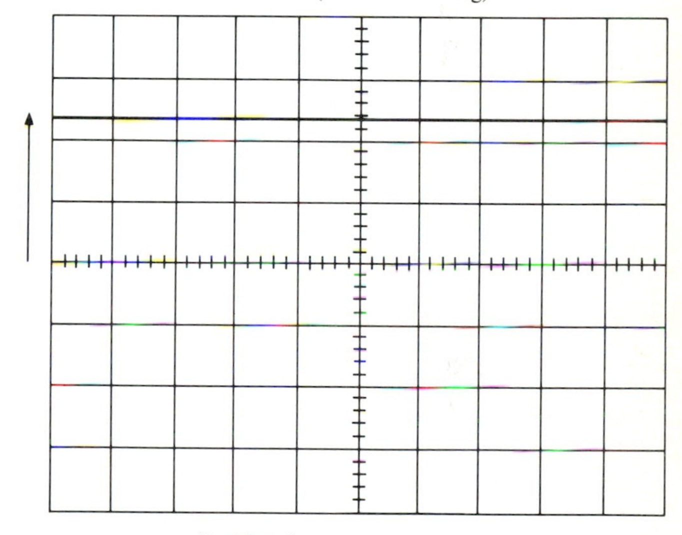

Positive dc trace moves upward.

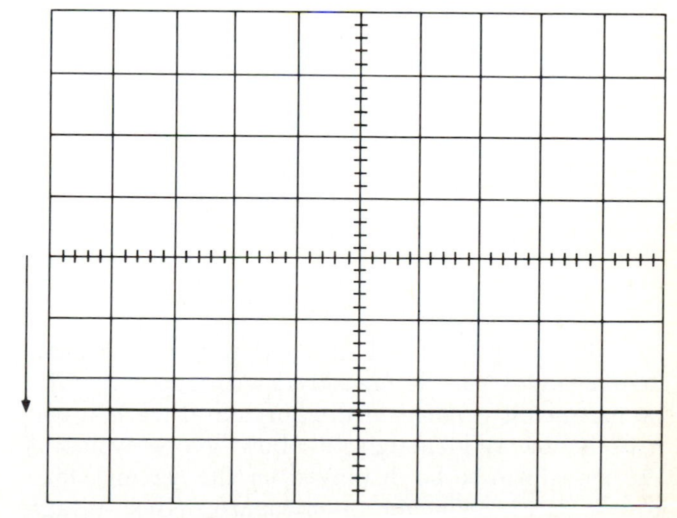

Negative dc trace moves downward.

FIGURE 8–4
Measuring dc voltage using an oscilloscope

The ac is applied to the vertical input. The selector switch should be moved to the ac position to block out any dc that may be part of the input signal and could thus interfere with the measurement.

EXAMPLE 8–2

In measuring the ac ripple voltage across the output of a 300-V power supply, the 300 V dc are not important and should be blocked out. If this is not done, it may be impossible to locate the ac voltage on the display (Figure 8–5).

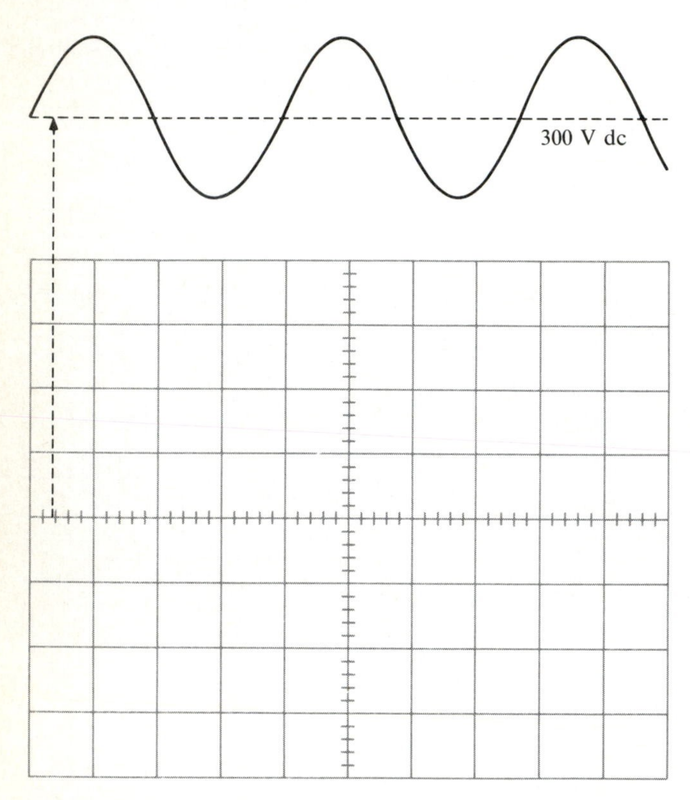

FIGURE 8–5
dc not blocked. When a large dc voltage is not blocked out, the ac voltage cannot be measured.

The trace is lined up so that the peak-peak displacement can be measured with the graticule. Sometimes it is easier if the horizontal trace sweep rate is lowered enough to allow many cycles of the waveform to be displayed on the screen. Once the vertical deflection is measured, the voltage can be calculated by multiplying the peak-peak displacement by the setting of the vertical attenuator.

EXAMPLE 8–3

Find the peak-peak voltage for the waveform in Figure 8–6.

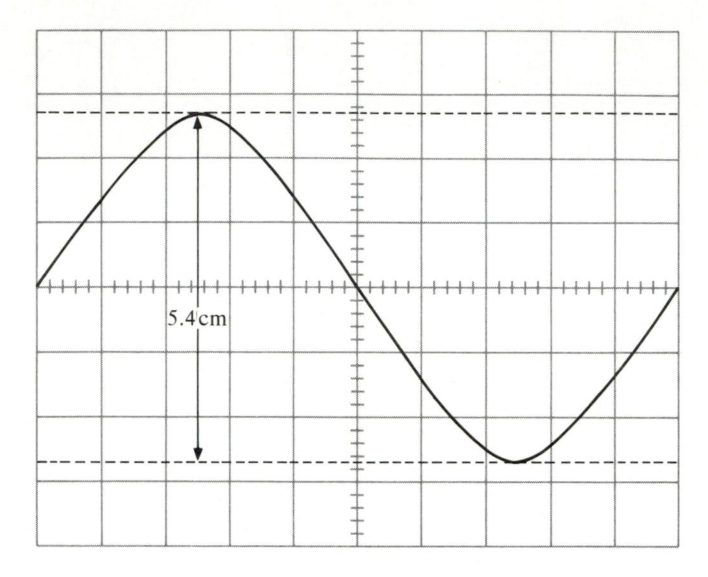

$5.4 \text{ cm} \times 5 \text{ V/cm} = 27 \text{ V peak-peak}$

FIGURE 8–6
Peak-peak measurement

Solution:
In the figure, the vertical attenuator is set at 5 V/cm, and the voltage therefore is

$$5.4 \text{ cm} \times 5 \text{ V/cm} = 27 \text{ V peak-peak} \quad \textbf{(8–2)}$$

If a ×10 probe is used, the display is 1/10 the actual voltage, and this reading will correspond to 270 V peak-peak.

Many oscilloscopes have a ×5 vertical magnifier, which magnifies the vertical displacement of the trace to five times its actual size. This capability is handy in measuring small voltages. Remember, however, that if the ×5 magnifier is used, the results must be divided by 5 to get the correct values.

EXAMPLE 8–4

The vertical displacement of a waveform (peak-peak) is measured to be 3.2 cm, the vertical attenuator is set at 0.01 V/cm, and a ×10 probe and ×5 magnifier are in use. What is the peak-peak voltage of the input waveform?

Solution:

$$V_{\text{pk-pk}} = 3.2 \text{ cm} \times 0.01 \text{ V/cm} = 0.032 \text{ V} \quad \textbf{(8–3)}$$

and because a $\times 10$ probe is used,

$$0.032 \text{ V} \times 10 = 0.32 \text{ V} \tag{8–4}$$

and because of the magnifier,

$$\frac{0.32 \text{ V}}{5} = 0.064 \text{ V} \tag{8–5}$$

In summary, when taking any ac voltage measurements:

1. Check the oscilloscope calibration.
2. Remember the $\times 10$ probe if one is used.
3. Check to see if the vertical has a magnifier and if it is engaged.
4. Make sure ac has been selected as the input mode.
5. Use as much of the screen as possible to minimize reading errors.
6. Multiply the vertical displacement by the vertical attenuator setting.
7. Multiply the result by 10 if a $\times 10$ probe is used.
8. Divide the result by any magnifying factor in the vertical section.

TIME AND FREQUENCY MEASUREMENTS

The horizontal oscillator in most oscilloscopes is calibrated against a time standard. The horizontal selector is marked in seconds, milliseconds, or microseconds per centimeter. When a waveform is applied to the vertical input and an appropriate vertical displacement has been selected, a horizontal setting is chosen to give the largest possible displacement in the horizontal direction. The large displacement reduces the possibility of error due to reading the oscilloscope. If several waveforms are very close together, the possible reading error is large compared to the actual reading. The same amount of reading error in waveforms spaced farther apart represents a smaller fraction of the actual value and thus a smaller percentage of error.

A displacement reading taken from the graticule should be taken from one point on a waveform to the corresponding point on the next waveform. Measurements taken between points on the signal where the steepest slope occurs will yield the most accurate readings. Good points are usually point-to-point, leading–edge-to-leading–edge or trailing–edge-to-trailing–edge (Figure 8–7). The reading is usually in centimeters and is multiplied by the setting on the horizontal time base. The result is the time of the waveform, not the frequency.

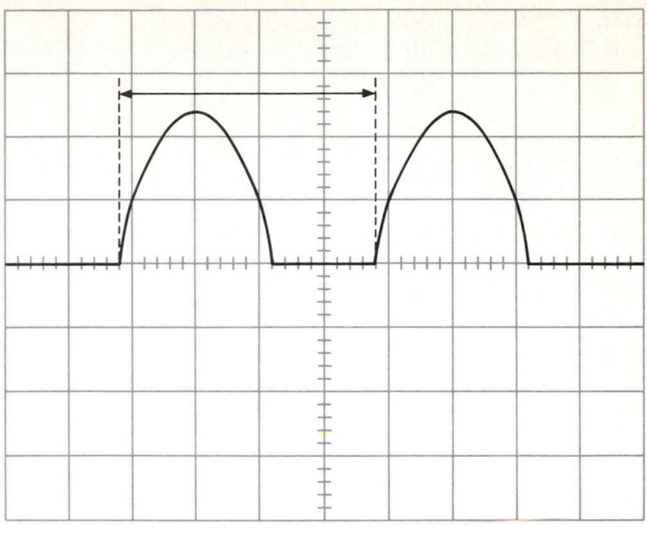

Measuring from one point on one wave to corresponding point on next wave

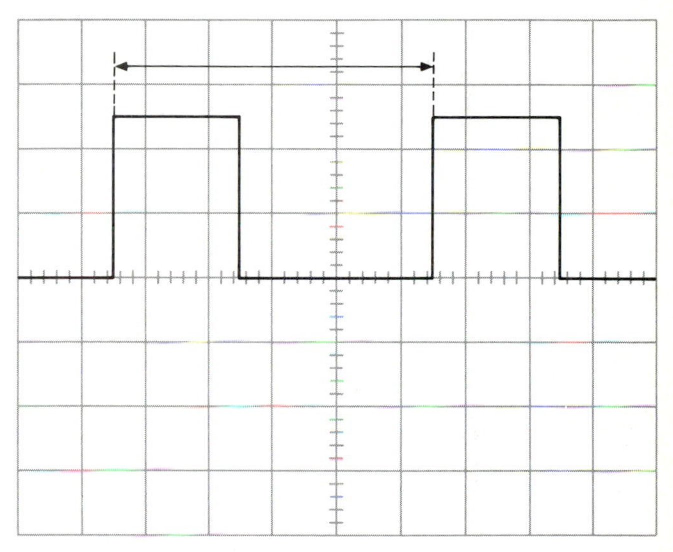

Measuring from leading edge to leading edge

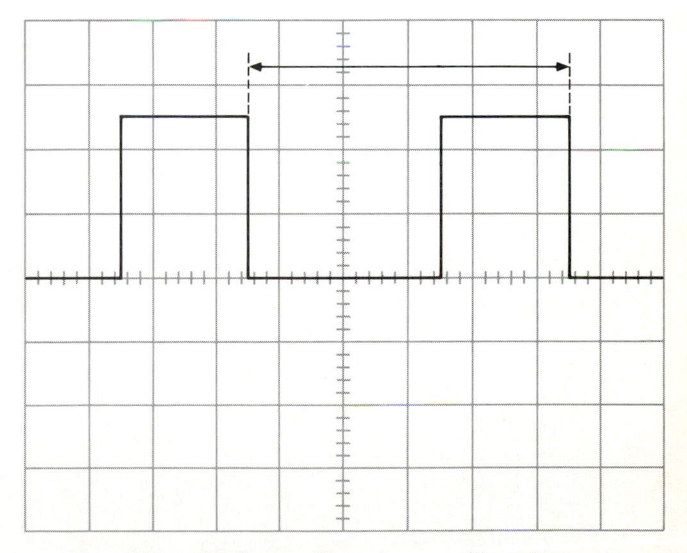

Measuring from trailing edge to trailing edge

FIGURE 8–7
Techniques for measuring time on an oscilloscope display

EXAMPLE 8–5
If the horizontal time base is set on 20 μs/cm and the horizontal displacement reading of one waveform is 7.8 cm, the time (T) of the waveform is

$$T = 7.8 \text{ cm} \times 20\ \mu\text{s/cm} = 156\ \mu\text{s} \qquad \textbf{(8–6)}$$

The frequency (f) of the waveform is the reciprocal of the time:

$$f = \frac{1}{T} \qquad \textbf{(8–7)}$$

where T is the period in seconds. Therefore,

$$f = \frac{1}{156\ \mu\text{s}} = 6410 \text{ Hz} \qquad \textbf{(8–8)}$$

A ×10 probe affects only the vertical section. However, there may be a ×10 magnifier on the oscilloscope. If the magnifier is in use, the figure seen is 10 times larger than normal. The true horizontal displacement is 1⁄10 the image on the CRT. The time measurement is also 1⁄10 of what has been calculated, and the frequency is 10 times greater than the value shown.

EXAMPLE 8–6
Find the frequency of the waveform in Figure 8–8 if the horizontal time base is set at 0.1 μs/cm and a ×10 magnifier is used in the horizontal section.

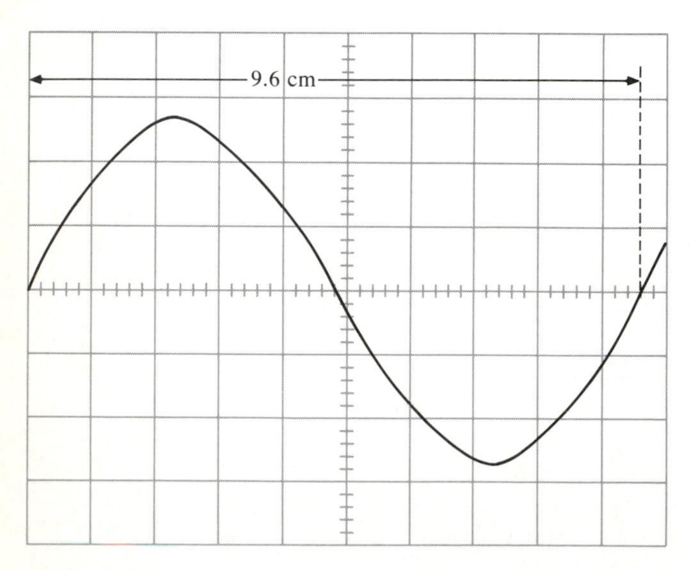

FIGURE 8–8

Solution:

$$T = \frac{9.6 \text{ cm} \times 0.1\ \mu\text{s/cm}}{10} = 0.096\ \mu\text{s} \qquad \textbf{(8–9)}$$

$$f = \frac{1}{0.096\ \mu\text{s}} = 10.417 \text{ MHz} \qquad \textbf{(8–10)}$$

or

$$T = 9.6 \text{ cm} \times 0.1\ \mu\text{s/cm} = 0.96\ \mu\text{s} \qquad \textbf{(8–11)}$$

$$f = \frac{10}{0.96\ \mu\text{s}} = 10.417 \text{ MHz} \qquad \textbf{(8–12)}$$

One of these corrections, but *not* both, must be made for the use of the ×10 magnifier!

DUAL-TRACE METHOD OF PHASE MEASUREMENT

If the phase difference between two signals is to be measured, each signal should be used as the input to a separate channel so that both signals can be displayed simultaneously. Use the chopped mode. A horizontal measurement between the two signals will then be possible.

In Figure 8–9, d is the horizontal measurement between the two signals. The period T of one complete cycle of a waveform is also taken. The ratio of d to T is the ratio of the phase angle to one complete cycle.

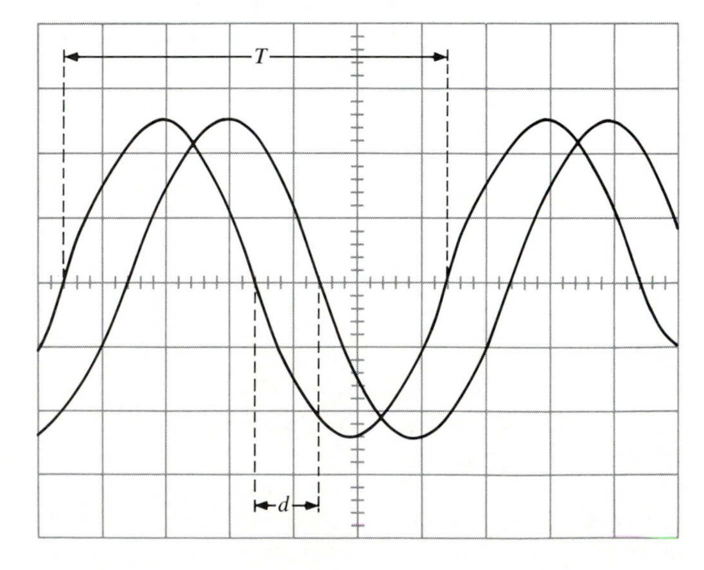

$d/T \times 360° =$ phase angle

FIGURE 8–9
Using two traces to measure phase angles

EXAMPLE 8–7
Find the phase angle between two signals if the horizontal measurement between the signals is 1 cm and the period of one waveform cycle is 6 cm.

Solution:

$$\frac{1\text{ cm}}{6\text{ cm}} \times 360° = 60° \qquad \textbf{(8–13)}$$

The phase angle between the two signals is 60°.

LISSAJOUS FIGURES

If the oscilloscope has the *x*-versus-*y* or A-versus-B capability, the technician can apply one signal to the vertical deflection plates while applying a second signal to the horizontal deflection plates. The horizontal sweep section is automatically disengaged at this time. The resulting waveform is called a *Lissajous figure*. This mode can be used to measure phase or frequency relationships between two signals.

Phase Measurements Using Lissajous Figures

If the signals forming a Lissajous figure are of equal frequency but there is a phase difference between them, the resulting waveform might look like Figure 8–10. A point-by-point plot may be taken to show how the figure is formed. Figure 8–11 represents a 45° phase difference between two signals of equal frequency. Figure 8–12 illustrates some other phase relationships shown by Lissajous figures.

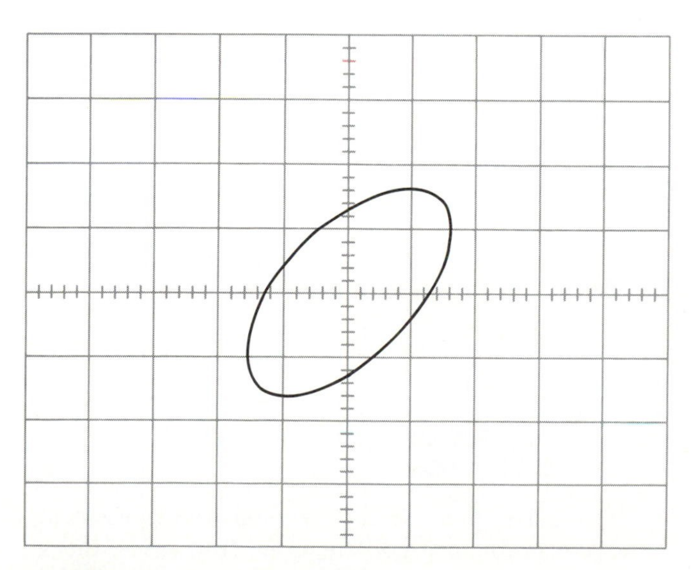

FIGURE 8–10
Lissajous figure resulting from two input signals out of phase but with the same frequency

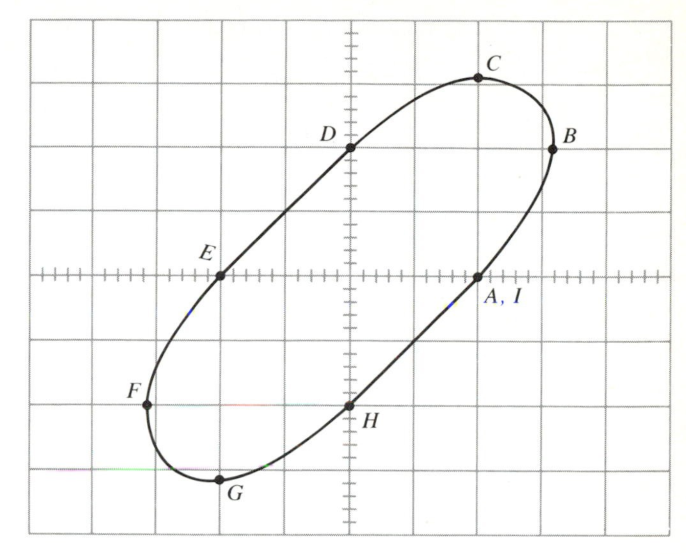

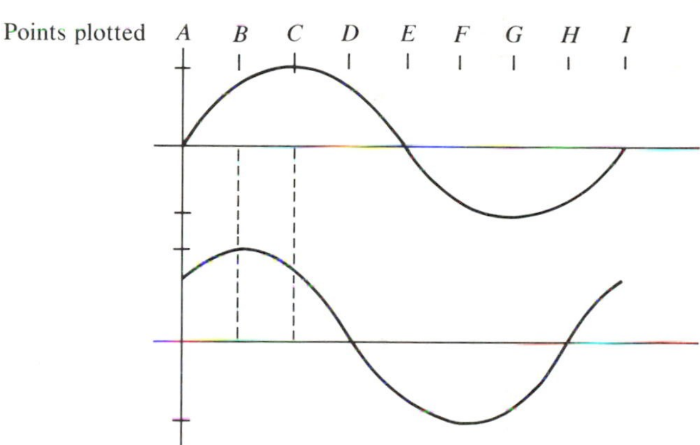

FIGURE 8–11
Plotting a Lissajous figure showing two signals 45° out of phase

If a stable pattern can be achieved, a reasonably accurate determination of the phase angle is possible. Only at 0° and 180° does the Lissajous figure give a unique answer. All other figures are associated with two phase-angle values. For example, the figure for 225° is the same as the figure for 135°. Thus, the figure gives an angle and its supplement.

In Figure 8–13,

$$\sin\theta = \frac{y_0}{y_{max}} \qquad \textbf{(8–14)}$$

$$\theta = \arcsin\frac{y_0}{y_{max}}$$

where y_0 is the displacement between the y-intercepts if the figure is centered, and y_{max} is the vertical peak-peak displacement. This yields θ, which is a reference angle and not necessarily the actual phase measurement.

EXAMPLE 8–8
Refer to Figure 8–14.

$$\sin \theta = \frac{2 \text{ cm}}{5 \text{ cm}} = 0.4 \qquad \textbf{(8–15)}$$
$$\theta = 23.6°$$

Angle θ = 23.6°, but this is a reference angle. Since the figure is nearly a line representing 180°, it represents either 180° + 23.6° = 203.6° or 180° − 23.6° = 156.4°.

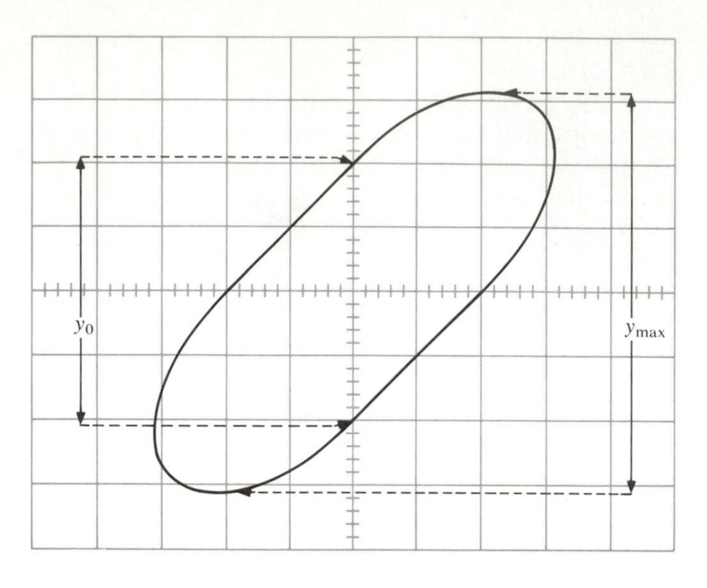

FIGURE 8–13
Phase-angle measurement using Lissajous figures

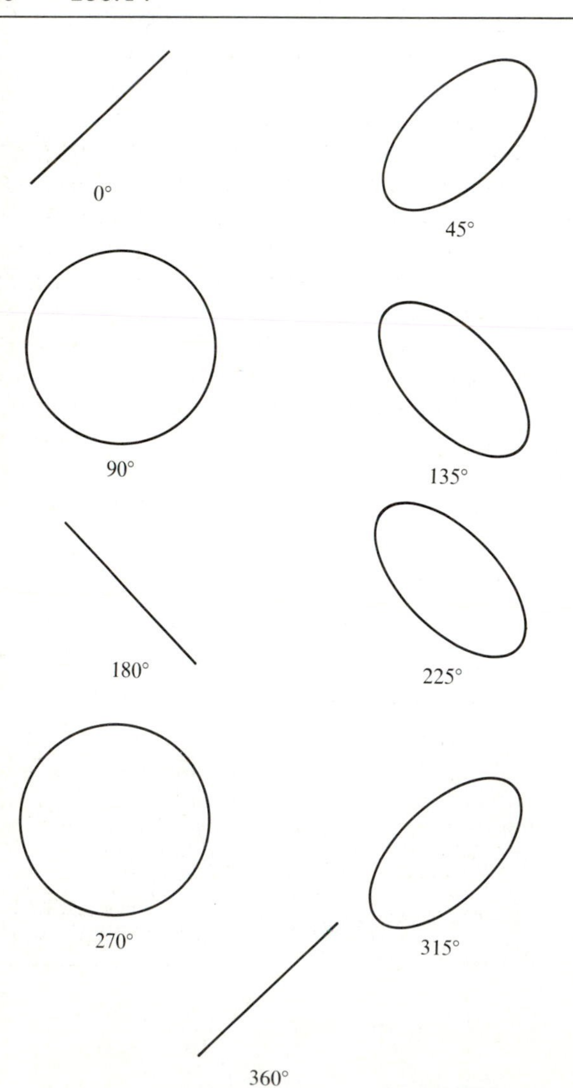

FIGURE 8–12
Phase relationships of Lissajous figures

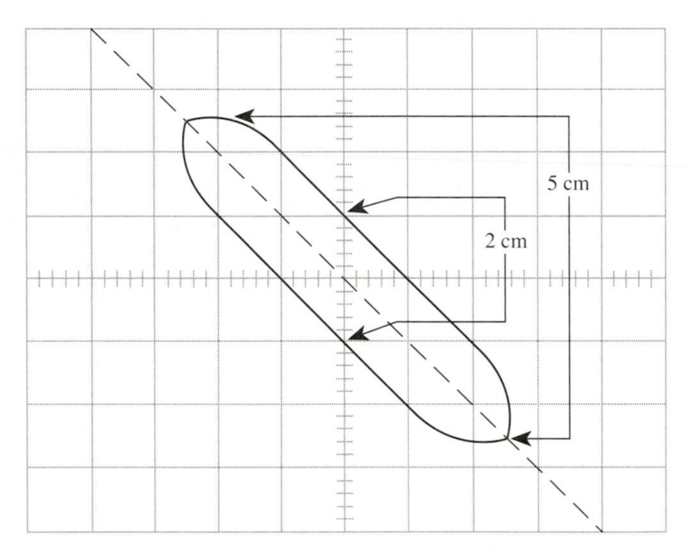

FIGURE 8–14

Frequency Measurements Using Lissajous Figures

If the signals input to the vertical and horizontal deflection plates are not the same frequency and a stable pattern can be established, the ratio of the frequencies can be determined. Figure 8–15 illustrates the formation of a Lissajous figure when the frequency ratio is 2:1. The horizontal trace moves

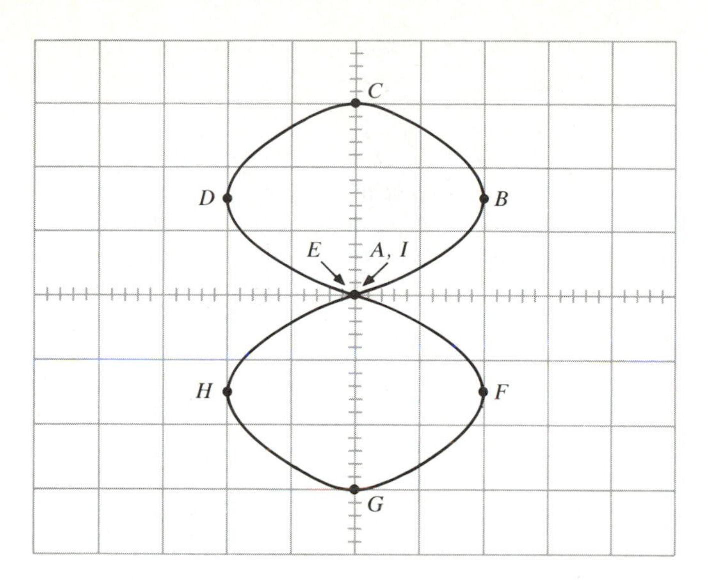

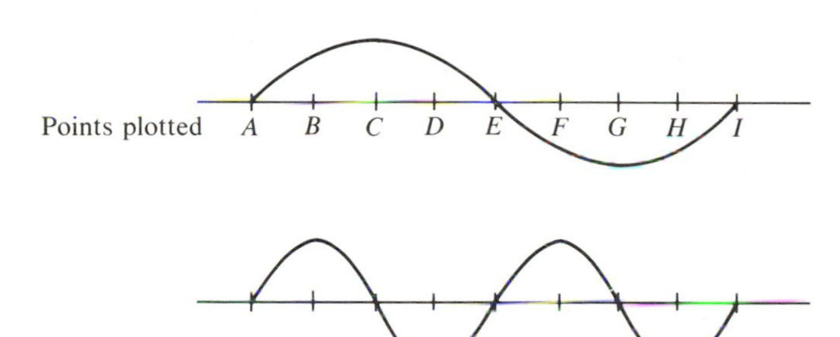

FIGURE 8–15
Plotting a Lissajous figure showing a 2:1 frequency ratio

back and forth twice while the vertical moves up and down once. Therefore, the frequency of the signal on the horizontal is twice the frequency of the signal on the vertical. Other frequency patterns appear in Figure 8–16.

As illustrated in Figure 8–16, the frequency ratio can be determined by the number of vertical peaks to the number of horizontal peaks. A pattern "open" at one end is considered to have one-half peak at that end (Figure 8–17).

If one of the signals used is a standard, the other frequency may be determined with considerable accuracy.

If the two signals have frequencies that differ by 1 Hz, the pattern will rotate at the rate of one cycle per second. (A 2-Hz difference represents two rotations per second.) Low frequencies are easier to measure than high frequencies because a stable pattern is easier to establish. With higher frequencies, stable patterns may be very difficult. It is virtually impossible to tune two 1-MHz generators to

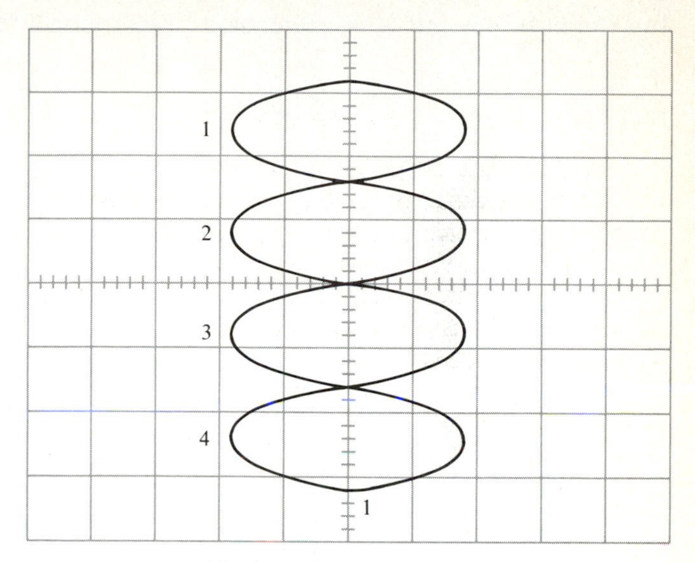

4 horizontal to 1 vertical 4:1

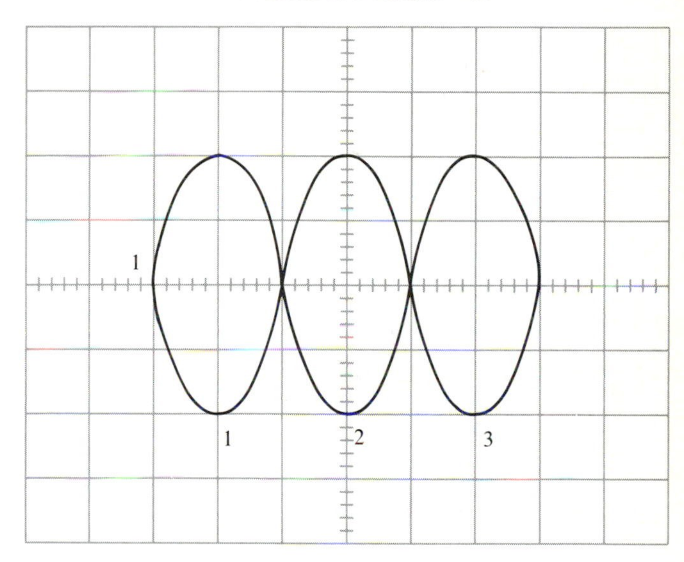

1 horizontal to 3 vertical 1:3

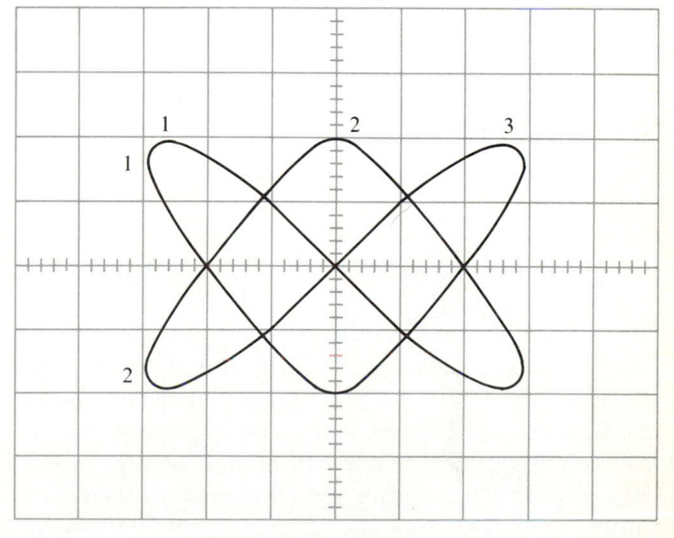

2 horizontal to 3 vertical 2:3

FIGURE 8–16
Lissajous patterns of some frequency ratios

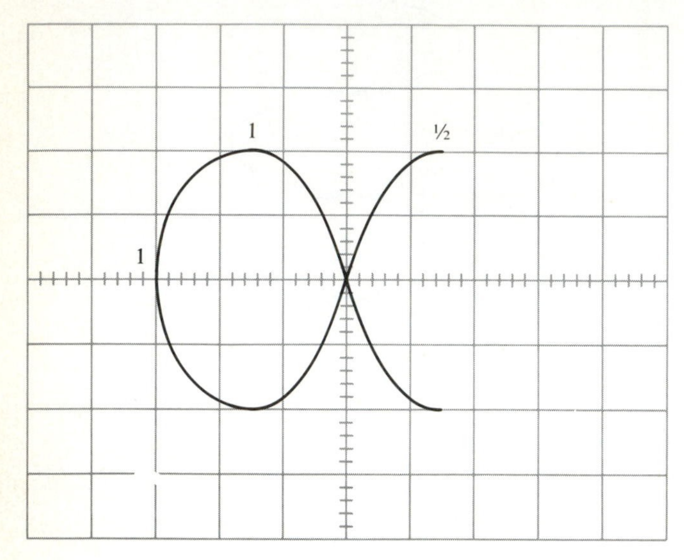

1.5:1 (same as 3:2)

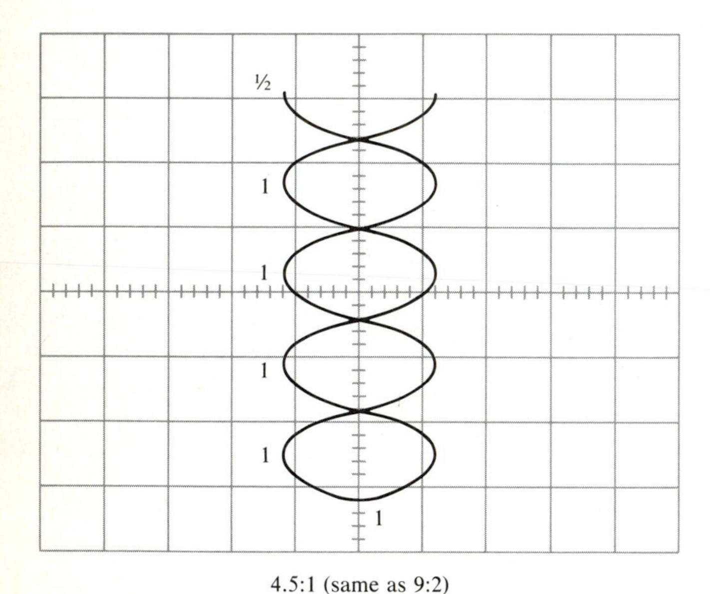

4.5:1 (same as 9:2)

FIGURE 8–17
Frequency ratios of open-ended figures

exactly 1,000,000 Hz or even close enough to get a stationary pattern.

At 200 Hz, the difference that produces a 1-Hz rotation in the pattern is 1 part in 200. At 1 MHz, the difference is 1 part in 1,000,000. Even under the best circumstances, it may be impossible to adjust the frequency sources to a high enough accuracy to produce a stationary Lissajous pattern. The following chart shows the accuracy required to produce a pattern that rotates at 1 Hz using both 200 Hz and 1 MHz as a reference:

Vertical	*Horizontal*	*Difference*	*Rotations*
200 Hz	199 Hz	0.5%	1/s
1,000,000 Hz	1,000,001 Hz	0.00001%	1/s

In many cases, even a glimpse of a circular pattern will suffice when taking a measurement. Although a circle shows a 90° phase difference between two signals, it also indicates that their frequencies are equal!

In calibrating audio frequency generators, WWV (National Bureau of Standards radio station) may be used as one input. WWV transmits 440 Hz, 500 Hz, and 600 Hz at different times during the hour. With the audio generator as the other input, patterns may be formed in a 1:1, 1:2, 1:3, 1:4, 2:1, 3:1, 4:1 sequence. Thus, several points can be plotted and a calibration curve drawn (Figure 8–18). Then, if the local oscillator is used, a reading taken from the dial can be compared to the standard frequency.

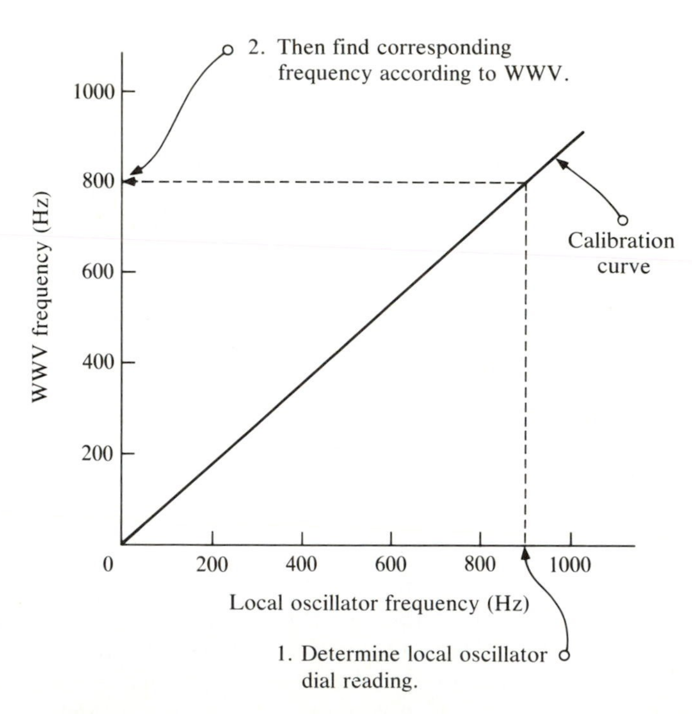

FIGURE 8–18
Using a calibration curve to determine the local oscillator frequency

A+B AND A−B

If an oscilloscope is equipped with an A+B function, two signals may be used as inputs and the A+B engaged to determine the sum of their amplitudes. If the oscilloscope has the capability to invert channel B, comparison measurements may be taken. One signal is input to channel A; the other signal is input to channel B and inverted. The A+B switch may then be used to determine the difference between the signals as follows:

$$A + (-B) = A - B \qquad \textbf{(8–16)}$$

USING THE DELAYED SWEEP MODE

Oscilloscopes with a delayed sweep mode usually have a cursor that can be used to locate the part of a waveform to be magnified. When the oscilloscope is used in the delayed sweep mode, it may be necessary to turn down the intensity of the main trace in order to locate the cursor. The cursor is a light beam on the main trace. It is moved along the main trace by means of a vernier adjustment. The difference between the main sweep rate and the delayed sweep rate determines the width of the cursor. If the delayed sweep rate is changed, the width of the cursor will also change. After the appropriate sweep rate has been selected and the cursor has been positioned, a change to mixed mode will cause the display to change. Part of the trace (up to the cursor) will be swept at the main trace sweep rate, and the rest of the trace will be swept at the delayed trace sweep rate (Figure 8–19).

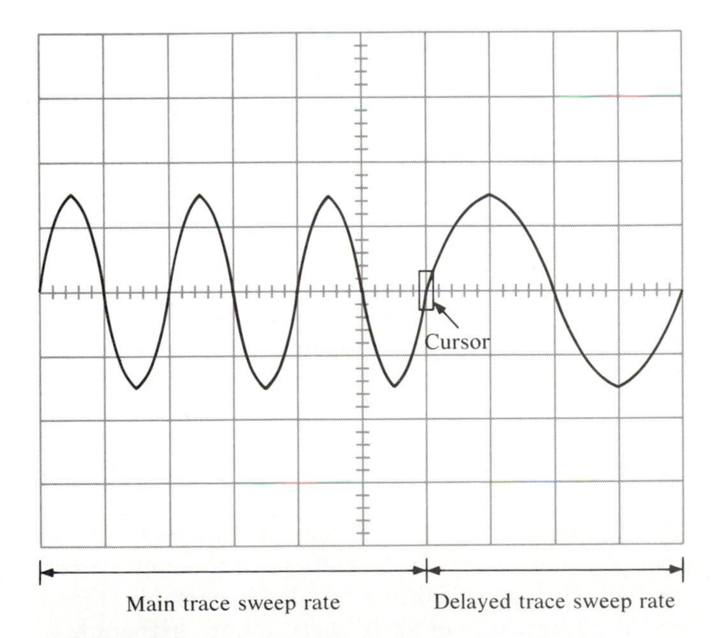

FIGURE 8–19
Illustration of mixed-mode trace rates

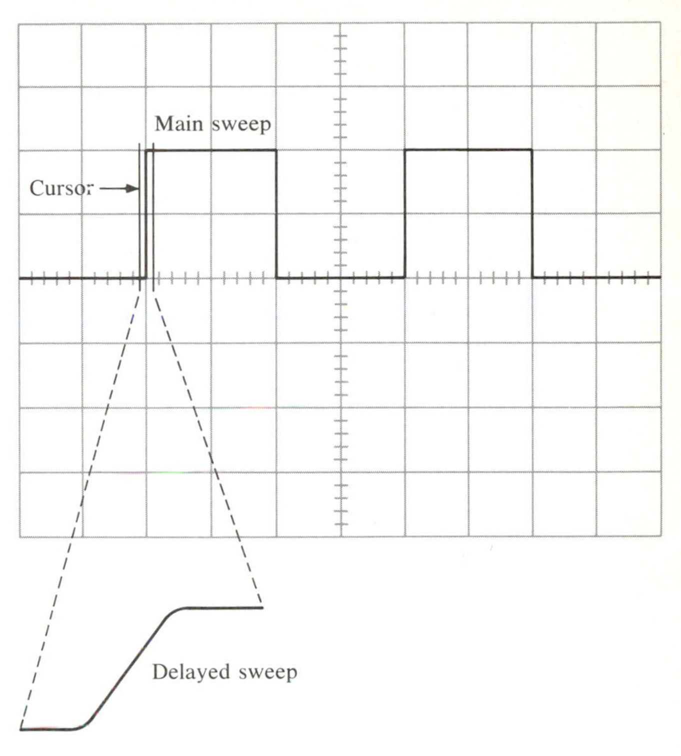

FIGURE 8–20
Magnified portion of a waveform

This mixed-mode setting can be used to better locate the part of the wave to be isolated by the delay function. When the oscilloscope is switched to the delay mode, the portion of the main trace isolated by the cursor appears to be magnified but will be displayed at the rate indicated by the delayed sweep rate selector (Figure 8–20). It is important to remember which mode the oscilloscope is in and to use the appropriate measuring factor.

EXAMPLE 8–9

The main sweep rate is set at 0.1 ms/cm, and the delayed trace rate is set at 0.1 μs/cm. What is the rise time of the trace in Figure 8–21?

Solution:

Rise time and fall time are measured from 10% to 90% of the amplitude. Since the part of the display to be measured (indicated by vertical lines on Figure 8–21) is approximately 2.3 cm in the horizontal direction, the measurement is

$$2.3 \text{ cm} \times 0.1 \ \mu\text{s/cm} = 0.23 \ \mu\text{s} \qquad \textbf{(8–17)}$$

Remember that if the ×10 horizontal multiplier is also used, the results must be adjusted accordingly.

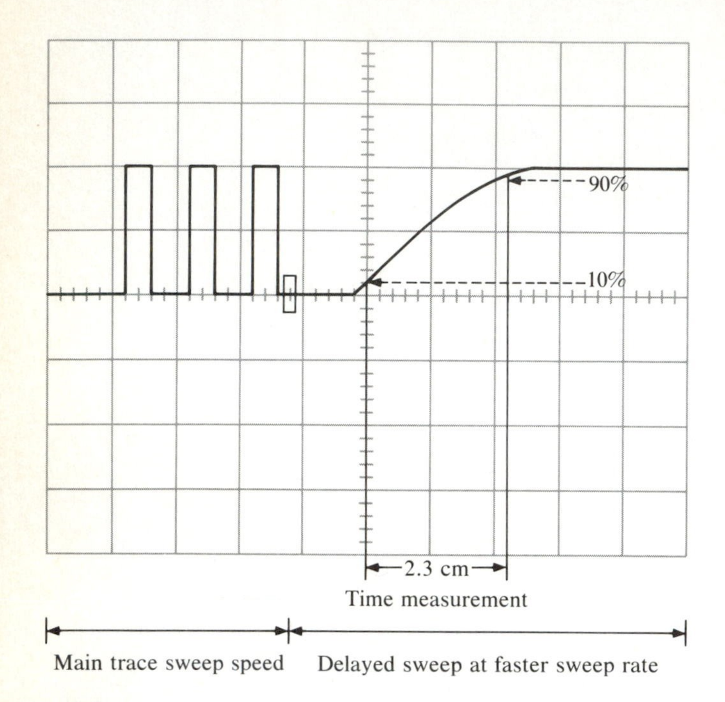

FIGURE 8–21
Mixed-mode sweeps at two different rates

z-AXIS MODULATION

z-axis modulation is an intensity modulation that allows the technician to externally blank out the main trace. It can be used to create interesting displays or for frequency measurements. A standing pattern is produced on the display when the signal to the z-axis input has an appropriate frequency and intensity (Figure 8–22).

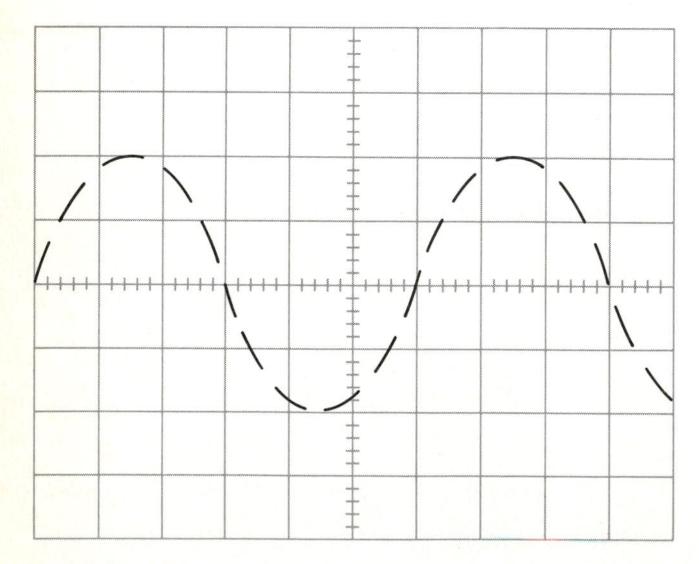

FIGURE 8–22
z-axis modulation

The number of blanks displayed can be used to determine the ratio of the frequency of the main trace to the frequency of the z-axis input signal. Ten blanks are shown for every one cycle of the main trace signal. If the frequency of the main trace or the z-axis input signal is known, the ratio of 10:1 will yield the frequency of the other.

NOTES ON FIELD SERVICE

The oscilloscope is a popular measuring tool. It can be used to quickly measure voltage, frequency, and phase by giving visual information. The numerical values are often not as informative as the visual display.

Oscilloscopes can be purchased for as little as $250 or as much as $20,000 depending on their capabilities.

Field service is usually restricted to fuse changes, power cord replacement, or probe replacement. If a probe is identified as not working, it is important to:

1. Replace the probe with one specifically designed for the make and model of oscilloscope, or
2. Understand the possible effects of using a probe not matched to the oscilloscope.

Impedance match changes and frequency considerations are the most important factors to keep in mind here. Since oscilloscope circuitry is so complex, it is usually not cost efficient to try other than the indicated field service repairs. Manufacturers have test setups to quickly pinpoint faulty components. The same procedure for the average technician, not specifically trained in troubleshooting this piece of equipment, takes a great deal of time. And even if the faulty component(s) are located, the parts will have to be ordered from the manufacturer. In most cases, therefore, it is recommended that nonoperating oscilloscopes be referred to a field service representative of the manufacturer.

SUMMARY

The proper use of the oscilloscope as a measuring tool is a necessary skill. Voltage, frequency, time, and phase relationships can be accurately measured using this versatile instrument. Although a one-time "hands-on" experience may demonstrate the oscilloscope's capabilities, hours of practice are usually required to perfect the techniques involved.

EXERCISES

1. If you ignore the power supply, an oscilloscope can be represented by four functional blocks: the vertical, horizontal, trigger, and display systems. Briefly describe the purposes of these four functional blocks.
2. Most test and measurement instruments are designed for one specific use. For example, voltmeters measure voltage, a Wheatstone bridge measures resistance, and frequency counters count frequency. But oscilloscope displays present more information than is available from most other measurement instruments. Explain why.
3. What are the two most common oscilloscope measurements?
4. What is the best technique for making amplitude measurements?
5. What is the best technique for making time measurements?
6. There are two common methods of determining the phase shift between two waveforms. Briefly describe both.
7. Are phase measurements the only use for A-versus-B measurements?
8. How can differential measurements be made?
9. What is a delayed sweep?
10. You have made a direct amplitude measurement of the sinewave representing household current, and the result is 330 V. Calculate the following measurements:
    a. Peak voltage   b. Average value   c. rms voltage
11. What is the phase angle between the two signals that form the Lissajous figure in Figure 8–23?

FIGURE 8–23
Diagram for Exercise 11

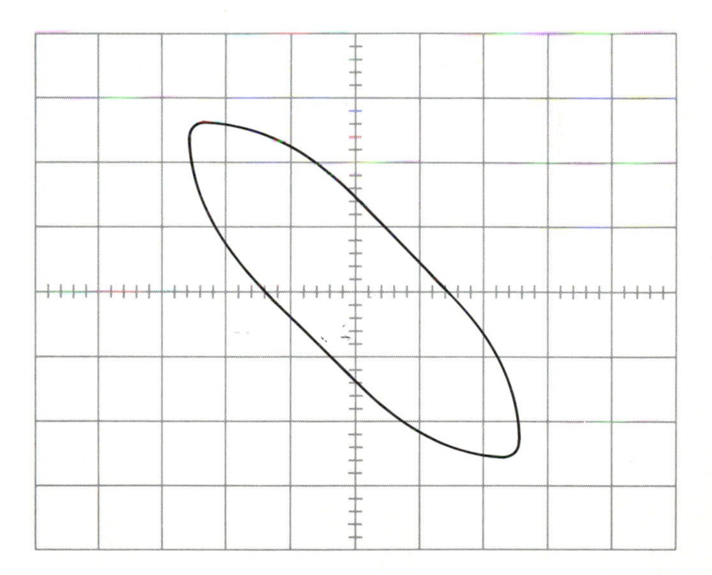

12. In Figure 8–24, if the frequency on the horizontal input is 1 kHz, what is the frequency of the signal on the vertical input?

FIGURE 8–24
Diagram for Exercise 12

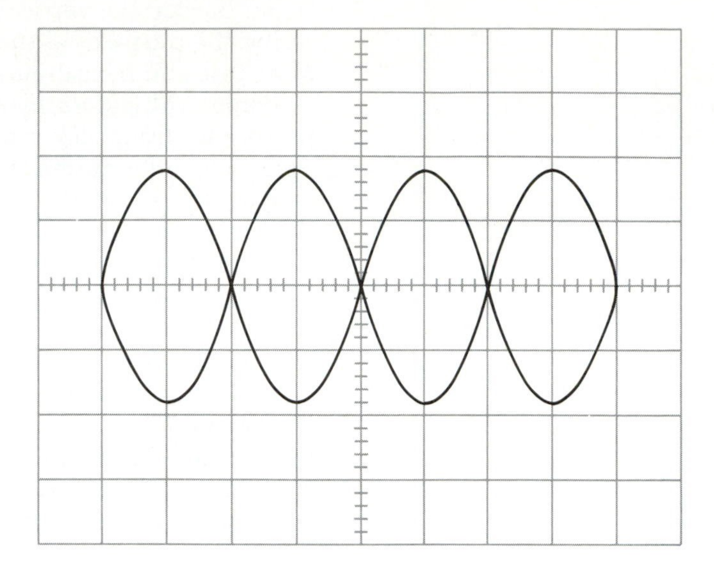

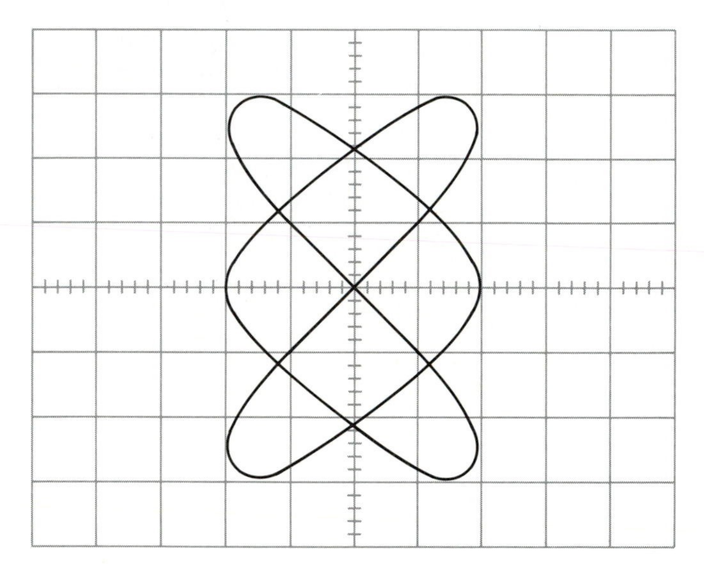

13. Figure 8–25 represents an oscilloscope display. The grid is in centimeters and a $\times 10$ probe is being used. Give the peak-peak amplitude and frequency at the input signal for oscilloscopes with each of the following settings:

a. Vertical—0.5 V/cm
Horizontal—0.1 ms/cm

b. Vertical—1 V/cm
Horizontal—20 μs/cm
(magnified $\times 10$)

c. Vertical—5 V/cm
(magnified $\times 5$)
Horizontal—0.5 μs/cm

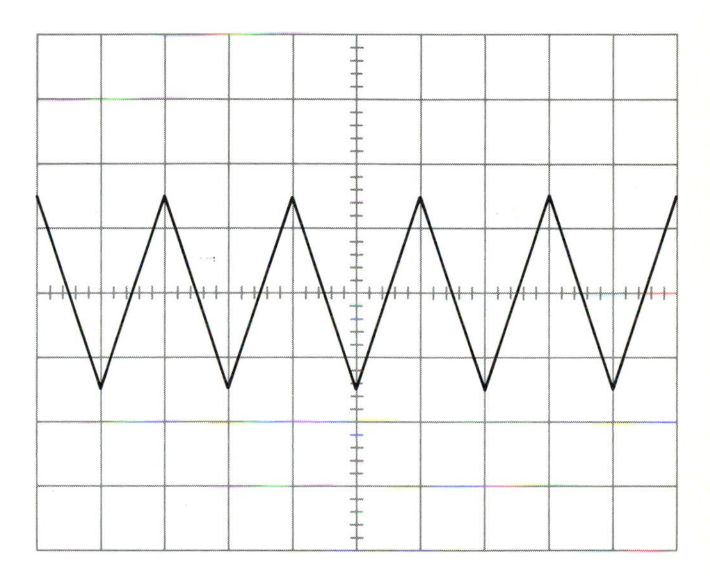

FIGURE 8–25
Diagram for Exercise 13

LAB PROBLEM 8–1
Oscilloscope Familiarization

Equipment

One general purpose oscilloscope (10 MHz); two function generators (1 Hz to 1 MHz); one DMM; one frequency counter.

Objective

To become familiar with the operation and use of an oscilloscope in measuring voltage and frequency.

Procedure

1. Plug in the oscilloscope (CRO).
2. Connect a probe to the channel A input (if dual-trace) or to the single vertical input (if single-channel).
3. Select channel A with the mode switch.
4. Turn on the CRO using the power switch.
5. Make sure the trigger select switch is in the automatic-mode position.
6. If the probe has a selector switch, set the switch to the ×1 position. Otherwise, remember that it is most likely a ×10 probe and that this fact will have to be considered later.
7. Set the dc-GND-ac switch to the ac position.
8. Set the trigger source switch to internal.
9. Connect the probe tip to the calibration output terminal if one is available. Otherwise, use a function generator.
10. Adjust the vertical attenuator until a 1-cm vertical deflection is observed.
11. Adjust the brightness until you see the trace clearly. It is IMPORTANT that the brightness be at a comfortable MINIMUM.
12. Adjust the focus until a sharp trace occurs.
13. Adjust the sweep time/cm switch (horizontal attenuator) until three or four waveforms appear on the display (probably about 5 ms/cm).
14. Rotate the vertical position switch. Locate the trace so that the bottom of the signal is on the horizontal axis of the graticule (the trace will usually be 1 cm high).
15. If the trace drifts across the screen, adjust the triggering level clockwise and then counterclockwise until the drifting stops.
16. Adjust (one at a time) the following controls fully clockwise and fully counterclockwise to see what effect (if any) they have on the trace. Include your observations in your report.
    a. Vertical position
    b. Horizontal position
    c. Intensity (brightness)
    d. Focus
    e. Vertical attenuator (volts/cm)
    f. Horizontal sweep (sweep time/cm)
17. Repeat steps 2 through 16 using channel B if available.
18. Many oscilloscopes have variable switches on the vertical and horizontal attenuators. These switches control the corresponding "sizes" of the waveforms between the calibrated values shown on the dials. They should be in the calibrated position if accurate voltage and time measurements are to be taken.

Plug in the function generator and connect its output to the CRO probe. Remember that there must be *two* connections from the CRO to any device if measurements are to be taken. Turn on the function generator. Select the 1-kHz range and the sinewave function. Set the frequency dial to 1, which will yield 1 × 1 kHz, or a 1-kHz signal. Adjust the amplitude until you get a 2-cm vertical display. Adjust the horizontal sweep on the CRO until you see about five sinewaves on the display. Since you have a 2-cm-high (peak-peak) vertical displacement, the reading is

$$2 \text{ cm} \times 1 \text{ V/cm (vertical attenuator)} = 2 \text{ V peak-peak}$$

If you are using a ×10 probe, the reading will be

$$2 \text{ V peak-peak} \times 10 = 20 \text{ V peak-peak}$$

Adjust the horizontal position until a peak is centered on the middle intersection of the graticule. There will be a 2-cm separation between the points of maximum slope. The time of one cycle (the period) is

$$2 \text{ cm} \times 0.5 \text{ ms/cm (horizontal sweep)} = 1 \text{ ms}$$

The frequency of the signal can be calculated using

$$f = \frac{1}{T} = \frac{1}{0.001 \text{ s}} = 1 \text{ kHz}$$

Use two function generators, one connected to channel A and one connected to channel B. Repeat the familiarization procedures outlined previously. Try the square and triangular signal output of the function generators. Try using a multimeter to measure the ac output of the function generator. Since the multimeter usually displays an rms reading, you might have to use the following relationships to calculate rms and peak-peak voltages:

$$v_{\text{rms}} = 0.707 v_{\text{pk}}$$
$$v_{\text{pk}} = .5 v_{\text{pk-pk}}$$

These relationships should always be remembered when a CRO and multimeter are being used.

Object 2

To measure frequency using Lissajous figures.

Procedure

Use two signal generators and consider the first generator (generator 1) as the known or standard frequency source. Assume that it has correct dial markings. Generator 2 will be the variable (unknown) source. Set generator 1 at 1 kHz. Vary the frequency of generator 2 until a circular pattern occurs on the scope. Record this pattern in Table 8–1. Record the number of vertical and horizontal tangencies and calculate the frequency of generator 2. (A vertical tangent is a point where the figure is tangent to the vertical axis.) Repeat this procedure for other convenient ratios and complete Table 8–1. Use at least one other standard frequency for the last three entries (e.g., 3 kHz).

TABLE 8–1
Complete the table.

Lissajous Figure	No. of Vertical Tangencies	No. of Horizontal Tangencies	Frequency of Generator 1	Frequency of Generator 2

Set the standard at 1 kHz. Adjust the unknown source until the circular pattern appears to revolve at one complete cycle per second. Use a frequency meter to measure the frequency of both generators. Enter the results in Table 8–2. Repeat this measuring procedure, setting the test generator at 2 kHz, 3 kHz, and 500 Hz. Use the appropriate Lissajous figures and measurements from the frequency counter.

TABLE 8–2
Complete the table.

Frequency of Generator 1	Frequency of Generator 2	Frequency Counter Measurement
1 kHz	1 kHz	
1 kHz	2 kHz	
1 kHz	3 kHz	
1 kHz	500 Hz	

Object 3

To use the oscilloscope and Lissajous figures for phase measurements.

Procedure

Wire the circuit in Figure 8–26.

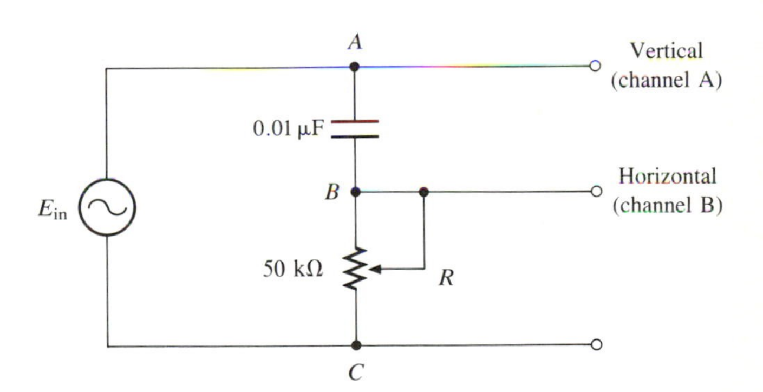

FIGURE 8–26
Circuit illustrating phase measurement

Set the frequency of E_{in} at 1 kHz. Set R at zero ohms. Set the signal voltage at 4 V peak-peak. Center the display. Change R to 10kΩ and record the pattern in Table 8–3. Follow the diagram in Figure 8–27 and record the measured and calculated values in Table 8–3. Repeat this procedure and complete the table. In your conclusions, outline the procedures followed and comment on the relative success of each.

TABLE 8–3
Complete the table.

Resistance (kΩ)	Lissajous Figure	Measured Values				Calculated Values	
		y_0	y_{max}	θ	ϕ	θ	ϕ
10							
15							
22							
30							
50							

$\tan \theta = X_C/R$
$\tan \phi = R/X_C$

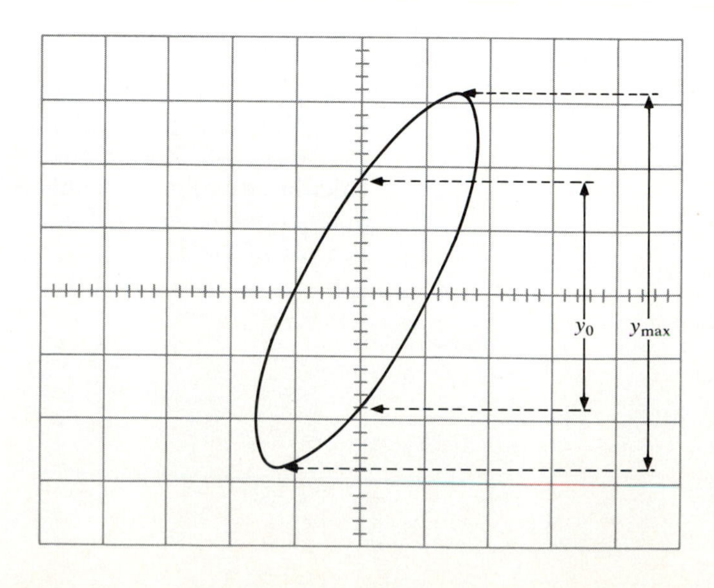

FIGURE 8–27
Diagram illustrating phase measurement

9 Potentiometers and Potentiometric Bridges

OUTLINE

OBJECTIVES

POTENTIOMETERS

- ☐ What Is a Potentiometer?
- ☐ Potentiometer across a Single Source of emf
- ☐ Thevenin Equivalent of a Potentiometer across a Single emf Source
- ☐ Potentiometer across Two emf Sources
- ☐ Thevenin Equivalent of a Potentiometer across Two emf Sources
- ☐ Slightly Unbalanced Bridge

POTENTIOMETRIC (SLIDE-WIRE) BRIDGES

- ☐ Slide-Wire Potentiometer
- ☐ Potentiometric Bridge
- ☐ Normalizing the Bridge
- ☐ The Sensitivity Control
- ☐ Measuring Potentials
- ☐ Thevenin Equivalent of the Bridge
- ☐ Methods of Using the Bridge
- ☐ Advantages and Disadvantages

SUMMARY

EXERCISES

LAB PROBLEM 9–1—Slightly Unbalanced Potentiometer

LAB PROBLEM 9–2—Potentiometric Bridges

OBJECTIVES

After completing this chapter, you will be able to:

- ☐ Discuss what a potentiometer is.
- ☐ Discuss what a potentiometer does.
- ☐ Develop the Thevenin equivalent of a potentiometer across a single source of emf.
- ☐ Develop the Thevenin equivalent of a potentiometer wired between two sources of emf.
- ☐ Give the characteristics of a slightly unbalanced potentiometric bridge.
- ☐ Develop the Thevenin equivalent of a slightly unbalanced potentiometric bridge.
- ☐ Give the applications of a slightly unbalanced potentiometric bridge.
- ☐ Discuss the characteristics of a slide-wire potentiometer.
- ☐ Normalize a potentiometric bridge.
- ☐ Describe the uses of a potentiometric bridge.
- ☐ Describe the measuring modes of a potentiometric bridge.
- ☐ Discuss the advantages and disadvantages of a potentiometric bridge as a measuring device.
- ☐ Extend the range of a potentiometric bridge.

Potentiometers have many uses. They can be used in variable-resistance configurations as "pots" or rheostats. They can also be used to measure resistances and potentials accurately. Electrical voltages and varying resistances are the products of devices called transducers. Transducers convert many different kinds of physical energy into electrical energy or varying electrical quantities. With potentiometers, the varying electrical quantities can be processed for measurement by the common meters discussed in previous chapters. This chapter presents an analysis of potentiometers and describes their various applications.

POTENTIOMETERS

What is a Potentiometer?

A potentiometer typically consists of a coil of high-resistance wire arranged to serve as a variable resistor. It usually has connections at either end of the coil and at the wiper (Figure 9–1). Figure 9–2 gives the schematic diagram for a potentiometer, and Figure 9–3 illustrates several potentiometer types.

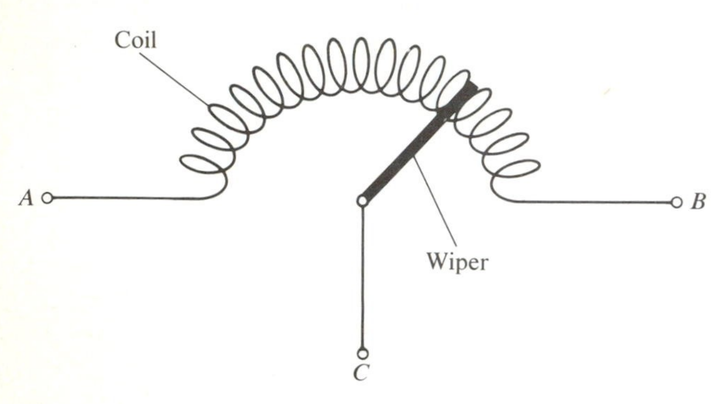

FIGURE 9–1
Inside a potentiometer

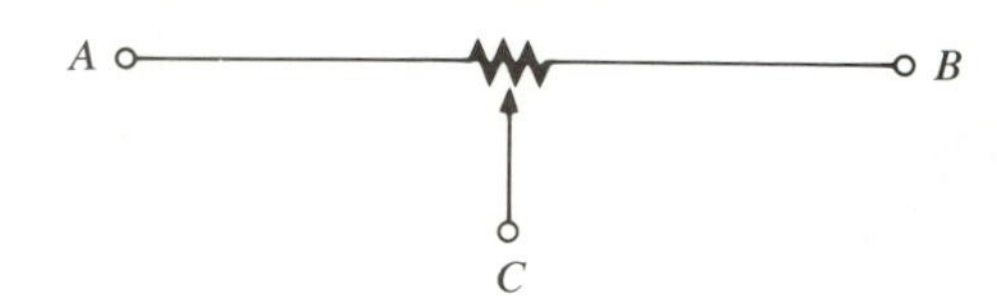

FIGURE 9–2
Schematic diagram of a potentiometer.

Potentiometer across a Single Source of emf

When the potentiometer is wired across a single source of emf, it serves as a voltage divider (Figure 9–4). If the wiper is set at some fractional portion

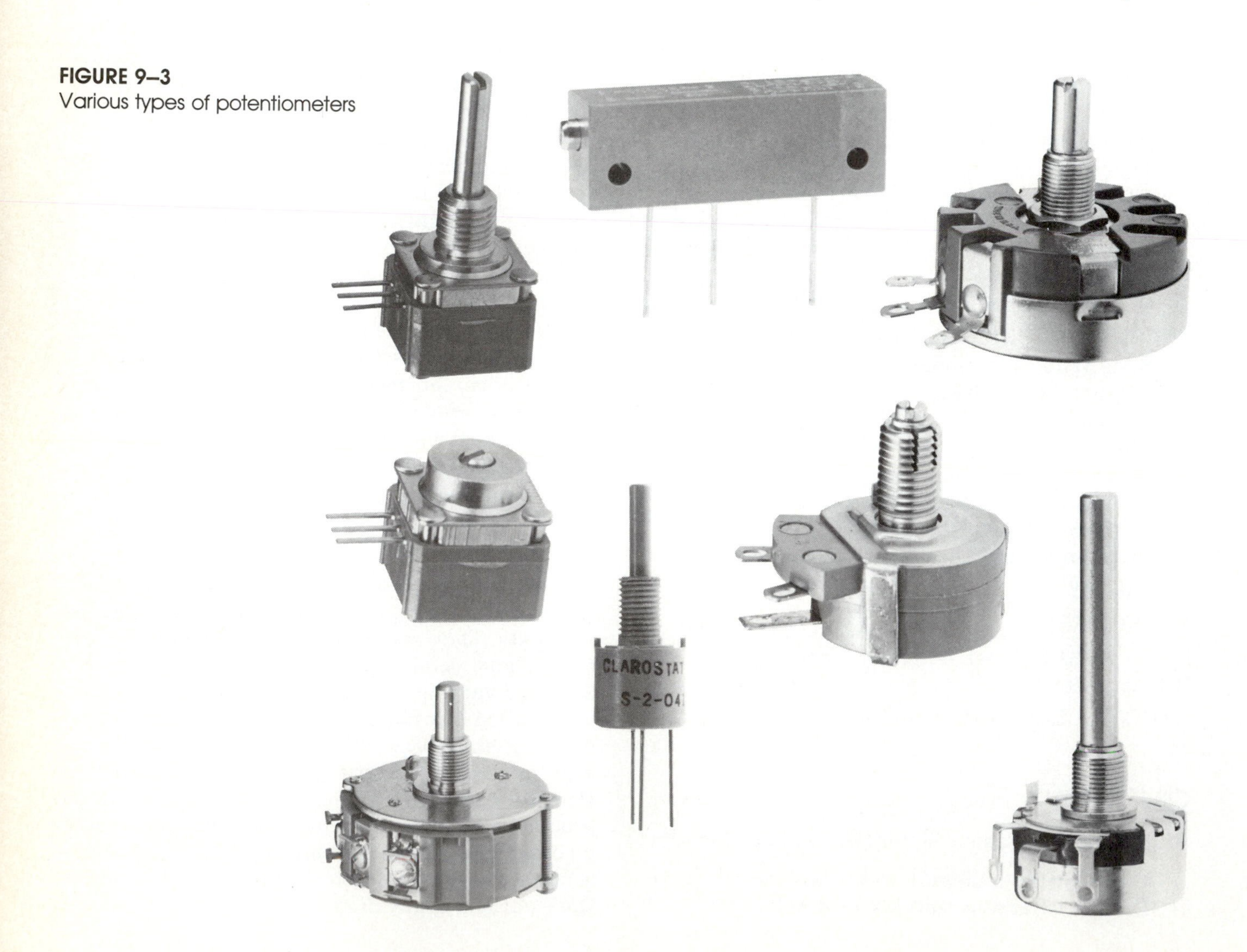

FIGURE 9–3
Various types of potentiometers

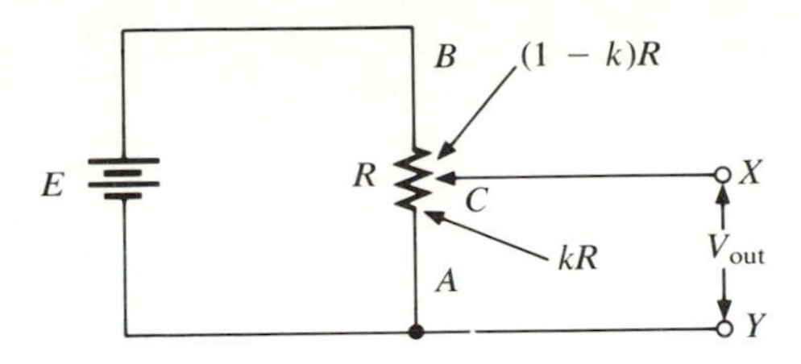

FIGURE 9–4
The potentiometer as a voltage divider

of the way from A to B, then R is divided into two parts, R_{AC} and R_{BC}, so that $R_{AC} = kR$ and $R_{BC} = (1 - k)R$ (k is a simple fraction, such as ½, ⅓, ¾, . . .). The purpose of this arrangement is to divide the emf into smaller parts.

Thevenin Equivalent of a Potentiometer across a Single emf Source

The Thevenin equivalent circuit as "seen" at the terminals X and Y can be developed as follows:

$$V_{Th} = E\left(\frac{kR}{kR + (1 - k)R}\right)$$

$$= E\frac{kR}{R} \tag{9–1}$$

$$= kE$$

Therefore, if $E = 10$ V and the wiper is set at $k = 1/4$ (one fourth of the way between A and B),

$$V_{Th} = (1/4)(10\text{ V}) = 2.5\text{ V} \tag{9–2}$$

Quantity R_{Th} is found by replacing E with its internal resistance (assume this to be zero ohms in this problem). The Thevenin resistance now occurs between X and Y (see Figure 9–5):

$$R_{Th} = kR \parallel (1 - k)R \tag{9–3}$$

$$R_{Th} = k(1 - k)R \tag{9–4}$$

which produces the circuit in Figure 9–6.

Notice that if $k = 1/2$,

$$R_{Th} = (1/2)(1 - 1/2)R = 1/4R \tag{9–5}$$

and if $k = 1/4$,

$$R_{Th} = (1/4)(1 - 1/4)R = 3/16R \tag{9–6}$$

and if $k = 3/4$,

$$R_{Th} = (3/4)(1 - 3/4)R = 3/16R \tag{9–7}$$

The effective resistance at the terminals plotted as a function of k (the position of the wiper) is shown in Figure 9–7. Notice that the maximum resistance is achieved when the wiper is in the middle ($k = 1/2$), and that this value is only $1/4R$.

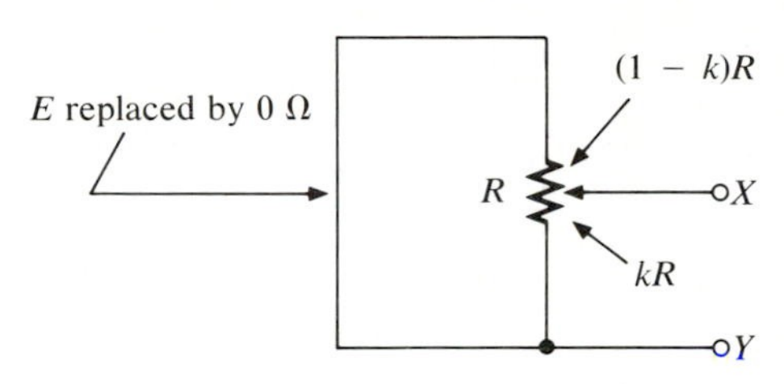

FIGURE 9–5
Finding the resistance at terminals X and Y

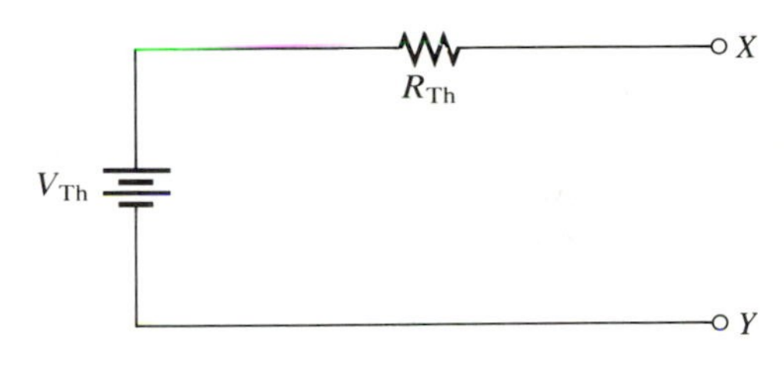

FIGURE 9–6
Thevenin equivalent of a potentiometer

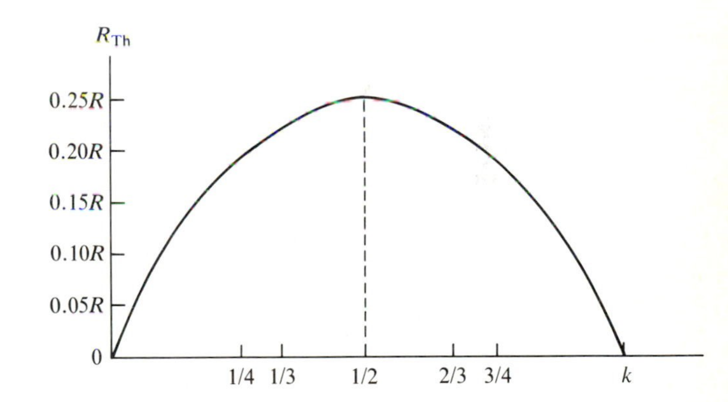

FIGURE 9–7
Resistance versus wiper position

EXAMPLE 9–1
Show the Thevenin equivalent of the potentiometric circuit in Figure 9–8. Assume that $k = 1/5$.

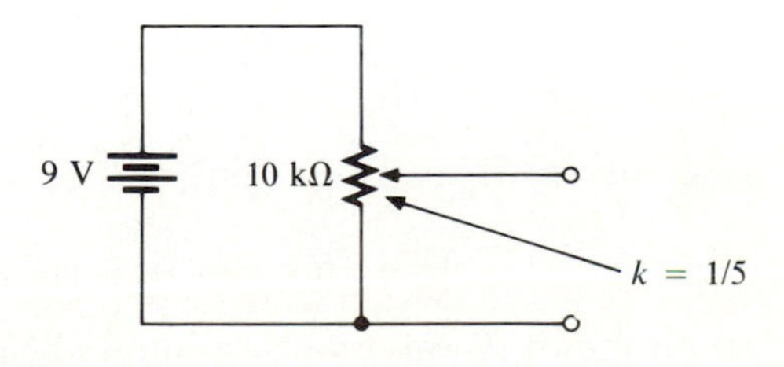

FIGURE 9–8

Solution:

$$V_{Th} = (9\text{ V})(1/5) \quad \textbf{(9–8)}$$
$$= 1.8\text{V}$$
$$R_{Th} = (1/5)(1 - 1/5)(10\text{ k}\Omega)$$
$$= (4/25)(10\text{ k}\Omega) \quad \textbf{(9–9)}$$
$$= 1.6\text{ k}\Omega$$

EXAMPLE 9–2
What voltage will be dropped across the output terminals if a 5-kΩ load is attached across the potentiometric circuit in Figure 9–8?

Solution:
Refer to Figure 9–9.

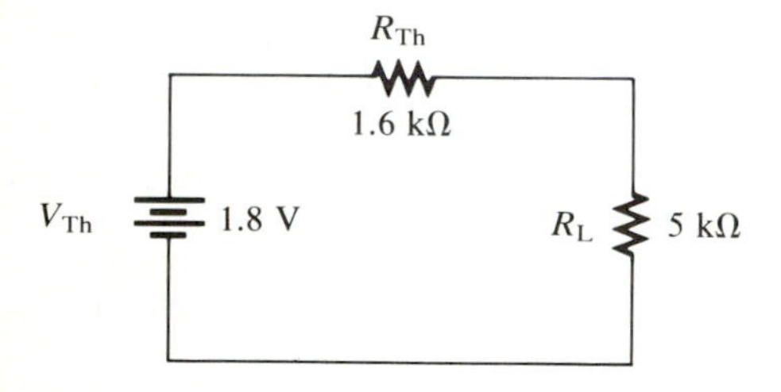

FIGURE 9–9

$$V_L = 1.8\text{ V}\left(\frac{5\text{ k}\Omega}{1.6\text{ k}\Omega + 5\text{ k}\Omega}\right) = 1.36\text{ V} \quad \textbf{(9–10)}$$

Thus, with the 1.8-V source, only 1.36 V are available at the load. This is an example of the loading effects caused by the *internal* resistance of a potentiometer (the Thevenin equivalent source resistance).

Potentiometer across Two emf Sources

Consider the circuit in Figure 9–10. The current flowing through R_1 and R_2 can be calculated using Ohm's law:

$$I = \frac{V_1 - (-V_2)}{R_1 + R_2} \quad \textbf{(9–11)}$$

The voltage across R_2 can be calculated from

$$V = IR \quad \textbf{(9–12)}$$

and the voltage at X relative to ground (Y) equals $-V_2$ plus the voltage across R_2.

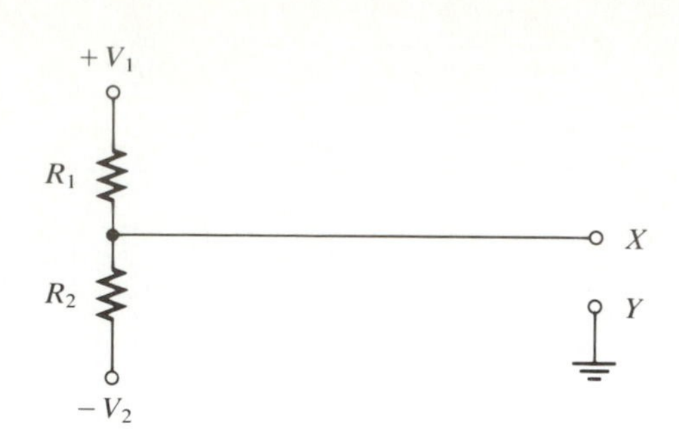

FIGURE 9–10
Potentiometer across two sources of emf

EXAMPLE 9–3
Let $V_1 = +15$ V, $V_2 = -10$ V, $R_1 = 5$ kΩ, and $R_2 = 2$ kΩ. The current through R_1 and R_2 is

$$I = \frac{15\text{ V} - (-10\text{ V})}{5\text{ k}\Omega + 2\text{ k}\Omega} \quad \textbf{(9–13)}$$
$$= 3.57\text{ mA}$$

The voltage across R_2 is (3.57 mA)(2 kΩ) = 7.14 V, and the potential at point X is -10 V + 7.14 V = -2.86 V.

Thevenin Equivalent of a Potentiometer across Two emf Sources

It is useful to develop the Thevenin equivalent of the circuit in Figure 9–10 as seen through terminals X and Y. To develop the Thevenin resistance, consider the circuit as redrawn in Figure 9–11. Replacing V_1 and V_2 with their internal resistances (assumed to be zero ohms) results in the circuit of Figure 9–12. The resistance between terminals X and Y is now R_1 in parallel with R_2. To develop the

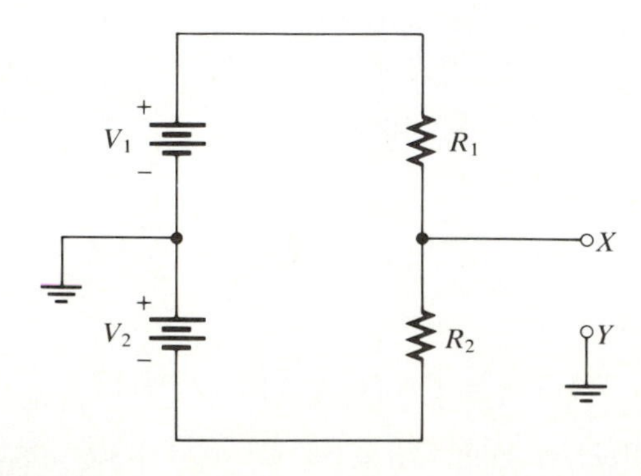

FIGURE 9–11
Potentiometer circuit redrawn

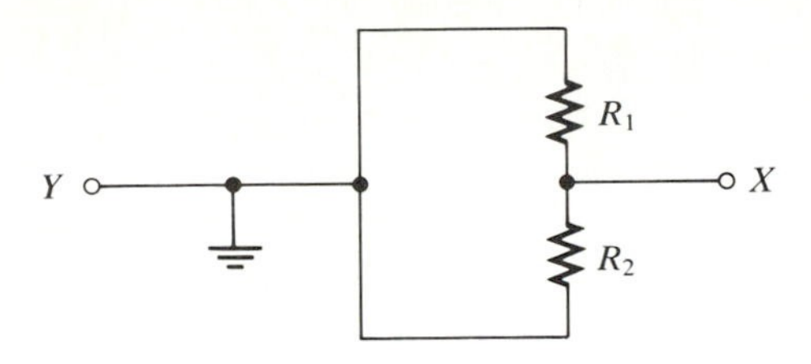

FIGURE 9–12
Finding the Thevenin resistance

Thevenin potential between X and Y in the circuit in Figure 9–10, begin at V_1 and subtract the voltage across R_1 to get the potential at X (V_X) relative to ground (terminal Y); in equation form,

$$V_1 - V_{R_1} = V_X \quad \textbf{(9–14)}$$

By voltage divison,

$$V_{R_1} = [V_1 - (-V_2)]\frac{R_1}{R_1 + R_2} \quad \textbf{(9–15)}$$

This simplifies to

$$V_{R_1} = (V_1 + V_2)\frac{R_1}{R_1 + R_2} \quad \textbf{(9–16)}$$

Therefore,

$$V_X = V_1 - (V_1 + V_2)\frac{R_1}{R_1 + R_2} \quad \textbf{(9–17)}$$

Collecting the fractions over the common denominator gives

$$\begin{aligned} V_X &= V_1 - \frac{(V_1 + V_2)R_1}{R_1 + R_2} \\ &= \frac{V_1R_1 + V_1R_2 - V_1R_1 - V_2R_1}{R_1 + R_2} \quad \textbf{(9–18)} \\ &= \frac{V_1R_2 - V_2R_1}{R_1 + R_2} \end{aligned}$$

The Thevenin equivalent circuit is shown in Figure 9–13.

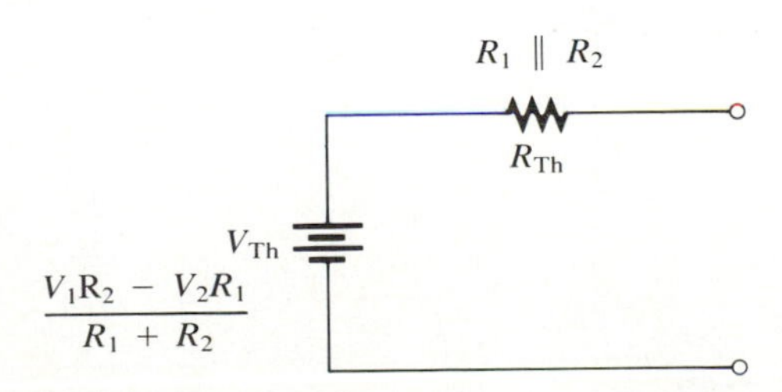

FIGURE 9–13
Thevenin equivalent of the circuit in Figure 9–10

This conclusion could have been reached using voltage division and superposition. With V_1,

$$V_{X_1} = V_1\left(\frac{R_2}{R_1 + R_2}\right) \quad \textbf{(9–19)}$$

With V_2,

$$V_{X_2} = -V_2\left(\frac{R_2}{R_1 + R_2}\right) \quad \textbf{(9–20)}$$

Superimposing yields

$$\begin{aligned} V_{X_1} + V_{X_2} &= V_1\left(\frac{R_2}{R_1 + R_2}\right) \\ &\quad - V_2\left(\frac{R_1}{R_1 + R_2}\right) \quad \textbf{(9–21)} \\ &= \frac{V_1R_2 - V_2R_1}{R_1 + R_2} \end{aligned}$$

EXAMPLE 9–4
Draw the Thevenin equivalent of the circuit in Figure 9–14.

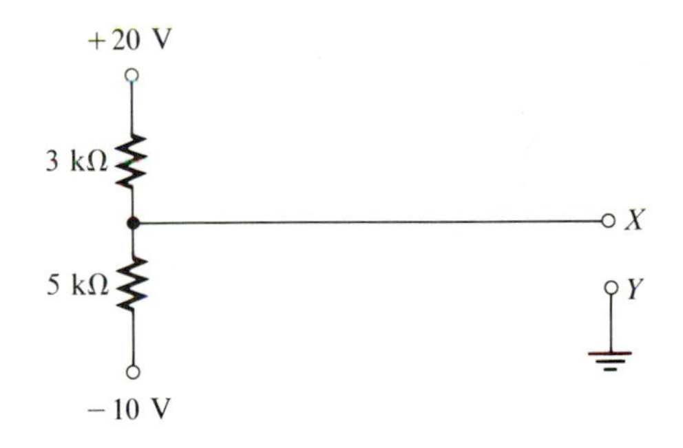

FIGURE 9–14

Solution:

$$R_{Th} = 3\text{ k}\Omega \parallel 5\text{ k}\Omega = 1875\ \Omega \quad \textbf{(9–22)}$$

$$\begin{aligned} V_{Th} &= \frac{(20\text{ V})(5\text{ k}\Omega) - (10\text{ V})(3\text{ k}\Omega)}{8\text{ k}\Omega} \quad \textbf{(9–23)} \\ &= 8.75\text{ V} \end{aligned}$$

(See Figure 9–15.)

This can be checked using equation (9–10):

$$\begin{aligned} I &= \frac{20\text{ V} - (-10\text{ V})}{8\text{ k}\Omega} \quad \textbf{(9–24)} \\ &= 3.75\text{ mA} \end{aligned}$$

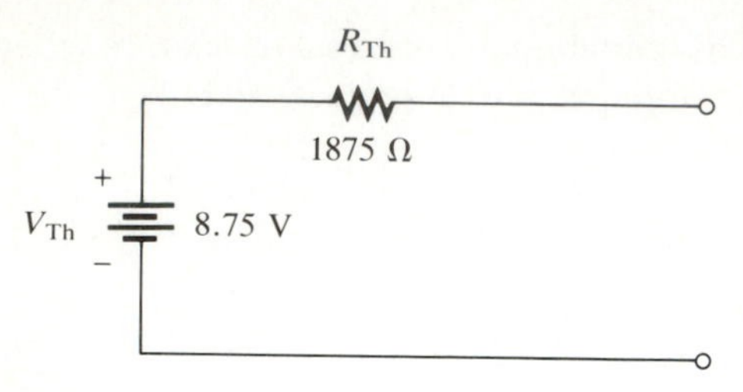

FIGURE 9–15

The voltage across $R_2 = (3.75 \text{ mA})(5 \text{ k}\Omega) = 18.75$ V. Therefore, the potential at point X is 18.75 V higher than -10, or 8.75 V.

EXAMPLE 9–5

What voltage is developed across a 5-kΩ resistive load connected across terminals X and Y in Figure 9–16?

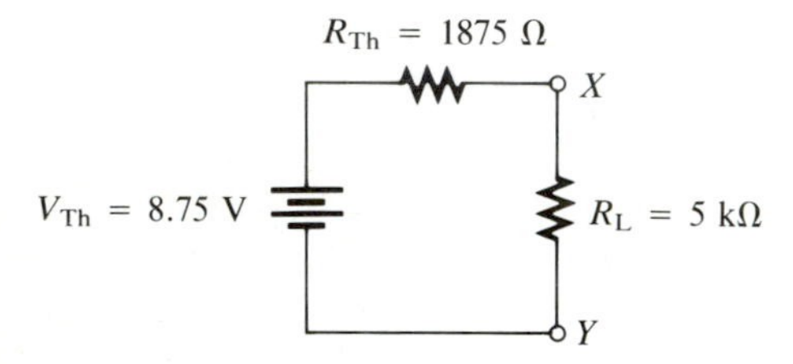

FIGURE 9–16

Solution:

$$V = 8.75 \text{ V}\left(\frac{5 \text{ k}\Omega}{6.875 \text{ k}\Omega}\right) = 6.36 \text{ V} \tag{9–25}$$

This is another example of the loading effect of a potentiometer. With an 8.75-V source, only 6.36 V are available at the load. The rest of the voltage is dropped on the 1875 Ω of internal resistance.

Slightly Unbalanced Bridge

Consider the special bridge in Figure 9–17A. Its Thevenin equivalent is shown in Figure 9–17B. This is an example of a balanced bridge since the voltage at X relative to ground is zero volts. Because this configuration is of little value at this time, consider the lower resistor changing by ΔR which is a small change in R ($\Delta R \ll R$) (Figure 9–18). (For a reasonable approximation, ΔR should be less than 10% of R.) The Thevenin equivalent becomes

$$R_{Th} = R \parallel (R + \Delta R) = \frac{R(R + \Delta R)}{R + (R + \Delta R)} \tag{9–26}$$

If ΔR is small,

$$R_{Th} = \frac{R^2}{2R} = \frac{R}{2} \tag{9–27}$$

as in the balanced bridge. And

$$V_{Th} = \frac{V(R + \Delta R) - VR}{R + R + \Delta R} = \frac{VR + V\Delta R - VR}{2R + \Delta R} \tag{9–28}$$

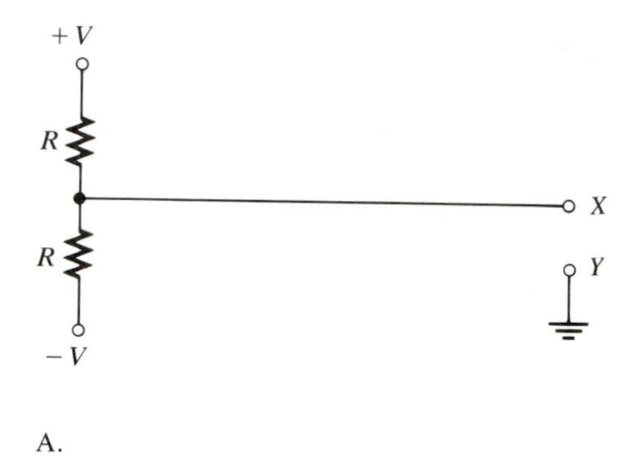

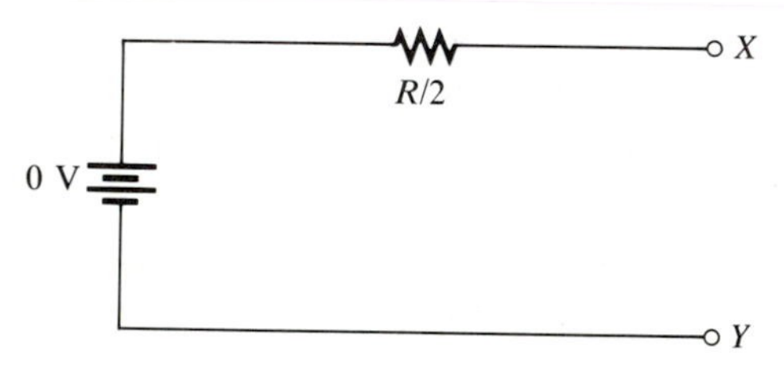

FIGURE 9–17
Balanced potentiometric bridge

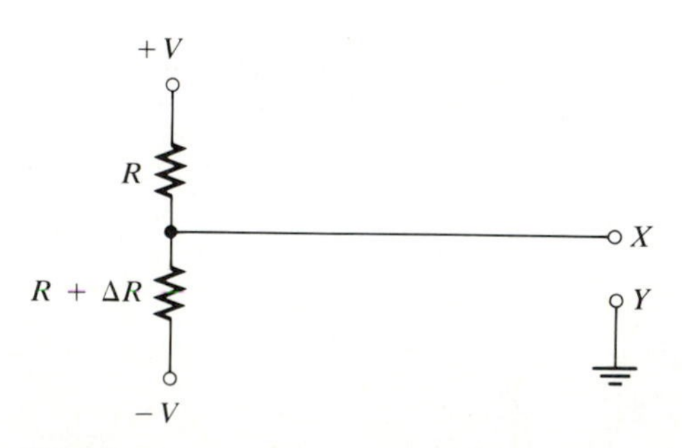

FIGURE 9–18
Slightly unbalanced potentiometric bridge

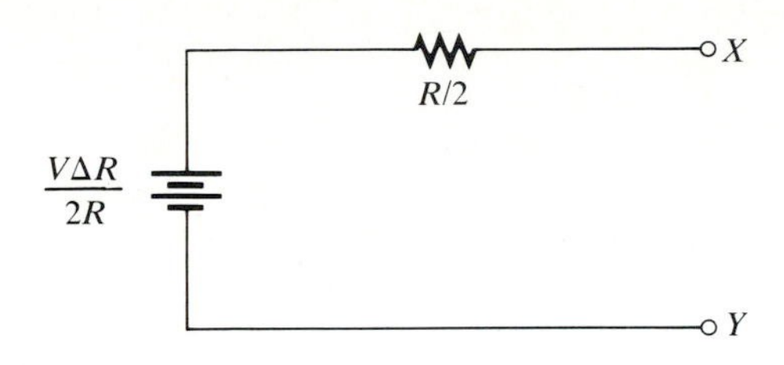

FIGURE 9–19
Thevenin equivalent of a slightly unbalanced potentiometric bridge

If $\Delta R \ll R$, then $\Delta R \ll 2R$ also. Therefore,

$$V_{Th} = \frac{V\Delta R}{2R} \tag{9–29}$$

(Figure 9–19). This is for a slightly unbalanced bridge.

EXAMPLE 9–6
What voltage is developed across $R_1 = 2\ \text{k}\Omega$ in the circuit shown in Figure 9–20?

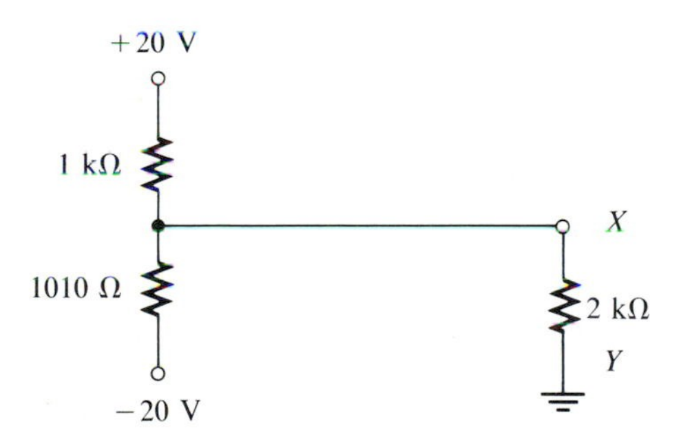

FIGURE 9–20

Solution:
The Thevenin equivalent appears in Figure 9–21.

$$V_{Th} = \frac{(20\ \text{V})(10\ \Omega)}{2(1\ \text{k}\Omega)} = 0.1\ \text{V} \tag{9–30}$$

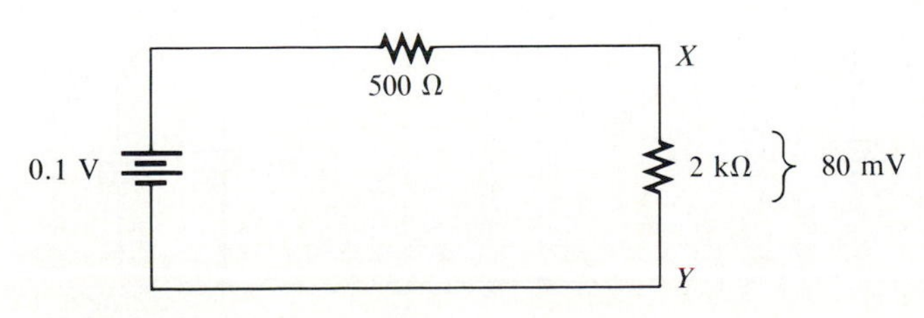

FIGURE 9–21

$$R_{Th} = \frac{1000\ \Omega}{2} = 500\ \Omega \tag{9–31}$$

The voltage across R_1 is

$$0.1\ \text{V}\left(\frac{2\ \text{k}\Omega}{2500\ \Omega}\right) = 80\ \text{mV} \tag{9–32}$$

If $R + \Delta R$ is a transducer such as a strain gage or a resistance thermometer whose resistance changes by 1% (e.g., $R = 1000\ \Omega$, $\Delta R = 10\ \Omega$), a voltage will be developed across X and Y which is measurable. Thus, a change in stress or temperature can be measured as an electric quantity.

POTENTIOMETRIC (SLIDE-WIRE) BRIDGES

Slide-Wire Potentiometer

Wire can be manufactured so that its resistance is a linear function of the length; for example, 3 cm = 3 Ω, 1 m = 100 Ω. In this case, a 2-m length of wire, often called a *slide wire,* will have a resistance of 200 Ω. This 2-m wire is connected in the circuit shown in Figure 9–22, where R_r is a rheostat. If the resistance of the wire is 200 Ω and has 2 V dropped across it,

$$I = \frac{V}{R} = \frac{2\ \text{V}}{200\ \Omega} = 10\ \text{mA} \tag{9–33}$$

will be flowing through the circuit. To provide this current, 4 V must be dropped across the rheostat. Using Ohm's law,

$$R_r = \frac{V}{I} = \frac{4\ \text{V}}{10\ \text{mA}} = 400\ \Omega \tag{9–34}$$

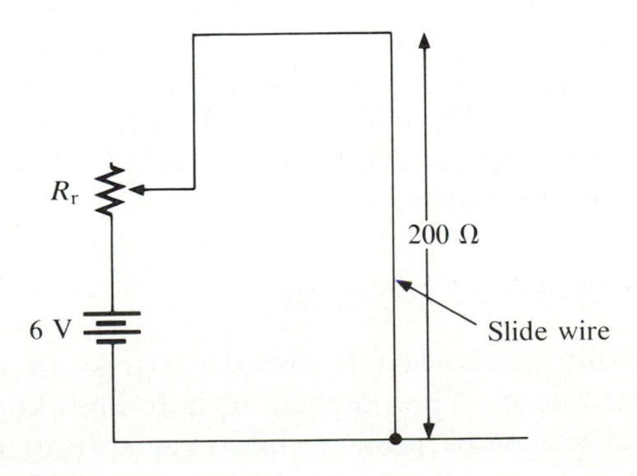

FIGURE 9–22
Slide-wire potentiometer

The variable resistance is used because the potential of the battery may not be exactly 6 V. With R_r set at 400 Ω, the voltage across the 2-m slide wire is 2 V and the voltage across a 1-mm length of the wire will be 1 mV. If a stick 2 m long, calibrated in millimeters, is arranged next to the wire, each millimeter mark will correspond to 1 mV. This process is called *normalizing* the bridge.

Potentiometric Bridge

If a standard cell E_s is attached to the circuit as shown in Figure 9–23 and a galvanometer is used as a detecting instrument, the resulting circuit is called a *potentiometric bridge*. The voltage of the standard cell is usually known accurately to five significant figures (e.g., 1.0195 V). If the pointer is set at the 1.0195-m mark on the meter stick, there should be equal potential on both sides of the meter, creating a potential difference of zero across the meter, which would indicate zero amperes flowing through the meter movement. This is a balanced bridge.

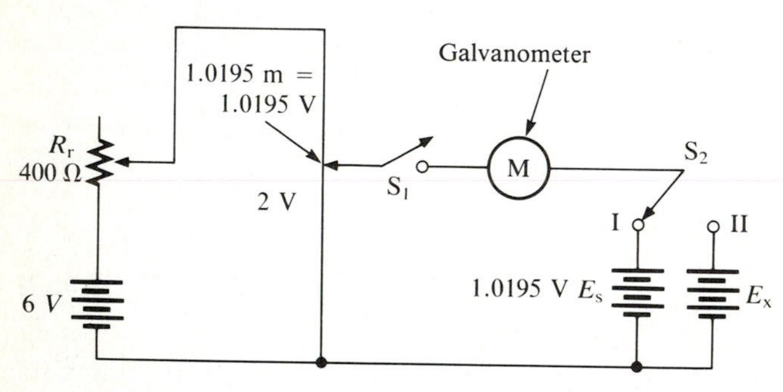

FIGURE 9–23
Slide-wire potentiometric bridge

Normalizing the Bridge

If current is indicated and the bridge is unbalanced, the rheostate is adjusted to bring the bridge back into balance. This normalizing process is a calibration technique. When the pointer is set to any place on the meter stick, the indicator will be a voltage reading in millivolts.

The Sensitivity Control

A variable resistance is usually wired in series with the meter. When a reading is to be taken, the sensitivity control (this variable resistance) is adjusted so that maximum resistance is in series with the meter. This configuration protects the meter from an overloaded condition (Figure 9–24). When

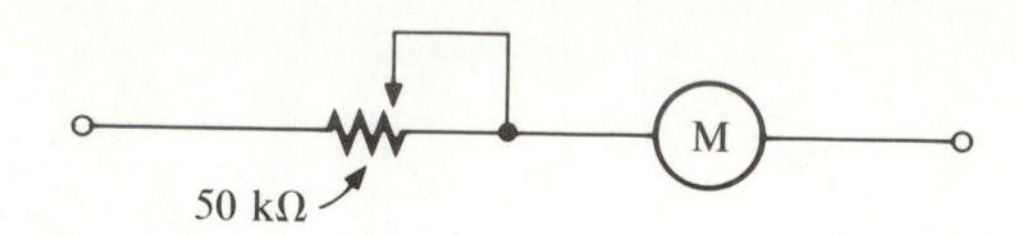

FIGURE 9–24
Maximum resistance, minimum sensitivity

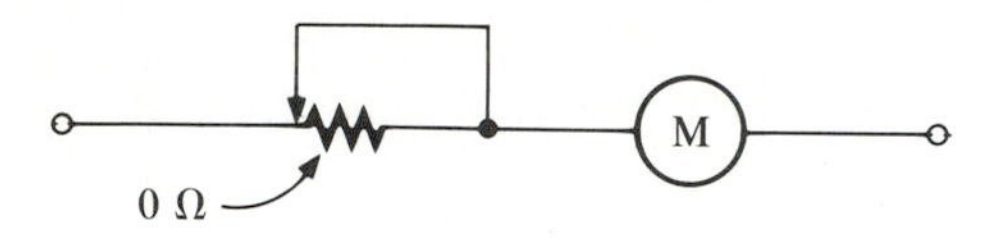

FIGURE 9–25
Minimum resistance, maximum sensitivity

the bridge is nearly balanced, the sensitivity control is adjusted so that minimum resistance is in series with the meter, thus increasing the meter's sensitivity (Figure 9–25).

Measuring Potentials

When the potentiometric bridge has been normalized and calibrated, the switch (S_2 in Figure 9–26) is moved to position II to measure an unknown potential. When the unknown potential is placed across the bridge, the sensitivity control is set at maximum resistance (minimum sensitivity) and the pointer is moved across the slide wire until the meter indicates zero current flow. The sensitivity control is then reduced toward zero as the bridge is more nearly balanced. When the sensitivity is at a maximum (resistance at a minimum) and the meter reads zero current flow, the bridge is balanced and the unknown potential can be read from the meter stick. (The length measurement on the meter stick corresponds to the voltage of the unknown potential.) This balanced condition (zero current) is also called the *null* condition.

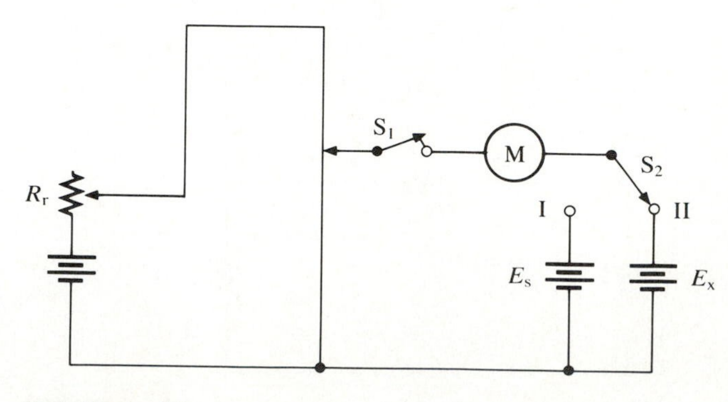

FIGURE 9–26
Measuring an unknown potential

Thevenin Equivalent of the Bridge

It is useful to develop the Thevenin equivalent of the slide-wire potentiometer (refer to Figure 9–27). As was shown earlier, loading effects are easily analyzed using this approach. If points X and Y are to be considered,

$$R_{\text{Th}} = R_1 \parallel [(R - R_1) + R_r] \qquad \textbf{(9–35)}$$

and V_{Th} is the voltage indicated by the meter stick.

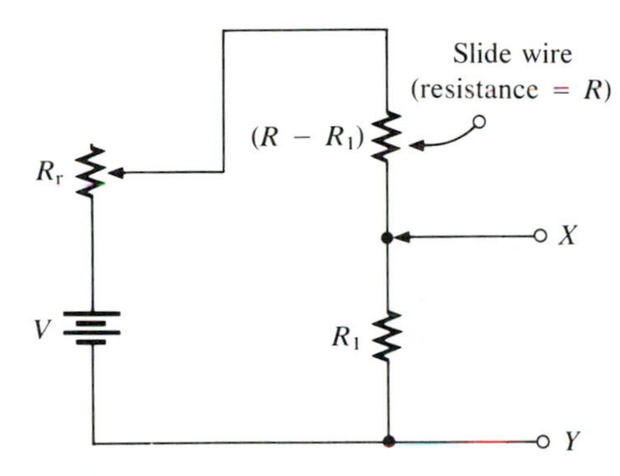

FIGURE 9–27
Deriving the Thevenin equivalent of the slide-wire potentiometer

---

EXAMPLE 9–7
Refer to Figure 9–28.

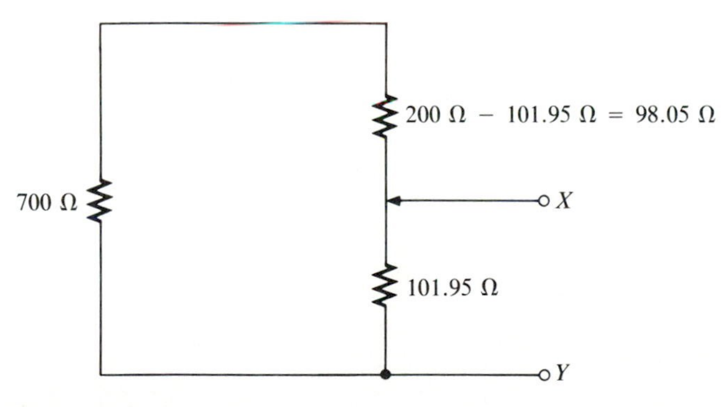

FIGURE 9–28

What will be the working current if this bridge is normalized using a standard cell? What is the value of R_r required to produce this condition? Assume that E_x is 1.0195 V, the internal resistance of the standard cell is 50 Ω, and a 50-μA meter movement with 2 kΩ of internal resistance is used and V = 9 V. (Consider the sensitivity resistance to be at minimum—zero ohms.)

Solution:

$$I_{\text{working}} = \frac{V}{R} = \frac{2\text{ V}}{200\ \Omega} = 10\text{ mA} \qquad \textbf{(9–36)}$$

$$V_{\text{rheostat}} = (9 - 2)\text{ V} = 7\text{ V} \qquad \textbf{(9–37)}$$

$$R_r = \frac{V}{I} = \frac{7\text{ V}}{10\text{ mA}} = 700\ \Omega \qquad \textbf{(9–38)}$$

EXAMPLE 9–8
When the switch in Figure 9–26 is moved to position II to measure E_x and the pointer is set at 1.6250 m, the potential of E_x is 1.6250 V. Before the pointer is moved from 1.0195 V to 1.6250 V, a potential will occur across the meter. Using the Thevenin equivalent, determine the current as seen looking back into terminals X and Y.

Solution:

$$R_{\text{Th}} = 101.95\ \Omega \parallel [(200\ \Omega - 101.95\ \Omega) + 700\ \Omega] \qquad \textbf{(9–39)}$$
$$= 90.4\ \Omega$$

And V_{Th} = 1.0195 V, yielding the Thevenin circuit in Figure 9–29.

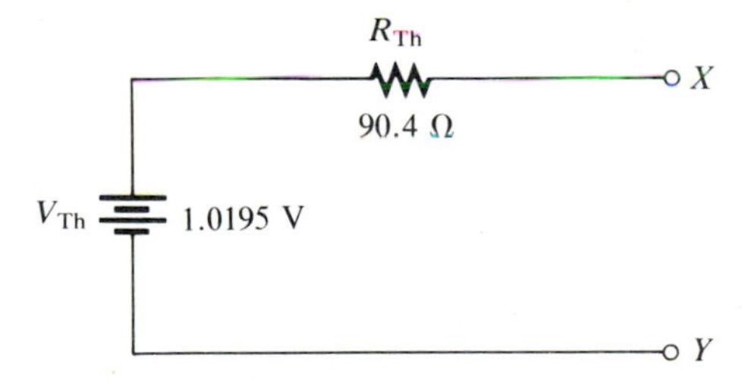

FIGURE 9–29

When the rest of the circuit is attached, assuming that the internal resistance of the unknown potential is zero, the voltage across the meter (2 kΩ) can be calculated using voltage division:

$$V = (1.6250\text{ V} - 1.0195\text{ V})\frac{2000\ \Omega}{2090.4\ \Omega} \qquad \textbf{(9–40)}$$
$$= 0.579\text{ V}$$

The current through the meter will be

$$I = \frac{V}{R} = \frac{0.579\text{ V}}{2000\ \Omega} = 289.6\ \mu\text{A} \qquad \textbf{(9–41)}$$

This illustrates the need for the sensitivity resistance. If the meter is left unprotected, damage might result. After the pointer is moved to the 1.6250-m mark, the sensitivity resistance can be reduced until the bridge is balanced.

Methods of Using the Bridge

The method just discussed is the null method. The measurement is taken when no current is flowing through the bridge. That is, the bridge is acting like an open circuit (infinite resistance). No current is drawn from the circuit under test; therefore, the potentiometric bridge does not load the circuit under test, making it an ideal test instrument.

The bridge can also be used in a mode where the meter is calibrated to indicate the potential (E_x) required to cause certain deflections. By finding the amount of deflection of the pointer on the galvanometer and knowing the sensitivity of the galvanometer, the current flowing through the meter can be found. This information can be used to determine the amount of imbalance in the bridge and thus the potential difference between the standardized bridge setting and the unknown potential. This latter method is rarely used since cumbersome calculations are required to find the unknown potential.

Advantages and Disadvantages

The potentiometric bridge has the advantage of infinite input impedance at the time of measurement. It yields a true reading of the emf of the circuit under test. The high degree of accuracy in taking potential measurements is another advantage. If the slide-wire pointer can be read to the nearest half of a millimeter, the reading can be taken to the nearest 0.0005 V!

The size and range of this instrument are definite disadvantages. Any instrument 2 m long is cumbersome, so the slide wire is wound into a linear precision potentiometer as shown in Figure 9–30. The bridge is limited to a very small range (the potentiometer in Figure 9–30 has a range of up to 2 V). This disadvantage can be overcome by a volt box which acts like the voltmeter multiplier resistors discussed in Chapter 4.

FIGURE 9–30
Linear precision potentiometer

SUMMARY

The use of potentiometers as variable resistances is universal. The use of potentiometers in potentiometric bridges is limited because of the availability of high-precision digital instruments that are portable, fast, accurate, and less expensive.

EXERCISES

1. A 10-kΩ potentiometer is connected across a 9-V battery (see Figure 9–31). What is the current drawn by a 2-kΩ load attached to the output terminals if the wiper is set at $k = 1/4$?

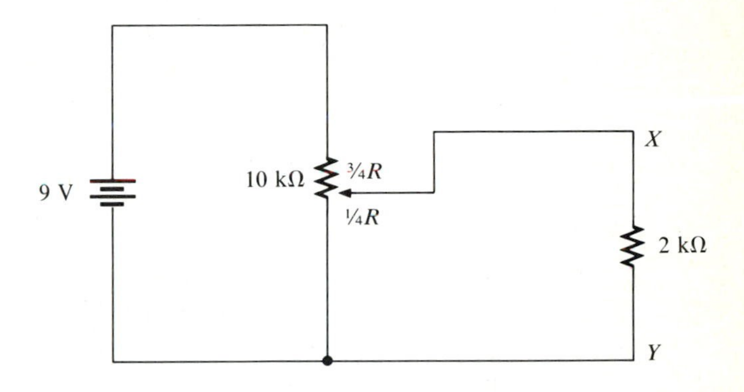

FIGURE 9–31
Circuit for Exercise 1

2. A 25-kΩ potentiometer is wired across a 32-V source (see Figure 9–32). The wiper is set at $k = 5/6$. What voltage is dropped across a 5-kΩ load wired to the output terminals?

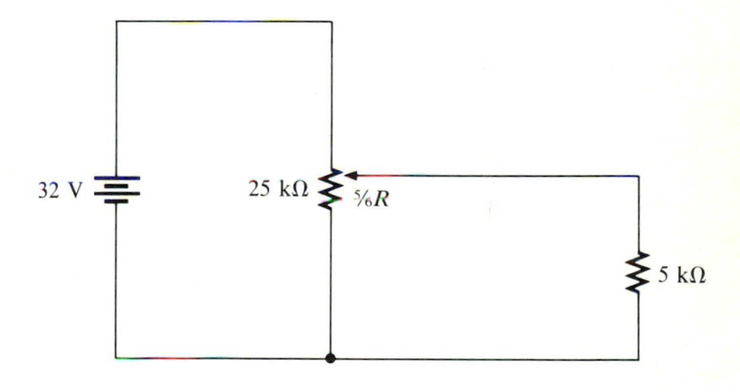

FIGURE 9–32
Circuit for Exercise 2

3. Find the Thevenin equivalent of the circuit in Figure 9–33.

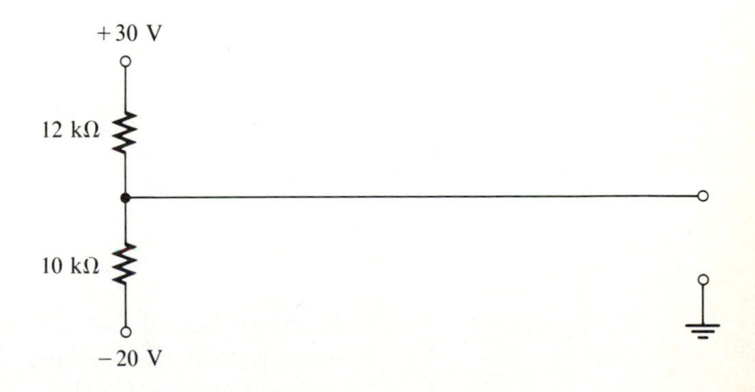

FIGURE 9–33
Circuit for Exercise 3

4. What current will the 10-kΩ load draw from the circuit of Figure 9–34?

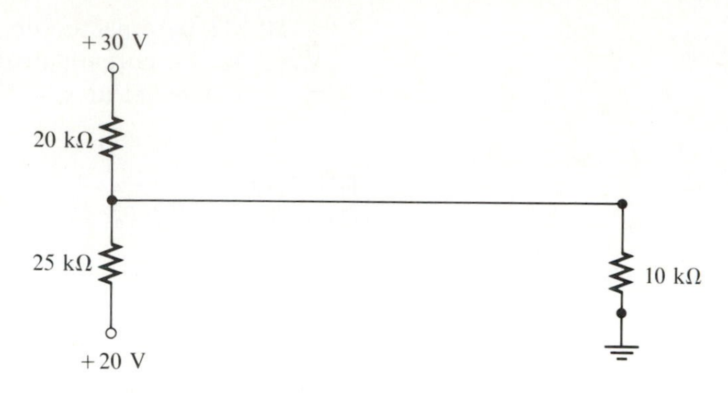

FIGURE 9–34
Circuit for Exercise 4

5. A strain gage has an ambient resistance of 1 kΩ at zero stress (see Figure 9–35). When a 500-lb force is applied to the gage, the resistance changes to 1050 Ω. What is the voltage developed across the output terminals of the circuit?

FIGURE 9–35
Circuit for Exercise 5

6. Refer to Figure 9–36. What value of E_s will produce a balanced condition (current through the galvanometer = 0 A)?

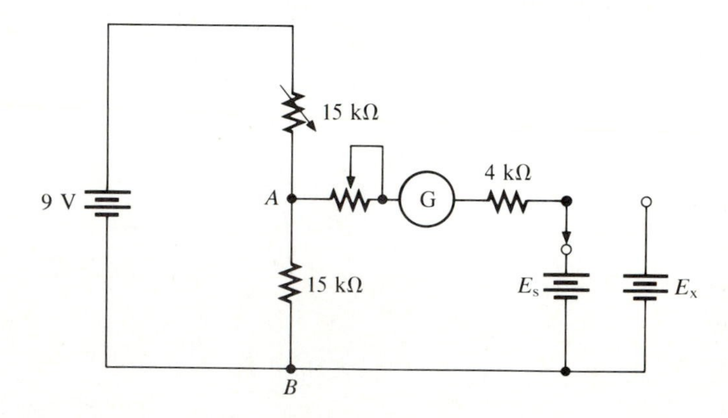

FIGURE 9–36
Circuit for Exercise 6

7. Refer to the slide-wire potentiometer in Figure 9–37. If the resistance of the wire is 200 Ω, the length of the wire is 2 m and the voltage (E) is 6 V. The voltage across the slide wire is 2 V. E_s equals 1.0195 V and E_x equals 1.6 V.
   a. Find the working current.
      If the bridge is standardized and the switch is moved to the unknown emf and the galvanometer indicates −25 μA flowing,
   b. Which has more positive potential, E_x or E_s?

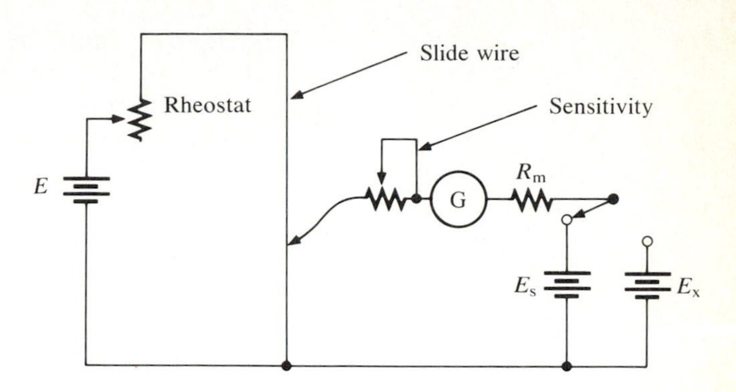

FIGURE 9–37
Circuit for Exercise 7

c. What must the pointer be set at to bring about balance? (Assume that the internal resistance of E_s is zero ohms.)
d. If E_x is accidentally wired in backwards (the polarity of the emf is reversed), how much current will flow through the galvanometer after the bridge is standardized?
e. What value of sensitivity resistance must be used to protect the galvanometer from overload if this error occurs? (Assume that the maximum E_x to be measured is 1.6 V).

LAB PROBLEM 9–1
Slightly Unbalanced Potentiometer

Equipment
One +5-V and one −5-V source (the sources should be electrically independent); one 2.2-kΩ and one 1.2-kΩ resistor; one 2-kΩ potentiometer; one DVM.

Object
To approximate the output of a slightly unbalanced bridge using a Thevenin equivalent.

Procedure
Wire the circuit shown in Figure 9–38.

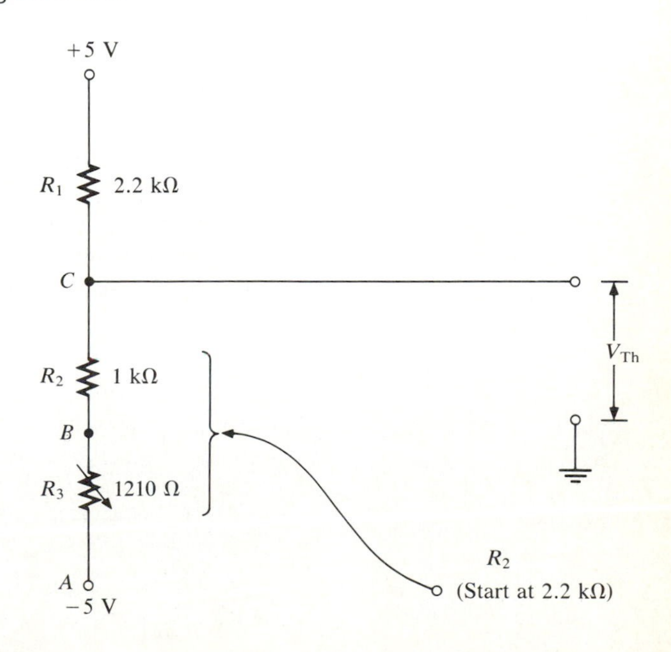

FIGURE 9–38
Circuit for Lab Problem 9–1

Vary R_3 by a small amount ($\Delta R_3 < 5\%$ of R_3, perhaps 25–50 Ω. Measure the resistance from A to C (remember to open the circuit).

Calculate the resistance and the output voltage for this slightly unbalanced bridge using the formulas

$$R_{\text{Th}} = R_1 \parallel (R_2 + R_3)$$

and

$$V_{\text{Th}} = \frac{V_1(R_2 + R_3) - V_2 R_1}{R_1 + R_2 + R_3}$$

The values can be applied to the Thevenin equivalent shown in Figure 9–39.

Measure the voltage. Does the approximation produce reasonably accurate results?

In your conclusion, comment on the effectiveness of using Thevenin's theorem to simulate the circuitry of a two-terminal output of any device with two output terminals.

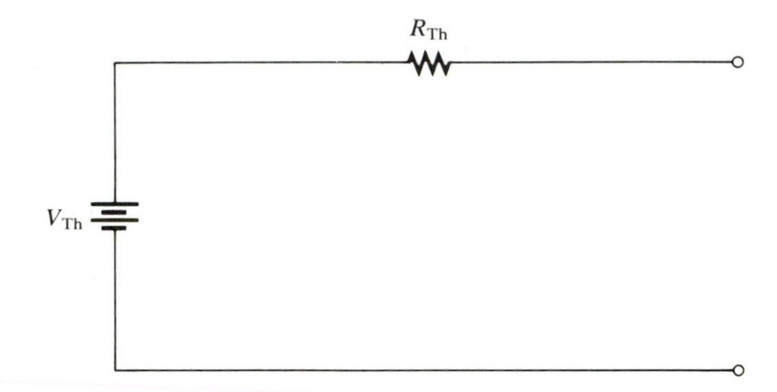

FIGURE 9–39
Thevenin equivalent circuit for Lab Problem 9–1

LAB PROBLEM 9–2
Potentiometric Bridges

Equipment

One calibrated potentiometer, one 10-turn linear potentiometer, or one slide-wire potentiometer; one dc power supply; one standard cell; one 10-kΩ variable resistor; some low-voltage sources to measure; one galvanometer.

Object

To use a potentiometric bridge to measure an unknown voltage using the null method of measurement. Since balance occurs when no current is being drawn from the source to be measured, the potentiometric bridge represents an infinite resistance and presents "no load" to the potential being measured. Because of the bridge's infinite resistance, the voltage measurement will be virtually perfect. A calibrated potentiometer, a precision potentiometer such as a 10-turn linear potentiometer, or a slide-wire potentiometer can be used. The accuracy required determines the choice of device. A voltage standard is also needed. This standard may be a secondary standard, or any lab standard if accuracy is not critical.

Procedure

Wire the circuit on the left of Figure 9–40.

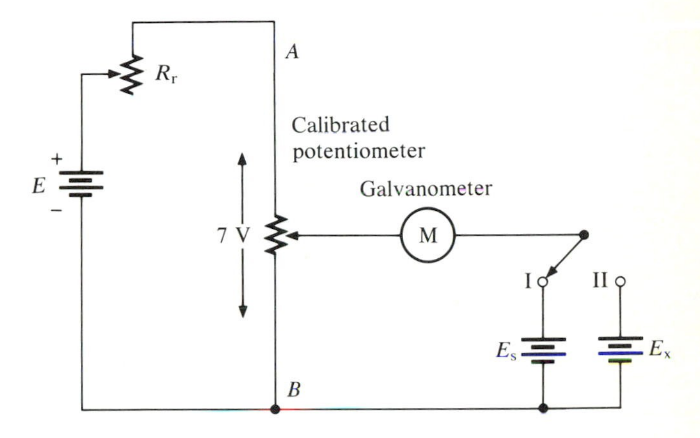

FIGURE 9–40
Circuit for Lab Problem 9–2

Set the potentiometer to yield 7 V across the entire potentiometer. This can be accomplished by setting R_r so that approximately 7 V occur across the bridge from point A to point B. Move the switch to position I (the top of the standard voltage source E_s). Move the wiper of the potentiometer until the galvanometer reads zero amperes. This position represents a potential equal to the standard voltage. When the switch is moved to position II (the top of the unknown voltage E_x) and the wiper of the potentiometer is moved to null the galvanometer, a ratio of resistances can be used to calculate the unknown voltage:

$$\frac{\text{resistance of } E_s}{\text{resistance of } E_x} = \frac{E_s}{E_x}$$

and so

$$E_x = E_s\left(\frac{\text{resistance of } E_x}{\text{resistance of } E_s}\right)$$

For example, if a 10-turn linear potentiometer is used, typical values might be

$$E_x = E_s\left(\frac{238\ \Omega}{331\ \Omega}\right) = 1.5\ \text{V}\left(\frac{238\ \Omega}{331\ \Omega}\right) = 1.07\ \text{V}$$

Record the necessary information in Table 9–1 and do the necessary calculations to produce an accurate voltage measurement.

TABLE 9–1
Complete the table.

Trial	E_s	Resistance of E_x	Resistance of E_s	E_x

Comment on the validity of your final results. Why is this method not in popular use if it yields such accurate results and represents zero loading error in measurements?

10 dc and ac Bridges

OUTLINE

OBJECTIVES

THE WHEATSTONE BRIDGE

THEVENIN EQUIVALENT OF A WHEATSTONE BRIDGE

- ☐ Unbalanced Bridge
- ☐ A Special Design
- ☐ Thevenin Equivalent of a Slightly Unbalanced Bridge

ac BRIDGES

SOME IMPEDANCE CONVERSION FORMULAS

OTHER BRIDGE CIRCUITS

IMPEDANCE BRIDGES

- ☐ Manual Impedance (*LRC*) Bridges
- ☐ Automatic Impedance (*LRC*) Bridges
- ☐ Theory of Operation of Automatic *LRC* Bridges

SUMMARY

EXERCISES

LAB PROBLEM 10–1—Impedance Bridge

LAB PROBLEM 10–2—Capacitance Comparison Bridge

OBJECTIVES

After completing this chapter, you will be able to:

- ☐ Recognize a Wheatstone bridge.
- ☐ Give the formula for a balanced Wheatstone bridge.
- ☐ Apply the null method of measuring resistance with a Wheatstone bridge.
- ☐ Derive the Thevenin equivalent of a Wheatstone bridge.
- ☐ Describe a special bridge design used with transducers.
- ☐ Recognize a slightly unbalanced bridge.
- ☐ Find the Thevenin equivalent of a slightly unbalanced Wheatstone bridge.
- ☐ Describe the methods of using a slightly unbalanced bridge.
- ☐ Describe a basic ac bridge.
- ☐ Give the formula for a balanced ac bridge.
- ☐ Convert parallel impedance to a series equivalent.
- ☐ Convert series impedance to a parallel equivalent.
- ☐ Describe a general impedance bridge, a manual impedance bridge, an automatic *LRC* bridge, and an automatic *LRC* comparator.

Dc and ac bridges are devices used to measure resistance, capacitance, inductance, voltage, and other electrical quantities. This chapter presents manual and automatic bridges, various bridge configurations, and different methods of using bridges.

THE WHEATSTONE BRIDGE

The *Wheatstone bridge* is a dc bridge composed of resistors that is used to measure potential differences. It is often employed to measure an unknown resistance using the *null* method of measurement.

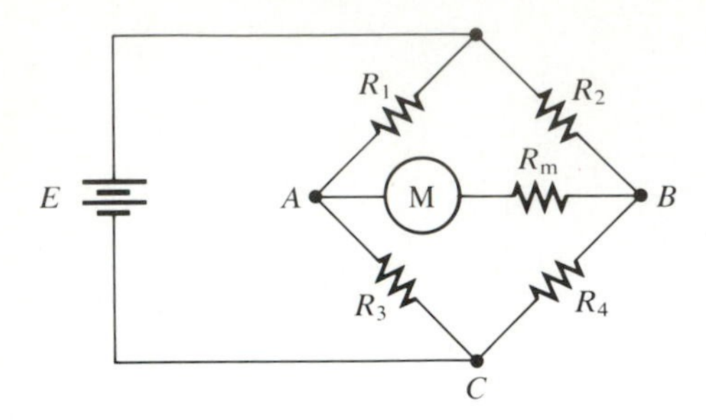

FIGURE 10–1
Wheatstone bridge

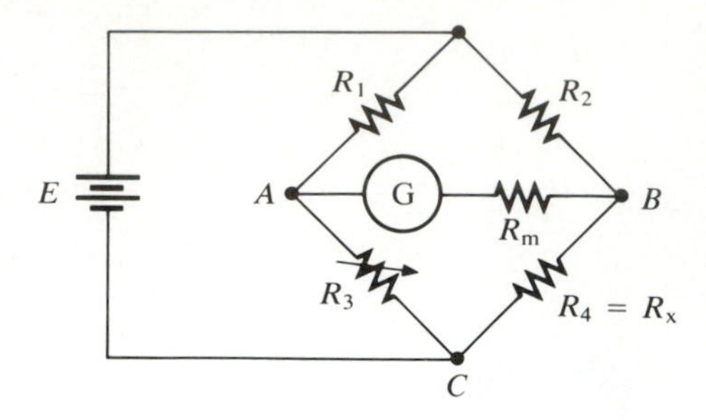

FIGURE 10–2
Practical Wheatstone bridge

This method is based on the concept that a bridge is balanced when the potential difference being measured is null, that is, zero. Figure 10–1 shows the circuit diagram of a Wheatstone bridge.

If the potential at point A equals the potential at point B,

$$V_{AC} = V_{BC} \tag{10–1}$$

and

$$V_{AC} - V_{BC} = 0 \tag{10–2}$$

By voltage division,

$$V_{AC} = E\left(\frac{R_3}{R_1 + R_3}\right) \tag{10–3}$$

and

$$V_{BC} = E\left(\frac{R_4}{R_2 + R_4}\right) \tag{10–4}$$

If $V_{AC} = V_{BC}$,

$$E\left(\frac{R_3}{R_1 + R_3}\right) = E\left(\frac{R_4}{R_2 + R_4}\right) \tag{10–5}$$

Dividing by E gives

$$\frac{R_3}{R_1 + R_3} = \frac{R_4}{R_2 + R_4} \tag{10–6}$$

Clearing of fractions results in

$$\begin{aligned} R_3(R_2 + R_4) &= R_4(R_1 + R_3) \\ R_2R_3 + R_3R_4 &= R_1R_4 + R_3R_4 \end{aligned} \tag{10–7}$$

Subtracting R_3R_4 from both sides yields

$$R_2R_3 = R_1R_4 \tag{10–8}$$

Typically, R_4 is the unknown branch. Therefore, the following is usually used:

$$R_4 = \frac{R_2R_3}{R_1} \tag{10–9}$$

Relationship (10–8) is independent of the applied voltage E. If three of the resistances in the bridge are known and the bridge is balanced, the fourth resistance can be found. A suitable meter can be connected across points A and B to determine the null condition. Usually a galvanometer (a center-deflection instrument) is used since the bridge may be unbalanced in a positive or negative direction (Figure 10–2).

THEVENIN EQUIVALENT OF A WHEATSTONE BRIDGE

Unbalanced Bridge

It is sometimes useful to have the Thevenin equivalent of a Wheatstone bridge to help in designing a measuring system for a particular application. To find the Thevenin equivalent, the open-circuit potential and internal resistance must be determined. To accomplish this, the power source is replaced by its internal resistance, R_S (assumed to be zero ohms), and the meter is removed to provide an open circuit, as shown in Figure 10–3A.

From Figure 10–3B,

$$R_{Th} = (R_1 \parallel R_3) + (R_2 \parallel R_4) \tag{10–10}$$

To determine the open-circuit potential at points A and B; the relationship

$$V_{AB} = V_{AC} - V_{BC} \tag{10–11}$$

is used. By voltage division,

$$V_{AC} = E\left(\frac{R_3}{R_1 + R_3}\right) \tag{10–12}$$

and

$$V_{BC} = E\left(\frac{R_4}{R_2 + R_4}\right) \quad \textbf{(10–13)}$$

Substitution and factoring gives

$$V_{Th} = V_{AB} = E\left(\frac{R_3}{R_1 + R_3} - \frac{R_4}{R_2 + R_4}\right) \quad \textbf{(10–14)}$$

(see Figure 10–4).

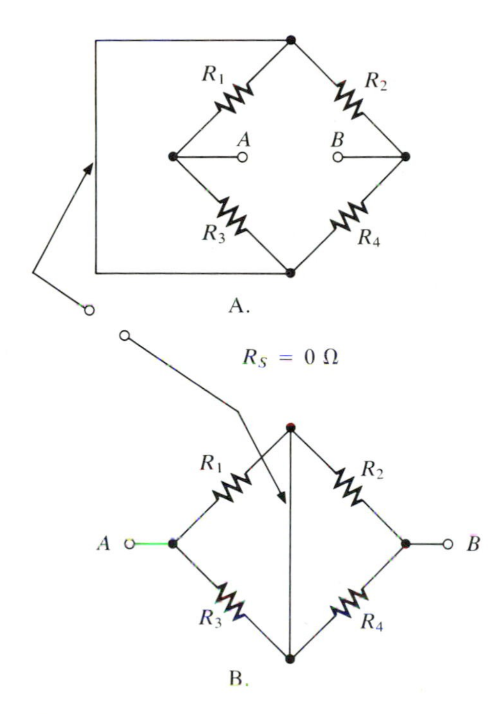

FIGURE 10–3
Thevenin equivalent of resistance

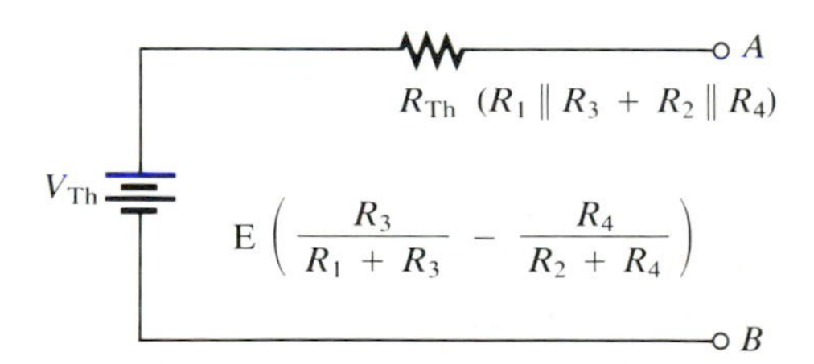

FIGURE 10–4
Thevenin equivalent of a Wheatstone bridge

EXAMPLE 10–1
Develop the Thevenin equivalent for the bridge in Figure 10–5A.

Solution:
"Looking into" terminals A and B and replacing the 9-V source with its internal resistance (assumed to be zero ohms—a short circuit), we see that the 3-kΩ and 6-kΩ resistors are in parallel, as are the

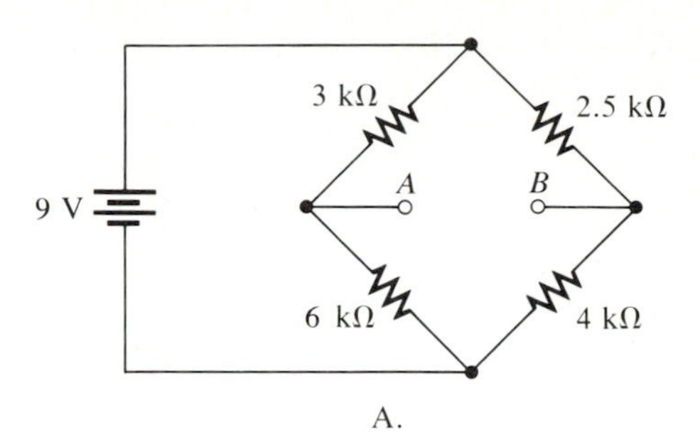

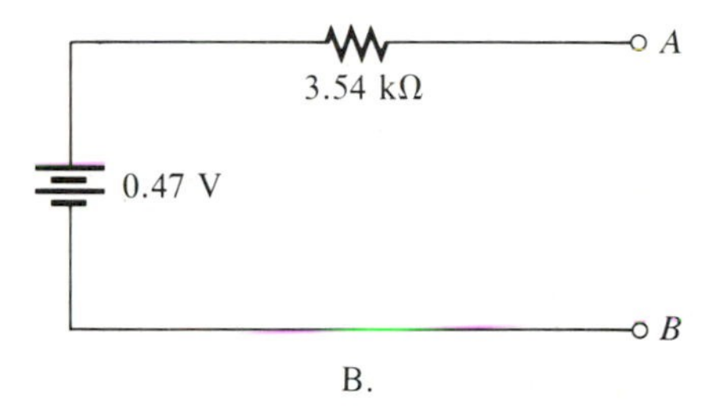

FIGURE 10–5

2.5-kΩ and 4-kΩ resistors. These two parallel combinations are in series. Thus,

$$\begin{aligned} R_{Th} &= (6\ \text{k}\Omega \parallel 3\ \text{k}\Omega) \\ &\quad + (2.5\ \text{k}\Omega \parallel 4\ \text{k}\Omega) \\ &= 3.54\ \text{k}\Omega \end{aligned} \quad \textbf{(10–15)}$$

Using equation (10–14),

$$\begin{aligned} V_{Th} &= 9\ \text{V}\left(\frac{6\ \text{k}\Omega}{3\ \text{k}\Omega + 6\ \text{k}\Omega} - \frac{4\ \text{k}\Omega}{2.5\ \text{k}\Omega + 4\ \text{k}\Omega}\right) \\ &= 0.46\ \text{V} \end{aligned} \quad \textbf{(10–16)}$$

(see Figure 10–5B). These are the voltage and resistance values needed before a measuring instrument is wired across terminals A and B.

EXAMPLE 10–2
The bridge in Figure 10–2 is balanced when the following values are used: $E = 10$ V, $R_1 = 10$ kΩ, $R_2 = 1$ kΩ, $R_3 = 2$ kΩ. Find R_x.

Solution:

$$\begin{aligned} R_4 &= R_x \\ R_x &= \frac{R_2 R_3}{R_1} \\ &= \frac{(1\ \text{k}\Omega)(2\ \text{k}\Omega)}{10\ \text{k}\Omega} \\ &= 200\ \Omega \end{aligned} \quad \textbf{(10–17)}$$

A Special Design

The resistances in the Wheatstone bridge circuit are referred to as *arms*. Arm R_3 is usually a variable resistance used to bring the bridge into balance. Arms R_1 and R_2 are called the *ratio* arms and are used to adjust the magnitude of resistances in the bridge to provide more suitable conditions for balancing the bridge.

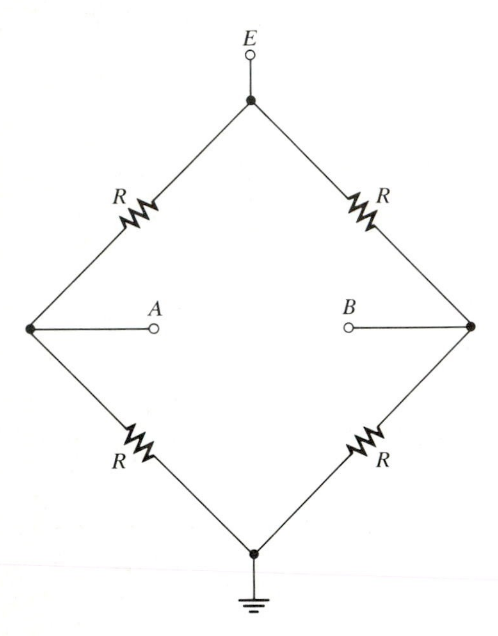

FIGURE 10–6
Special bridge design

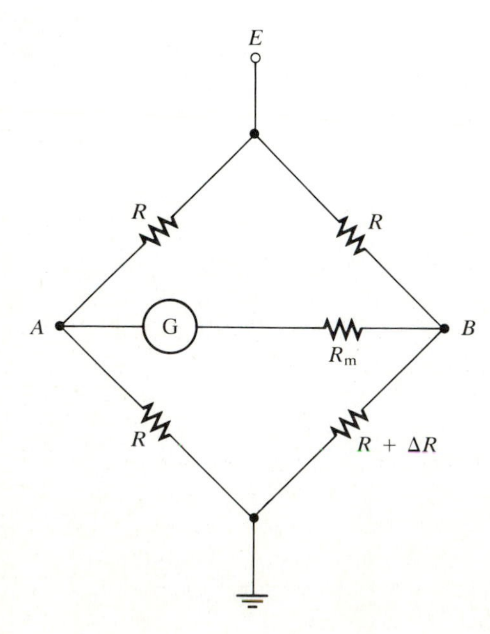

FIGURE 10–7
Slightly unbalanced Wheatstone bridge

Consider the bridge circuit in Figure 10–6. This bridge is balanced. If one branch were a transducer such as a strain gage or an optical transducer with resistance R, the bridge would still be at balance. If conditions were to change, causing the resistance of the transducer to vary by a small amount, ΔR, then the bridge would appear as shown in Figure 10–7. In this case, the bridge would be unbalanced. The amount of imbalance can be used to determine the variation in the resistance of the transducer. The meter can then be calibrated to indicate this variation in degrees, tons of force, lumens, and so on, depending on the type of transducer used.

Thevenin Equivalent of a Slightly Unbalanced Bridge

The Thevenin equivalent is useful in determining the amount of imbalance of a slightly unbalanced bridge. To develop the equivalent, the meter (load) is removed and the power source is replaced by the value of its internal resistance (here assumed to be zero), as shown in Figure 10–8A. The circuit is redrawn in Figure 10–8B. Redrawing the node at point C gives a clearer picture of the relationship between corresponding resistors (see Figure 10–8C). The circuit can be reduced to the simplified equivalent circuit in Figure 10–8D. The Thevenin equivalent of the circuit resistance appears in Figure 10–8E.

If $\Delta R \ll R$, then $R + \Delta R \cong R$. Therefore, the Thevenin resistance is R ohms.

Voltage division is used to evaluate the Thevenin (open-circuit) potential:

$$\begin{aligned} V_{BA} &= V_{BC} - V_{AC} \\ &= E\left(\frac{R + \Delta R}{R + R + \Delta R}\right) - E\left(\frac{R}{R + R}\right) \\ &= E\left(\frac{R + \Delta R}{2R + \Delta R} - \frac{1}{2}\right) \\ &= E\left(\frac{2R + 2\Delta R - 2R - \Delta R}{2(2R + \Delta R)}\right) \end{aligned} \qquad (10\text{–}18)$$

Since ΔR is small compared to $2R$, it can be ignored, yielding

$$V_{BA} = \frac{E\Delta R}{4R} \qquad (10\text{–}19)$$

If ΔR is positive, V_B is more positive than V_A. If ΔR is negative, V_B is more negative than V_A.

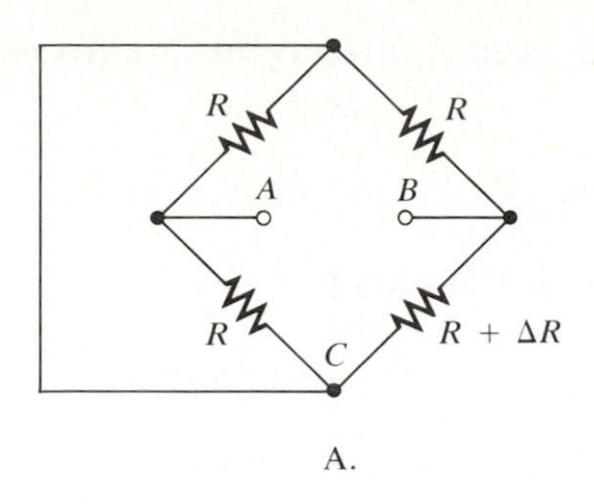

A.

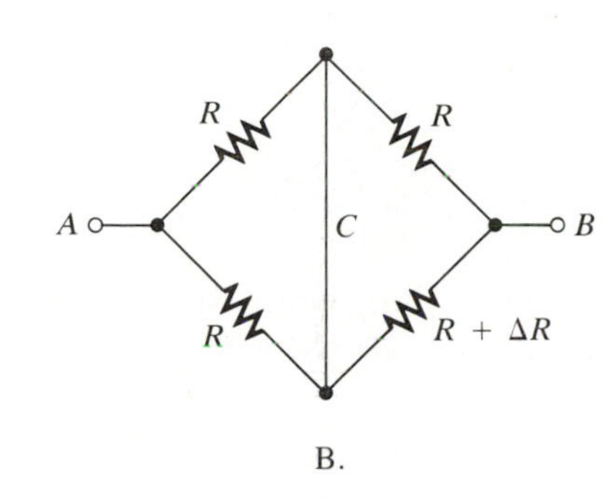

B.

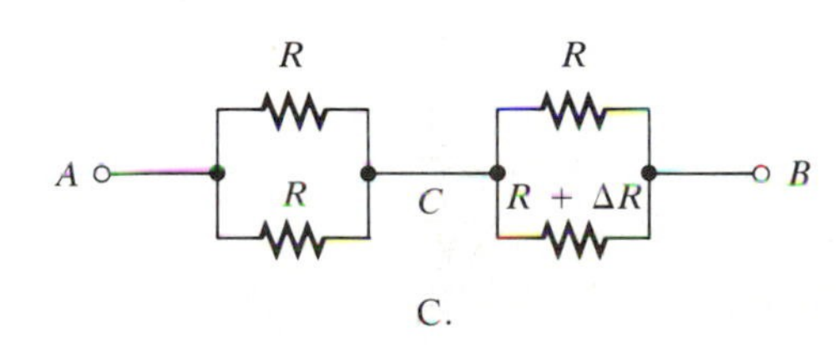

C.

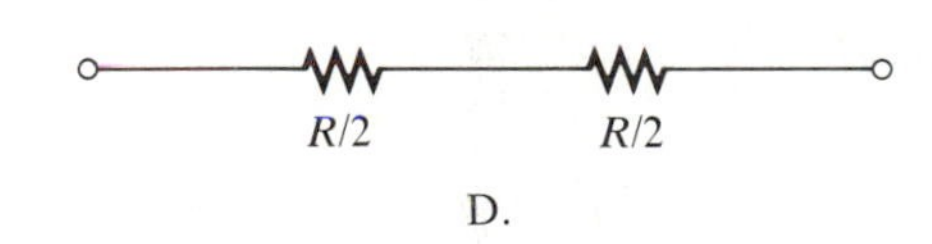

D.

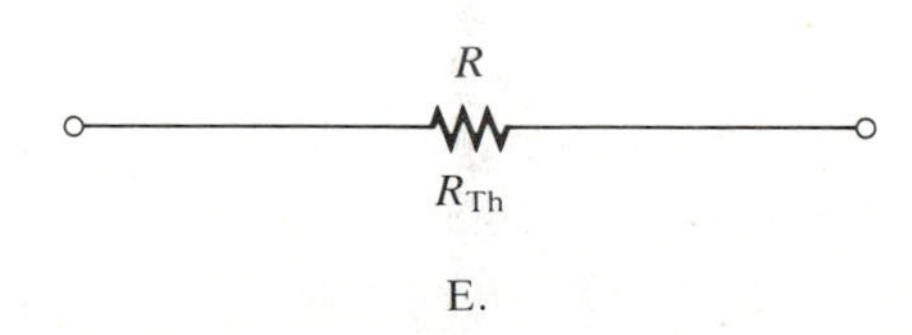

E.

FIGURE 10–8
Thevenin resistance of a slightly unbalanced bridge

EXAMPLE 10–3

Find the Thevenin equivalent for the bridge in Figure 10–9A. Notice that ΔR is very small compared to R.

Solution:

$$R_{\text{Th}} = 2\ \text{k}\Omega \tag{10–20}$$

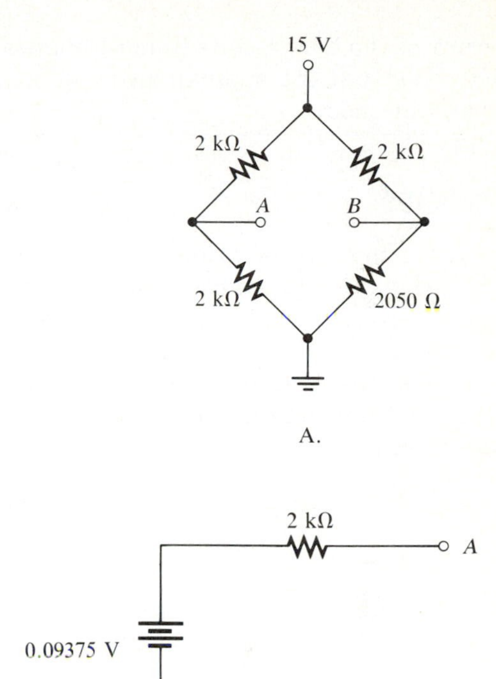

FIGURE 10–9

$$V_{\text{Th}} = \frac{(15\ \text{V})(50\ \Omega)}{4(2000\ \Omega)} = 0.09375\ \text{V} \tag{10–21}$$

(see Figure 10–9B).

EXAMPLE 10–4

Compare the results in Example 10–3 to the more accurate data resulting from the method that does not use the approximation ($\Delta R \ll R$).

Solution:
The alternative method gives

$$R_{\text{Th}} = (2\ \text{k}\Omega \parallel 2\ \text{k}\Omega) + (2\ \text{k}\Omega \parallel 2050\ \Omega) = 2012\ \Omega \tag{10–22}$$

This represents an error of 0.62%. And

$$V_{\text{Th}} = 15\ \text{V}\left(\frac{2050\ \Omega}{4050\ \Omega} - \frac{2\ \text{k}\Omega}{4\ \text{k}\Omega}\right) = 0.09259\ \text{V} \tag{10–23}$$

This represents an error of 1.25%. The results obtained with the approximation method (and the simpler formulas) are within an acceptable percentage of error of the true results in most applications, demonstrating that the approximation using $\Delta R \ll R$ is a valid method.

ac BRIDGES

Any or all arms of an ac bridge may be composed of combinations of inductances, capacitances, and resistances powered by an ac source. When the bridge is brought into balance—or nearly so—it can be used to measure capacitance, inductance, and series or parallel impedance (see Figure 10–10).

At balance,

$$\mathbf{v}_{AC} = \mathbf{v}_{BC} \qquad (10\text{–}24)$$

where **v** denotes a vector. And

$$\mathbf{v}_{AC} = v\left(\frac{\mathbf{Z}_3}{\mathbf{Z}_1 + \mathbf{Z}_3}\right) \qquad (10\text{–}25)$$

$$\mathbf{v}_{BC} = v\left(\frac{\mathbf{Z}_4}{\mathbf{Z}_2 + \mathbf{Z}_4}\right) \qquad (10\text{–}26)$$

Therefore,

$$v\left(\frac{\mathbf{Z}_3}{\mathbf{Z}_1 + \mathbf{Z}_3}\right) = v\left(\frac{\mathbf{Z}_4}{\mathbf{Z}_2 + \mathbf{Z}_4}\right) \qquad (10\text{–}27)$$

yielding

$$\frac{\mathbf{Z}_3}{\mathbf{Z}_1 + \mathbf{Z}_3} = \frac{\mathbf{Z}_4}{\mathbf{Z}_2 + \mathbf{Z}_4} \qquad (10\text{–}28)$$

Multiplication gives

$$\mathbf{Z}_2\,\mathbf{Z}_3 + \mathbf{Z}_3\,\mathbf{Z}_4 = \mathbf{Z}_1\,\mathbf{Z}_4 + \mathbf{Z}_3\,\mathbf{Z}_4 \qquad (10\text{–}29)$$

which reduces to

$$\mathbf{Z}_2\,\mathbf{Z}_3 = \mathbf{Z}_1\,\mathbf{Z}_4 \qquad (10\text{–}30)$$

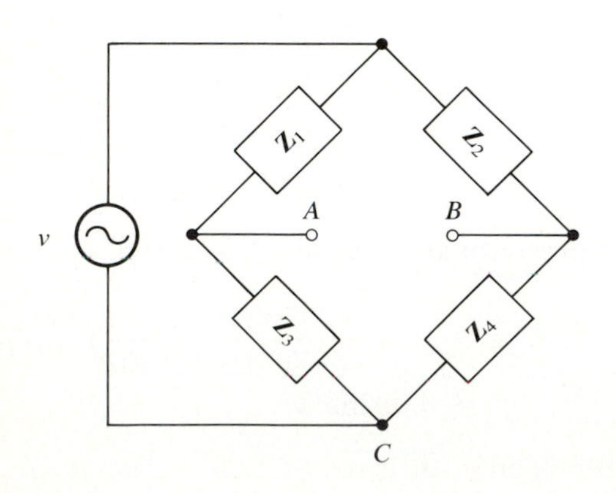

FIGURE 10–10
ac bridge

But $\mathbf{Z}_1$, $\mathbf{Z}_2$, $\mathbf{Z}_3$, and $\mathbf{Z}_4$ are vector quantities. Therefore,

$$(Z_2 \angle\theta_2)(Z_3 \angle\theta_3) = (Z_1 \angle\theta_1)(Z_4 \angle\theta_4) \qquad (10\text{–}31)$$

Vector multiplication gives

$$Z_2\,Z_3 \angle\theta_2 + \theta_3 = Z_1\,Z_4 \angle\theta_1 + \theta_4 \qquad (10\text{–}32)$$

at balance. Both conditions must be met for balance; that is,

$$Z_2\,Z_3 = Z_1\,Z_4 \qquad (10\text{–}33)$$

and

$$\angle\theta_2 + \angle\theta_3 = \angle\theta_1 + \angle\theta_4 \qquad (10\text{–}34)$$

(see Figure 10–11). This means that some ac bridge configurations cannot be balanced.

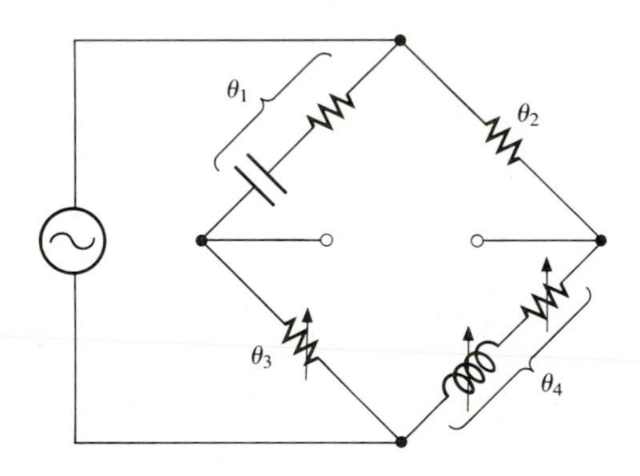

FIGURE 10–11

EXAMPLE 10–5
Demonstrate that the bridge in Figure 10–11 can be balanced.

Solution:
Since

$$-90° \leq \angle\theta_1 \leq 0° \qquad (10\text{–}35)$$

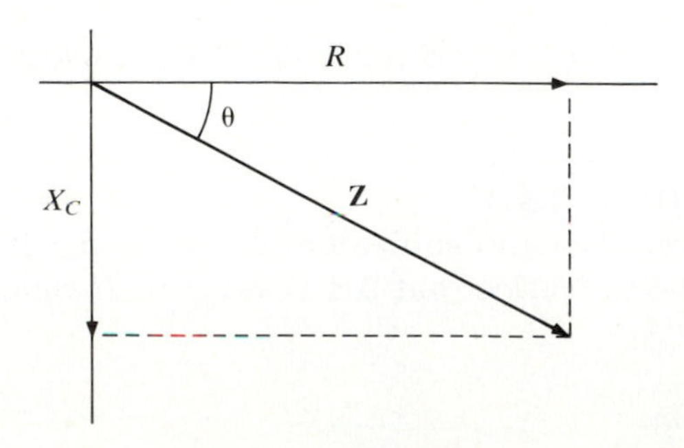

$$0° \leq \angle\theta_4 \leq +90° \qquad (10\text{–}36)$$

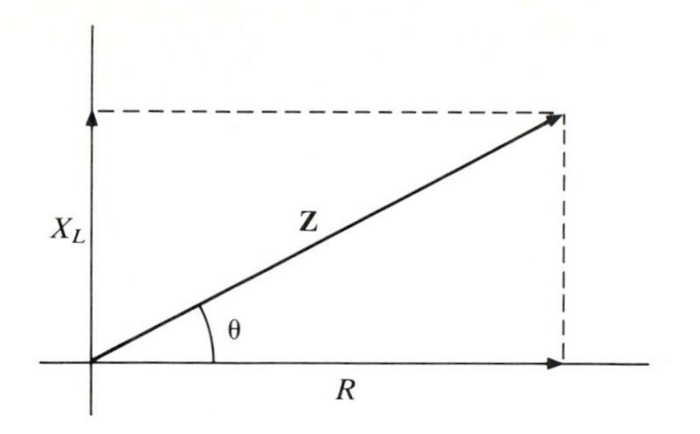

then

$$\underline{/\theta_1} + \underline{/\theta_4} \text{ could equal } 0° \qquad (10\text{–}37)$$

and

$$\underline{/\theta_3} + \underline{/\theta_2} = 0° \qquad (10\text{–}38)$$

Therefore, the bridge can be balanced.

EXAMPLE 10–6
Show that the bridge in Figure 10–12 cannot be balanced.

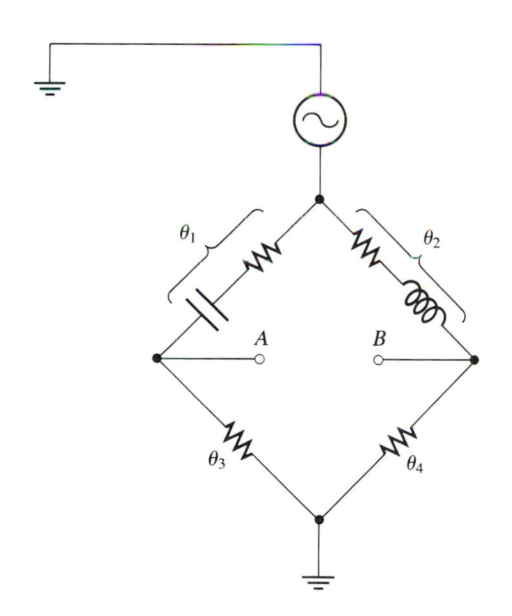

FIGURE 10–12

Solution:
Since $\mathbf{Z}_1$ is capacitive,

$$-90° \leq \underline{/\theta_1} \leq 0° \qquad (10\text{–}39)$$

and since $\mathbf{Z}_4$ is resistive,

$$\underline{/\theta_4} = 0° \qquad (10\text{–}40)$$

The sum of these angles results in a negative angle:

$$-90° \leq \underline{/\theta_1} + \underline{/\theta_4} \leq 0° \qquad (10\text{–}41)$$

$\mathbf{Z}_2$ is inductive. Its phase angle is therefore

$$0° \leq \underline{/\theta_2} \leq 90° \qquad (10\text{–}42)$$

Since $\mathbf{Z}_3$ is resistive,

$$\underline{/\theta_3} = 0° \qquad (10\text{–}43)$$

Therefore,

$$0° \leq \underline{/\theta_2} + \underline{/\theta_3} \leq 90° \qquad (10\text{–}44)$$

Because the sum of $\underline{/\theta_1}$ and $\underline{/\theta_4}$ must be between 0° and −90°, and the sum of $\underline{/\theta_2}$ and $\underline{/\theta_3}$ must be between 0° and +90°, the bridge cannot be balanced.

EXAMPLE 10–7
The bridge in Figure 10–13 is a capacitance comparison bridge. What values of C_x and R_x will cause a balanced condition?

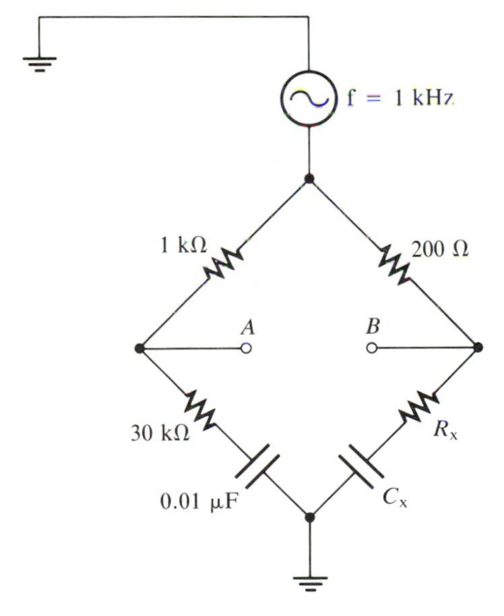

FIGURE 10–13
Capacitance comparison bridge

Solution:

$$1\text{ k}\Omega\left(R_x - \frac{j}{2\pi f C_x}\right)$$
$$= 200\ \Omega\left(30\text{ k}\Omega - \frac{j}{2\pi f(0.01 \times 10^{-6}\text{ F})}\right) \qquad (10\text{–}45)$$
$$(1\text{ k}\Omega)(R_x) - 1\text{ k}\Omega\left(\frac{j}{2\pi f C_x}\right)$$
$$= (200\ \Omega)(30\text{ k}\Omega) - 200\ \Omega\left(\frac{j}{2\pi f(0.01 \times 10^{-6}\text{ F})}\right)$$

Since the real numbers must be equal,

$$(1\text{k}\Omega)(R_x) = (200\ \Omega)(30\ \text{k}\Omega) \quad \textbf{(10–46)}$$
$$R_x = 6\ \text{k}\Omega$$

Because the imaginary numbers must be equal,

$$-j\left(\frac{1\ \text{k}\Omega}{2\pi f C_x}\right) = -j\left(\frac{200\ \Omega}{2\pi f(0.01 \times 10^{-6}\ \text{F})}\right) \quad \textbf{(10–47)}$$

Cancellation gives

$$\frac{1\ \text{k}\Omega}{C_x} = \frac{200\ \Omega}{0.01 \times 10^{-6}\ \text{F}} \quad \textbf{(10–48)}$$
$$C_x = 0.05 \times 10^{-6}\ \text{F} = 0.05\ \mu\text{F}$$

SOME IMPEDANCE CONVERSION FORMULAS

Any resistance can be considered the equivalent of two parallel resistances. Consider the 100-Ω resistance in Figure 10–14A. The 75-Ω and 25-Ω series combination in Figure 10–14B is its equivalent, as are the two parallel 200-Ω resistances in Figure 10–14C. The same relationship holds for impedances. Sometimes series or parallel equivalences are useful in analyzing impedance circuits. The conversion formulas will be developed in this section.

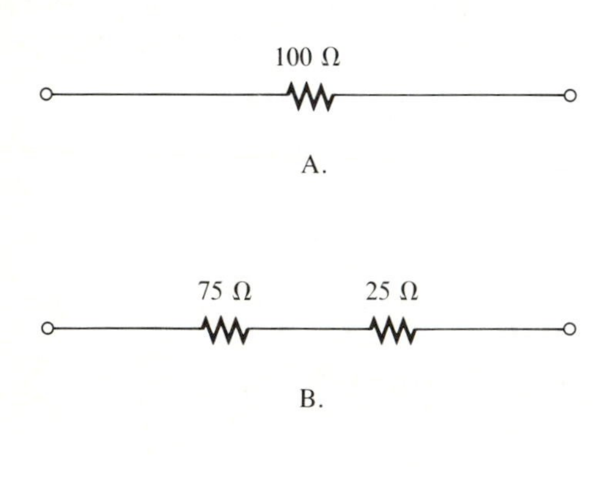

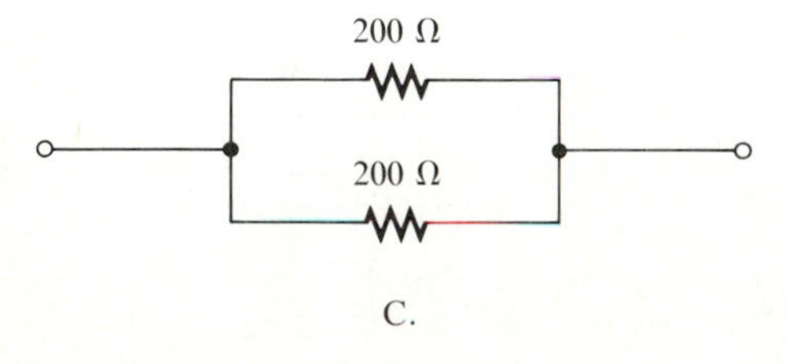

FIGURE 10–14
Different configurations of 100 Ω

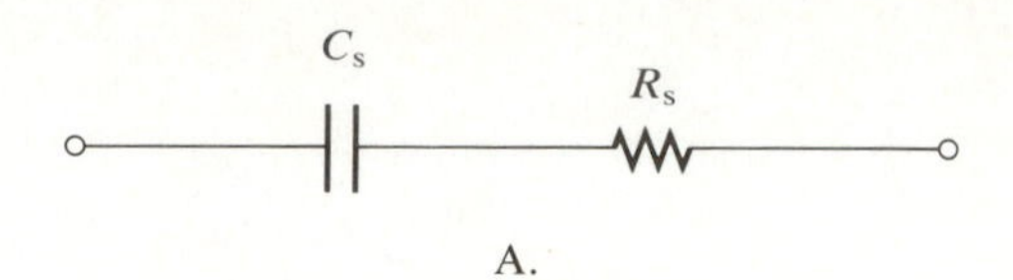

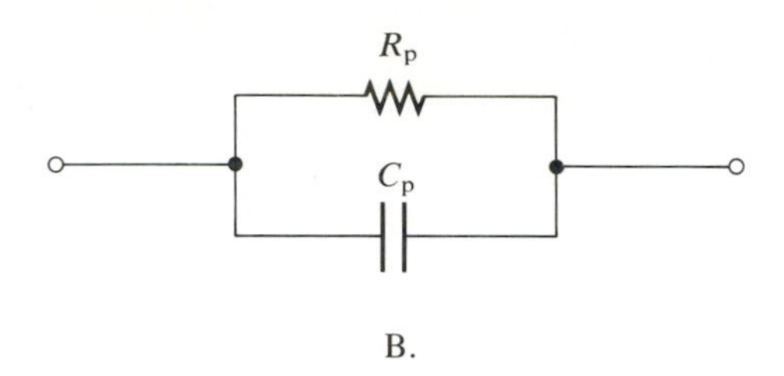

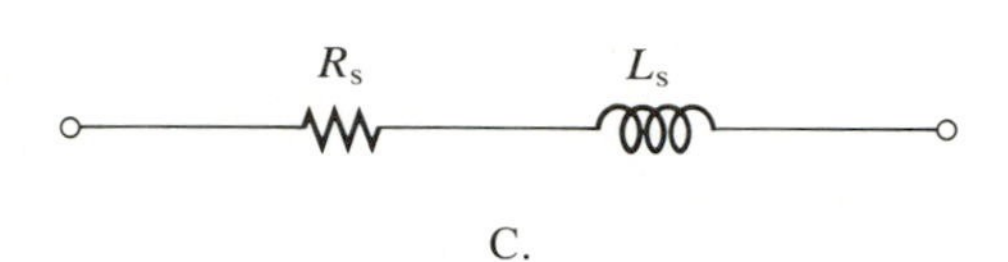

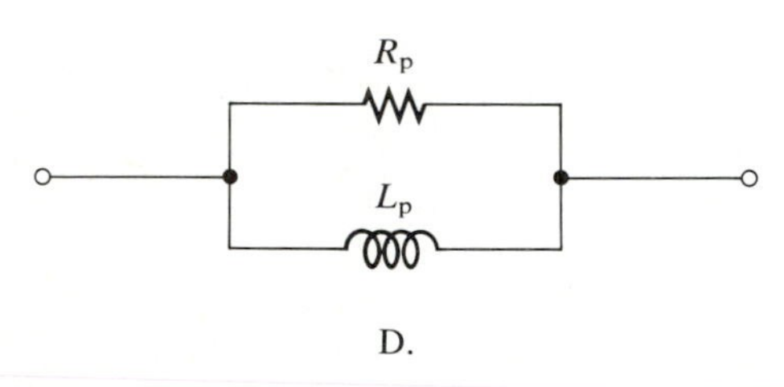

FIGURE 10–15
Series and parallel equivalences

Figure 10–15 shows the series and parallel equivalent circuits of a capacitance and a resistance (Figures 10–15A and B, respectively) and the series and parallel equivalent circuits of an inductance and a resistance (Figures 10–15C and D, respectively).

EXAMPLE 10–8
Connect a circuit consisting of a 100-Ω resistance in series with a 100-Ω inductive reactance that can be converted to an equivalent parallel circuit with a resistance in parallel with an inductive reactance (see Figure 10–16).

Solution:
Remember that impedances consist of a real or resistive component and an imaginary or reactive

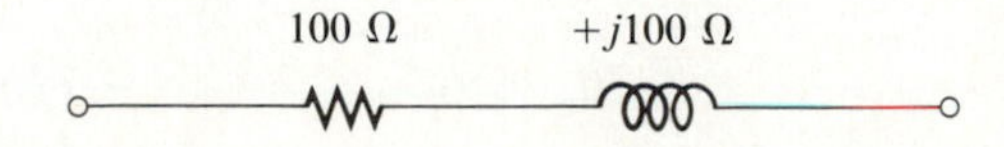

FIGURE 10–16

component. In the present example, the expression is

$$\mathbf{Z} = 100\ \Omega + j100\ \Omega \qquad \textbf{(10–49)}$$

Taking the inverse gives

$$\mathbf{Y} = \frac{1}{\mathbf{Z}} = \frac{1}{100\ \Omega + j100\ \Omega} \qquad \textbf{(10–50)}$$

where **Y** is the admittance. (It is useful to remember that the inverse of impedance [1/**Z**] is admittance [**Y**], that the inverse of resistance [$1/R$] is conductance [G], and that the inverse of reactance [$1/X$] is susceptance [B].) Multiplying both the numerator and the denominator by the complex conjugate of the denominator gives

$$\mathbf{Y} = \left(\frac{1}{100\ \Omega + j100\ \Omega}\right)\left(\frac{100\ \Omega - j100\ \Omega}{100\ \Omega - j100\ \Omega}\right) \qquad \textbf{(10–51)}$$

Since

$$(j \times j) = -1 \qquad \textbf{(10–52)}$$

it follows that

$$\mathbf{Y} = \frac{100\ \Omega}{(100\ \Omega)^2 + (100\ \Omega)^2} - j\frac{100\ \Omega}{(100\ \Omega)^2 + (100\ \Omega)^2} \qquad \textbf{(10–53)}$$

This expression reduces to

$$\mathbf{Y} = \frac{1}{200\ \Omega} - j\frac{1}{200\ \Omega} = G - jB \qquad \textbf{(10–54)}$$

The parallel equivalent of $100\ \Omega + j100\ \Omega$ is

$$200\ \Omega - \frac{200\ \Omega}{j} \qquad \textbf{(10–55)}$$

Multiplying the imaginary component by j/j gives

$$200\ \Omega + j200\ \Omega \qquad \textbf{(10–56)}$$

(see Figure 10–17).

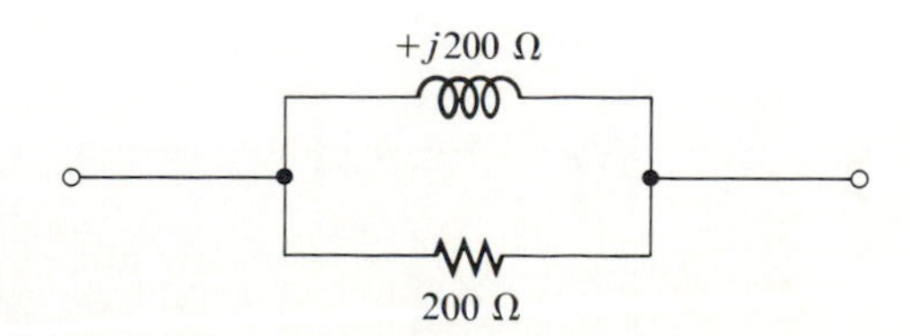

FIGURE 10–17

FIGURE 10–18
Series impedance

Notice that the parallel equivalent of a series combination of an inductive reactance and a resistance is another resistance and inductive reactance. Likewise, the equivalent of a capacitive reactance and a resistance is another capacitive reactance and resistance. The circuit in Figure 10–16 acts electrically the same as the circuit in Figure 10–17 for one frequency. Since X depends on frequency, if the frequency changes, X changes and the parallel equivalent must also change.

Two formulas for the conversion of series impedance to a parallel equivalent are now developed (see Figure 10–18). A series impedance can be expressed as

$$\mathbf{Z}_s = R_s \pm jX_s \qquad \textbf{(10–57)}$$

The fraction representing the admittance can now be rationalized:

$$\mathbf{Y} = \frac{1}{\mathbf{Z}} = \left(\frac{1}{R_s \pm jX_s}\right)\left(\frac{R_s \mp jX_s}{R_s \mp jX_s}\right) \qquad \textbf{(10–58)}$$

Expressing these values as parallel components yields

$$\mathbf{Y} = G + jB = \frac{1}{R_p} - \frac{1}{jX_p} \qquad \textbf{(10–59)}$$

which, written in terms of series component values, is

$$\frac{R_s}{R_s^2 + X_s^2} \mp j\frac{X_s}{R_s^2 + X_s^2} \qquad \textbf{(10–60)}$$

giving

$$R_p = \frac{R_s^2 + X_s^2}{R_s} \qquad \textbf{(10–61)}$$

and

$$X_p = \pm\frac{R_s^2 + X_s^2}{X_s} \qquad \textbf{(10–62)}$$

EXAMPLE 10–9
Find the parallel equivalent of the circuit in Figure 10–19A.

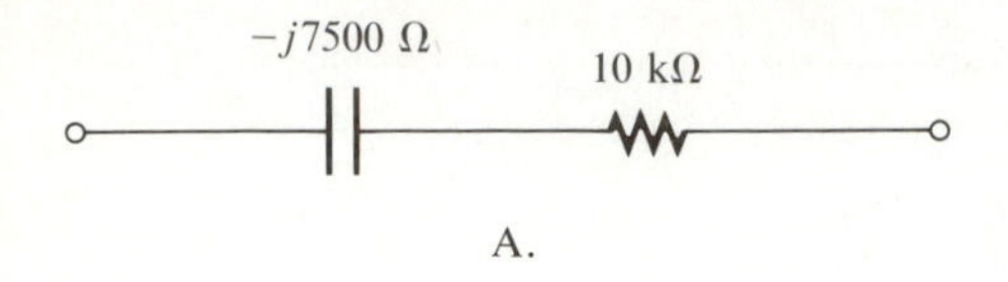

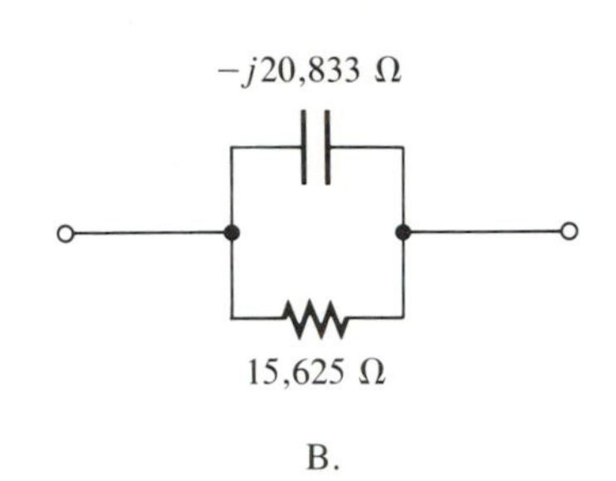

FIGURE 10–19
Series-to-parallel conversion

Solution:

$$R_p = \frac{R_s^2 + X_s^2}{R_s} \tag{10–63}$$

$$= \frac{(10\ \text{k}\Omega)^2 + (7500\ \Omega)^2}{10\ \text{k}\Omega}$$

$$= 15{,}625\ \Omega$$

$$X_p = -j\frac{(10\ \text{k}\Omega)^2 + (7500\ \Omega)^2}{7500\ \Omega} \tag{10–64}$$

$$= -j20{,}833\ \Omega$$

(see Figure 10–19B).

Of equal importance is the conversion from parallel circuits to series equivalent circuits. This change may simplify impedance arms of bridges for subsequent calculation and analysis. Using the formula

$$R_p = \frac{R_s^2 + X_s^2}{R_s} \tag{10–65}$$

and multiplying both the numerator and the denominator by $1/R_s^2$ gives

$$R_p = \frac{1 + X_s^2/R_s^2}{1/R_s} \tag{10–66}$$

The *quality factor* (Q) of a series circuit is defined as the ratio of reactance to resistance:

$$\frac{X_s}{R_s} = Q_s \tag{10–67}$$

Therefore,

$$R_p = (1 + Q^2)\,R_s \tag{10–68}$$

or

$$R_s = \frac{R_p}{1 + Q^2} \tag{10–69}$$

Using the second derived equation,

$$X_p = \frac{R_s^2 + X_s^2}{X_s} \tag{10–70}$$

and multiplying the numerator and the denominator by $1/X_s^2$ yields

$$X_p = \frac{R_s^2/X_s^2 + 1}{1/X_s} \tag{10–71}$$

$$= \left(\frac{1}{Q^2} + 1\right)X_s$$

$$X_s = \frac{X_p}{1 + 1/Q^2} \tag{10–72}$$

In formulas of this nature, remember that if *series* component values are being used,

$$Q = Q_s = \frac{X_s}{R_s} \tag{10–73}$$

and if *parallel* component values are being used,

$$Q = Q_p = \frac{R_p}{X_p} \tag{10–74}$$

EXAMPLE 10–10
Find the series equivalent of the circuit in Figure 10–20A.

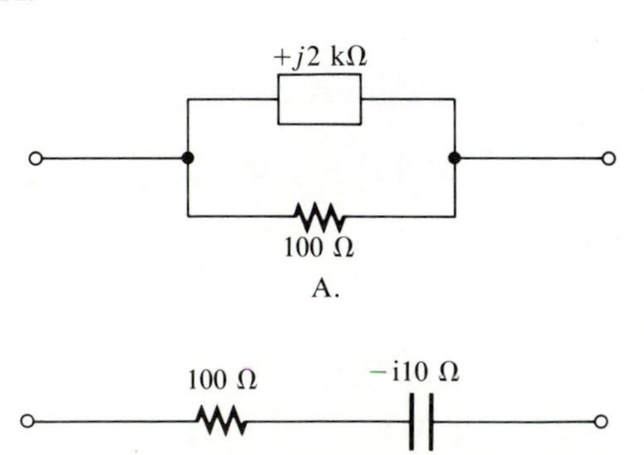

FIGURE 10–20

Solution:

$$Q = \frac{R}{X} = \frac{100\ \Omega}{2000\ \Omega} = 0.05 \qquad \textbf{(10–75)}$$

$$R_s = \frac{100\ \Omega}{1 + 0.05^2} \cong 100\ \Omega \qquad \textbf{(10–76)}$$

$$X_s = \frac{2\ \text{k}\Omega}{1 + 20^2} \cong +j5\ \Omega \qquad \textbf{(10–77)}$$

(see Figure 10–20B).

In high-Q circuits, X_p and X_s are approximately equal, while in low-Q circuits, R_s and R_p are approximately equal.

EXAMPLE 10–11
Find the series equivalent of the circuit in Figure 10–21A.

Solution:

$$Q = \frac{R}{X} = \frac{100\ \Omega}{2000\ \Omega} = 0.05 \qquad \textbf{(10–75)}$$

$$R_s = \frac{100\ \Omega}{1 + 0.05^2} \cong 100\ \Omega \qquad \textbf{(10–76)}$$

$$X_s = \frac{2\ \text{k}\Omega}{1 + 20^2} \cong +j5\ \Omega \qquad \textbf{(10–77)}$$

(see Figure 10–21B).

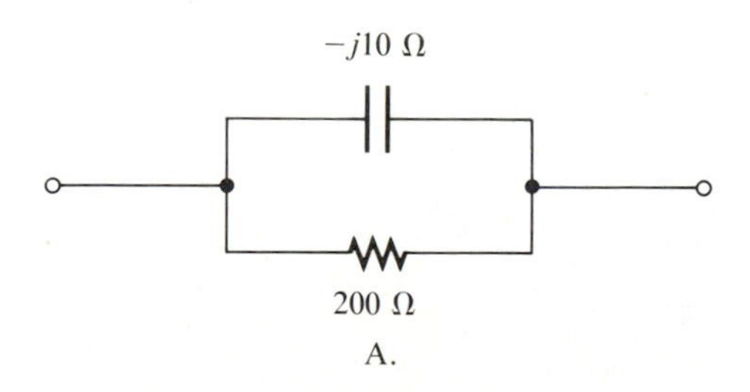

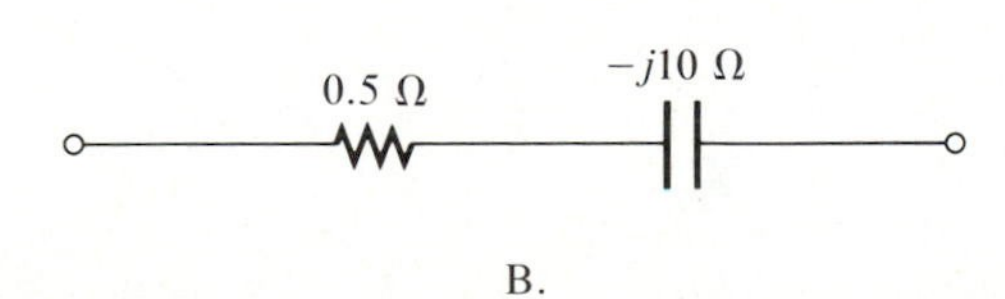

FIGURE 10–21

EXAMPLE 10–12
Find the unknown impedance of the *Maxwell bridge* in Figure 10–22.

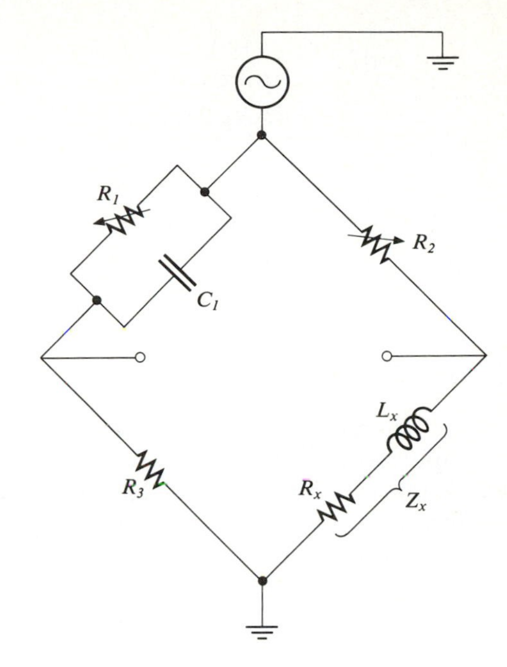

FIGURE 10–22
Maxwell bridge

Solution:
Since

$$\mathbf{Z}_1\,\mathbf{Z}_4 = \mathbf{Z}_2\,\mathbf{Z}_3 \qquad \textbf{(10–81)}$$

or

$$\mathbf{Z}_4 = \mathbf{Z}_2\,\mathbf{Z}_3\,\mathbf{Y}_1 \qquad \textbf{(10–82)}$$

then

$$R_x + jX_{L_x} = R_2\,R_3\left(\frac{1}{R_1} + \frac{1}{-jX_{C_1}}\right) \qquad \textbf{(10–83)}$$

Rationalizing gives

$$\begin{aligned} R_x + jX_{L_x} &= R_2\,R_3\left(\frac{1}{R_1} + j\,\frac{1}{X_{C_1}}\right) \\ &= \left(\frac{R_2R_3}{R_1} + j\,\frac{R_2R_3}{X_{C_1}}\right) \end{aligned} \qquad \textbf{(10–84)}$$

Therefore,

$$R_x = \frac{R_2R_3}{R_1} \qquad \textbf{(10–85)}$$

and

$$2\pi f L_x = \frac{R_2 R_3}{1/(2\pi f C_1)} = R_2 R_3\, 2\pi f C_1 \quad \textbf{(10–86)}$$

or

$$L_x = R_2\, R_3\, C_1 \quad \textbf{(10–87)}$$

OTHER BRIDGE CIRCUITS

Capacitance comparison *bridges* are used to measure capacitance. Figure 10–23 illustrates a *series* capacitance comparison bridge, and Figure 10–24 illustrates a *parallel* capacitance comparison bridge. Capacitances with a high Q factor ($Q = X/R$) are more easily measured using the series capacitance comparison bridge, whereas capacitances with a low Q factor are usually measured using a parallel capacitance comparison bridge.

Two bridge circuit configurations are used to measure inductance. Inductances with a low Q are usually measured using a Maxwell bridge (see Figure 10–22). For measuring inductances with a high value of Q, the *Hay bridge* is used (see Figure 10–25).

A *Wien bridge* circuit configuration is shown in Figure 10–26. This bridge is not normally used to measure impedance but finds extensive application where oscillator circuits and filters are required. The Wien bridge oscillator shown in Figure 10–27 is the most widely used circuit in commercial signal generators operating in the 5-Hz to 500-kHz frequency range. If R_1 is almost equal to R_2, the bridge is almost balanced, the circuit oscillates at

$$f = \frac{1}{2\pi RC} \quad \textbf{(10–88)}$$

and the output is larger than the input.

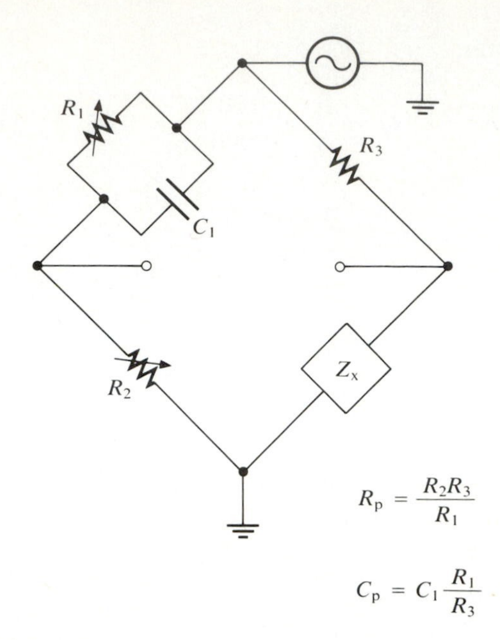

FIGURE 10–24
Parallel capacitance comparison bridge

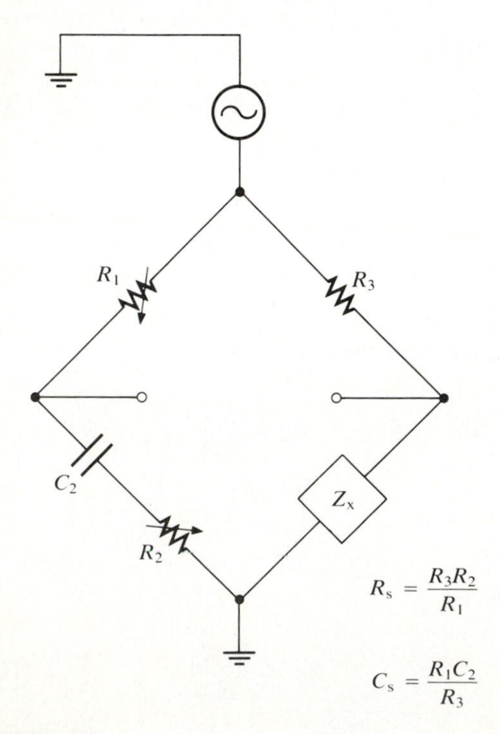

FIGURE 10–23
Series capacitance comparison bridge

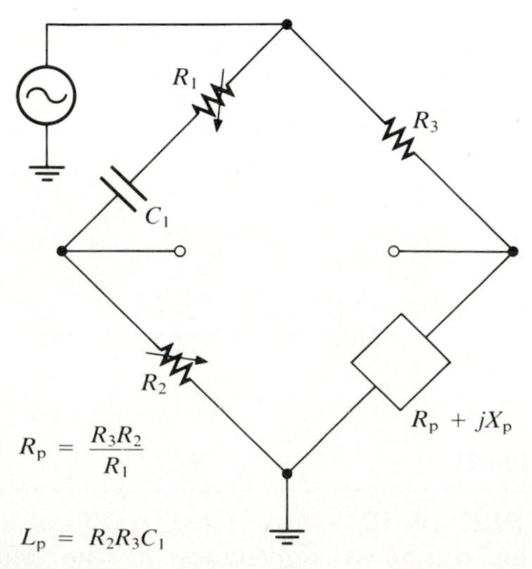

FIGURE 10–25
Hay bridge

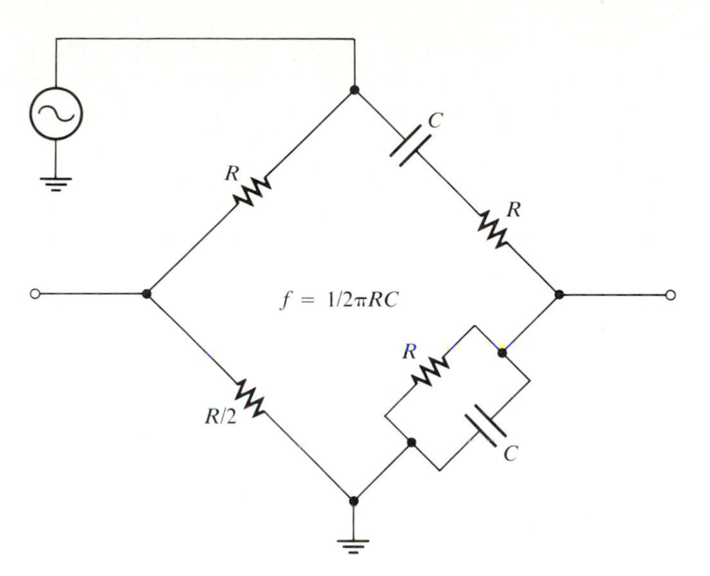

FIGURE 10–26
Wien bridge

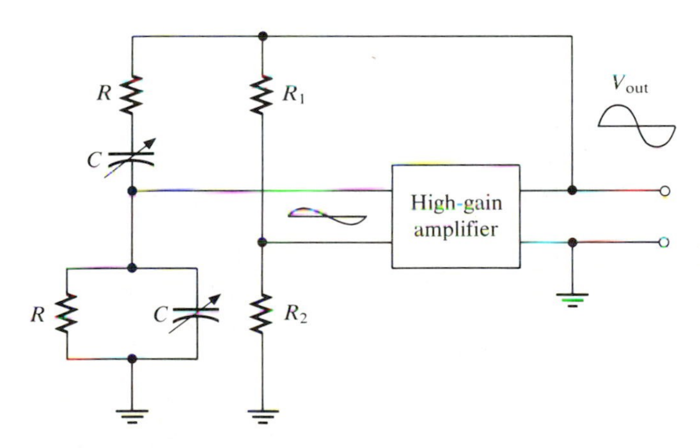

FIGURE 10–27
The Wien bridge as an oscillator

IMPEDANCE BRIDGES

An *impedance bridge* is a device used to measure inductance (L), resistance (R), capacitance (C), and the *dissipation factor* ($D = 1/Q$). Many commonly used bridges are manual bridges. Newer technology has led to the development of the automatic *LRC* bridge. Both bridge types will be discussed.

Manual Impedance (*LRC*) Bridges

With the manual bridge, the technician must select the parameter to be measured (L, R, C, or D). The technician must also choose the configuration that gives the best accuracy (series resistance and reactance or parallel resistance and reactance) and the range of the measurement. Most bridges have a single oscillator with a frequency of 1 kHz for reactance measurements and a dc source for resistance measurements. A typical bridge has a galvanometer and sensitivity adjustment for null methods of balancing. Figure 10–28 is an example of a manual impedance bridge.

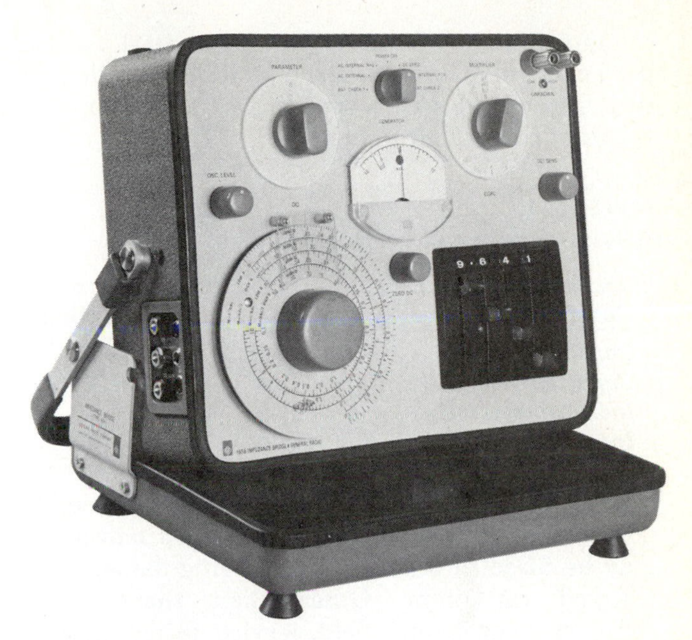

FIGURE 10–28
Manual impedance (*LRC*) bridge

Resistance measurements are best taken with a dc source when the circuit selected by the parameter dial is a Wheatstone bridge. The galvanometer needle crosses from one deflection through zero to another deflection as balance is approached. Finer adjustments of the variable resistances allow the needle to be brought to rest at zero deflection (or very near it). When zero deflection is achieved, the instrument indicates the resistance in ohms. This value is usually accurate to four significant figures.

When the C_s, C_p, L_s, or L_p parameter is selected, a different ac bridge configuration is chosen. Each configuration gives the most accurate measurement for the specific component. When taking C or L readings, it is important to realize that the meter may not zero. The technician may have to settle for a minimum of imbalance. For example, the needle may indicate a deflection of eight divisions for one setting, reduce to five divisions for the next setting, and then move back to eight divisions for the following setting. In this case, five divisions is as close to a null reading as possible with the configuration. When the bridge meter reaches equilibrium, it will indicate a value of L or C.

High-Q (low-D) components—$Q = X_s/R_s = R_p/X_p$—are more accurately measured using the configuration in Table 10–1. Table 10–1 can be used as a guide for selecting a series or parallel setting of the bridge corresponding to an appropriate range at component values.

TABLE 10–1
Settings for ranges of component values

Component Value	Q	Setting
Capacitor $< 1\ \mu F$	High	C_p
Capacitor $> 1\ \mu F$	Low	C_s
Inductance $< 1\ H$	Low	L_s
Inductance $> 1\ H$	High	L_p

The bridge dc source and oscillator are usually powered by batteries, and the meter movement may be fused to protect it from being damaged by charged capacitors. Other than these easily changed components, field service is not usually recommended because of the close tolerance ranges involved.

Automatic Impedance (*LRC*) Bridges

The relatively low cost of digital components, processors, and memories has prompted their use in the manufacture of automatic *LRC* bridges (Figure 10–29). An automatic impedance bridge is usually capable of measuring the *L, R, C,* and *D* parameters accurately and with greater speed than a manual bridge. Ranges from a few microhenrys to thousands of henrys and zero ohms to 100 megohms are available. The *LRC* bridge automatically decides whether the component under test is an inductance or a capacitance and determines the range of measurement to be taken. For larger values of inductance and capacitance, *Q* is usually large and a lower frequency is more likely to give accurate results. Included here are large-value electrolytic capacitors most commonly used as power supply filters. These capacitors should be tested at 120 Hz. Most automatic *LRC* bridges can operate the internal oscillator at 120 Hz and 1 kHz. Since these bridges are capable of indicating that an inappropriate frequency has been chosen, measurement error is unlikely.

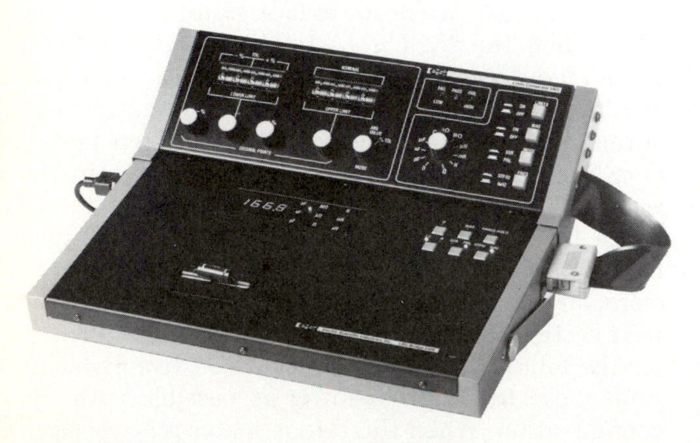

FIGURE 10–29
Automatic impedance (*LRC*) bridge

As with manual bridges, automatic bridges are capable of measuring series and parallel equivalents of capacitive or inductive reactance and resistance. The automatic *LRC* bridge can indicate whether the selected equivalent (series or parallel) will give accuracy within the prescribed limits.

The following are some advantages of the automatic *LRC* bridge:

1. Indicates the appropriate range for measurement (autoranging).
2. Indicates the appropriate parameter (automatically determines what component is being measured).
3. Indicates the appropriate series or parallel equivalent.
4. Indicates the appropriate measuring frequency.
5. Does not depend on the slow nulling method of the manual bridge.
6. Provides basic accuracy of ±0.25% with a four-digit display.

Theory of Operation of Automatic *LRC* Bridges

The automatic *LRC* bridge measures the voltage across and the current through a component. The bridge's microprocessor then calculates the value of the component under test. A sinewave is generated using a crystal oscillator with an accurately known frequency. Two reference square waves are produced at the same frequency with a phase relationship between them of exactly 90°. The voltage across and the current through the component are each measured relative to these two reference signals, providing in effect, two phase references in quadrature with each other. Once the four voltage and current values have been determined, the equivalent series or parallel reactance or resistance is calculated by the processor. The polarity of the reactance term indicates whether the component is capacitive or inductive. Since the frequency is accurately known, the value of the component is then computed by the processor.

Automatic *LRC* comparators are available. They will indicate if a component is within the prescribed tolerances entered. These bridges are particularly useful for testing large numbers of components for quality control.

SUMMARY

Dc and ac bridges are valuable measuring devices. Although manual bridges are still popular in industry, automatic *LRC* bridges are entering the market strongly. The technician should know the characteristics and operation of both bridge types.

EXERCISES

1. For the circuit in Figure 10–30, if $R_1 = 2$ kΩ, $R_2 = 3$ kΩ, and $R_3 = 5$ kΩ, calcuate R_x.
2. For the circuit in Figure 10–30, if $R_1 = 100$ Ω, $R_2 = 500$ Ω, and $R_3 = 300$ Ω, calcuate R_x.
3. For the circuit in Figure 10–30, $R_1 = 10$ kΩ, $R_2 = 5$ kΩ, and R_3 has a range from 1 kΩ to 10 kΩ. What range of resistances can be measured with this bridge?
4. Refer to Figure 10–30. It is found that $R_x = 1.5$ MΩ balances a Wheatstone bridge in which $R_2 = 750$ kΩ and $R_3 = 4$ MΩ. What value of R_1 is in the bridge?

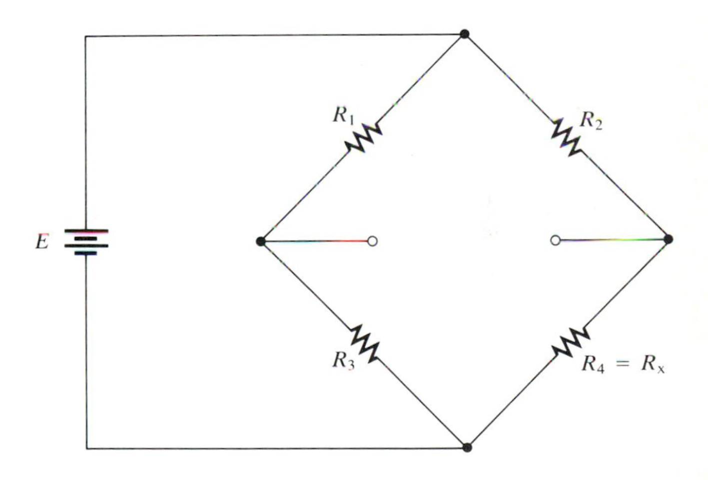

FIGURE 10–30
Circuit for Exercises 1 through 4

5. Derive the Thevenin equivalent for the following unbalanced Wheatstone bridge: $R_1 = 5$ kΩ, $R_2 = 7$ kΩ, $R_3 = 3$ kΩ, $R_4 = 9$ kΩ, $E = 9$ V.
6. Derive the Thevenin equivalent for the following unbalanced Wheatstone bridge: $R_1 = 1062$ Ω, $R_2 = 2352$ Ω, $R_3 = 5010$ Ω, $R_4 = 3$ kΩ, $E = 15$ V.
7. Find the parallel equivalent circuit for the series combination of $X_s = +j10$ kΩ, $R_s = 250$ kΩ.
8. Using the formula for a slightly unbalanced Wheatstone bridge, calculate the current drawn from each of the following bridges through a 50-μA, 2-kΩ meter:
   a. $R_1 = 5$ kΩ, $R_2 = 5$ kΩ, $R_3 = 5$ kΩ, $R_x = 5100$ Ω, $E = 12$ V
   b. $R_1 = 1$ kΩ, $R_2 = 1$ kΩ, $R_3 = 1$ kΩ, $R_x = 1050$ Ω, $E = 32$ V
   c. $R_1 = 10$ kΩ, $R_2 = 10$ kΩ, $R_3 = 10$ kΩ, $R_x = 10{,}075$ Ω, $E = 9$ V

9. For the circuit in Figure 10–31, calculate the unknown impedance as a series combination of R_x and L_x.

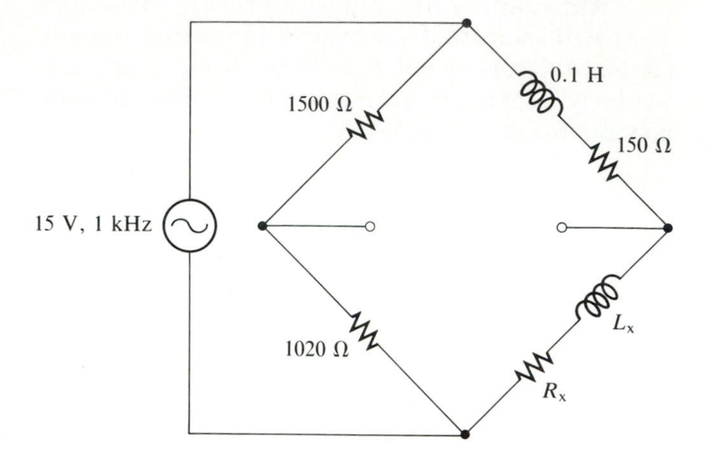

FIGURE 10–31
Circuit for Exercise 9

10. Calculate the unknown impedance for the circuit in Figure 10–32.

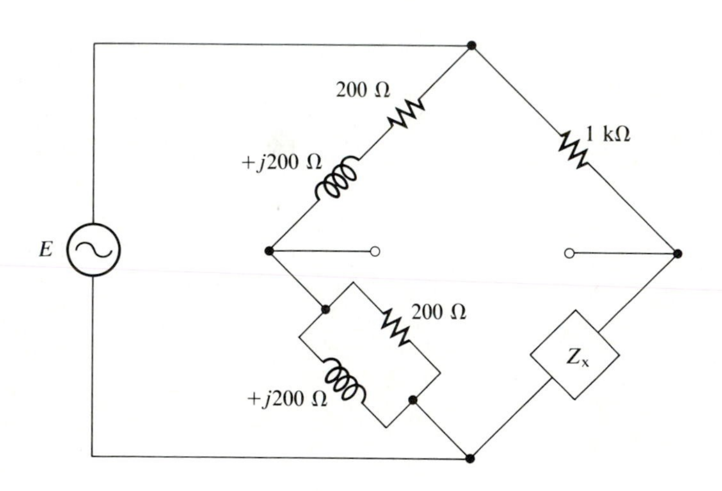

FIGURE 10–32
Circuit for Exercise 10

LAB PROBLEM 10–1
Impedance Bridge

Equipment

One manual impedance bridge; an assortment of resistors, capacitors, and inductors to measure; one-wire-wound resistor.

Object

To become familiar with the operation of an impedance bridge.

Procedure

Select several values of resistance, capacitance, and inductance. Measure the resistances using an impedance bridge and compare the measured values to rated values. Record the results in Table 10–2.

TABLE 10–2
Complete the table.

$R_{measured}$	R_{rated}

Measure C_s and R_s, or C_p and R_p, or both if possible. Enter the results in Table 10–3. Using either the results for C_s and R_s or C_p and R_p, calculate the other values (if using series values, calculate parallel values; if using parallel values, calculate series equivalent values). Indicate which values were used to begin the calculations and enter the results in Table 10–3. Compare, if possible, the results of the measurements and the calculations.

TABLE 10–3
Complete the table.

Measured				**Calculated**	
Series		*Parallel*		*(Series from Parallel or Parallel from Series)*	
C_s	R_s	C_p	R_p	$C_{s\text{ or }p}$	$R_{s\text{ or }p}$

Repeat this procedure by measuring L_s and R_s or L_p and R_p. Enter the measured and calculated results in Table 10–4. Repeat the procedure using inductances. Record the results in Table 10–4.

TABLE 10–4
Complete the table.

Measured				Calculated	
Series		Parallel		(Series from Parallel or Parallel from Series)	
L_s	R_s	L_p	R_p	$L_{s \text{ or } p}$	$R_{s \text{ or } p}$

Use clip leads about 1 ft or more long. Measure the resistance of the leads using the Wheatstone bridge. Measure the resistance and inductance of a wire-wound resistor.

LAB PROBLEM 10–2
Capacitance Comparison Bridge

Equipment
One 1.8-kΩ, one 5.6-kΩ, and one 10-kΩ resistor; one 0.1-μF capacitor; one function generator.

Object
To become familiar with the operation of a capacitance comparison bridge.

Procedure
Derive the formulas and use them to calculate the unknown values in the circuit of Figure 10–33.

FIGURE 10–33

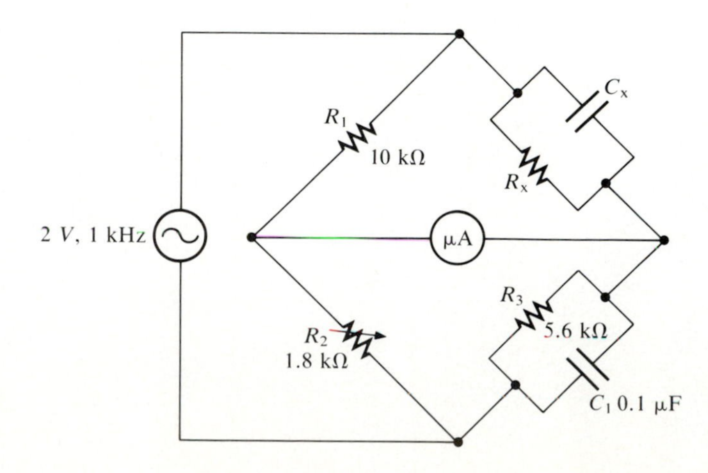

Wire the circuit using the calculated values. Adjust R_2. Record the results if R_2 is increased and decreased. Is the bridge balanced? Does the general formula hold?

Explain why the bridge in Figure 10–34 cannot be balanced.

FIGURE 10–34

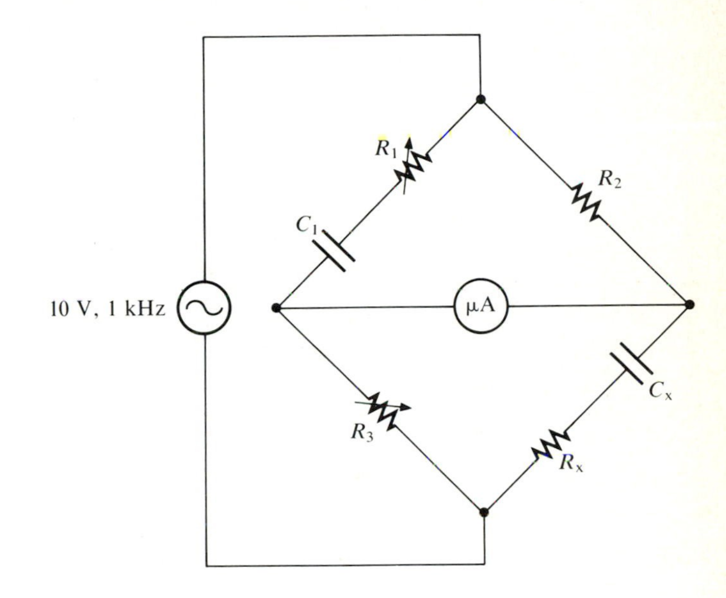

11 Signal-Processing Circuits

OUTLINE

OBJECTIVES
CLIPPERS AND CLAMPERS
AMPLIFIERS
ATTENUATORS
FREQUENCY MULTIPLIERS
FREQUENCY DIVIDERS
ANALOG-TO-DIGITAL CONVERTERS
DIGITAL-TO-ANALOG CONVERTERS
MODULATION TECHNIQUES
- Amplitude Modulation
- Frequency and Phase Modulation
- Pulse Modulation

SEPARATION OF TWO SIGNALS
SUMMARY
EXERCISES
LAB PROBLEM 11–1—D/A Converters
LAB PROBLEM 11–2—Input Attenuator

OBJECTIVES

After completing this chapter, you will be able to:

- State why signals must be processed.
- Describe several basic wave-shaping techniques.
- Describe how various ranges of signals can be changed to a common output level.
- Change the frequency of sinewave signals.
- Convert analog signals to corresponding digital signals.
- Convert digital signals to corresponding analog signals.
- Combine two or more signals.
- Separate combined signals.

Most measuring systems consist of three basic components: a signal source, processing circuits, and some kind of measuring device. Transducers and detectors provide the input signal. The processing circuits condition the signal for measurement, observation, or use. The processor may shape, increase, or reduce a signal, or change its frequency or phase. It may also convert the signal from analog format to digital or from digital to analog, combine signals, or shape signals for further processing.

This chapter considers some basic techniques for performing these changes. The discussion is not intended to be a complete presentation of processing circuits since the list is very long. Rather, several relatively simple representative networks will be presented to demonstrate how the changes are produced.

CLIPPERS AND CLAMPERS

Sometimes the amplitude of a signal must be limited to prevent the signal from driving an amplifier into saturation. This can be done by using a properly biased diode or transistor. (See Figure 11–1). In this case, when the pulse is positive, the voltage rises until it reaches 5+ V. A voltage of 5+ V is used because the potential barrier of the diode must be exceeded before conduction can take place. At

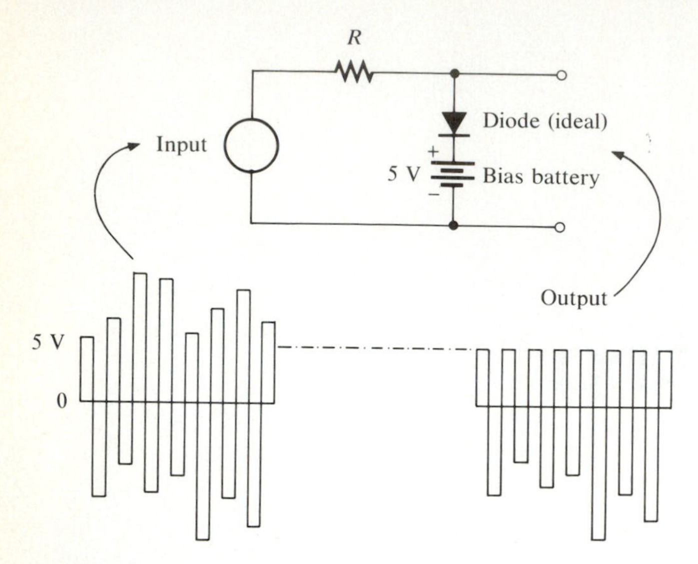

FIGURE 11–1
Diode-clipping circuit

this point, the diode is forward biased and allows only the 5 V of the battery to reach the output. Excess voltage is dropped across resistance R. In this example, the positive amplitude of the output signal cannot exceed 5 V. When the pulses are negative, the result is the unchanged signal because the diode is reverse biased.

Similarly, a transistor can be biased to clip both the positive and negative peaks of an input signal (see Figure 11–2). When the input signal exceeds the biased saturation point of the transistor, it drives the transistor into saturation. Because the output cannot exceed the saturation level, the waveform is clipped at this level. When the input waveform drives the transistor below cutoff, the same result occurs, and the waveform is clipped at the cutoff level.

The same results can be obtained using two diodes (in different directions) with batteries to bias

FIGURE 11–2
Biasing a transistor as a clipper

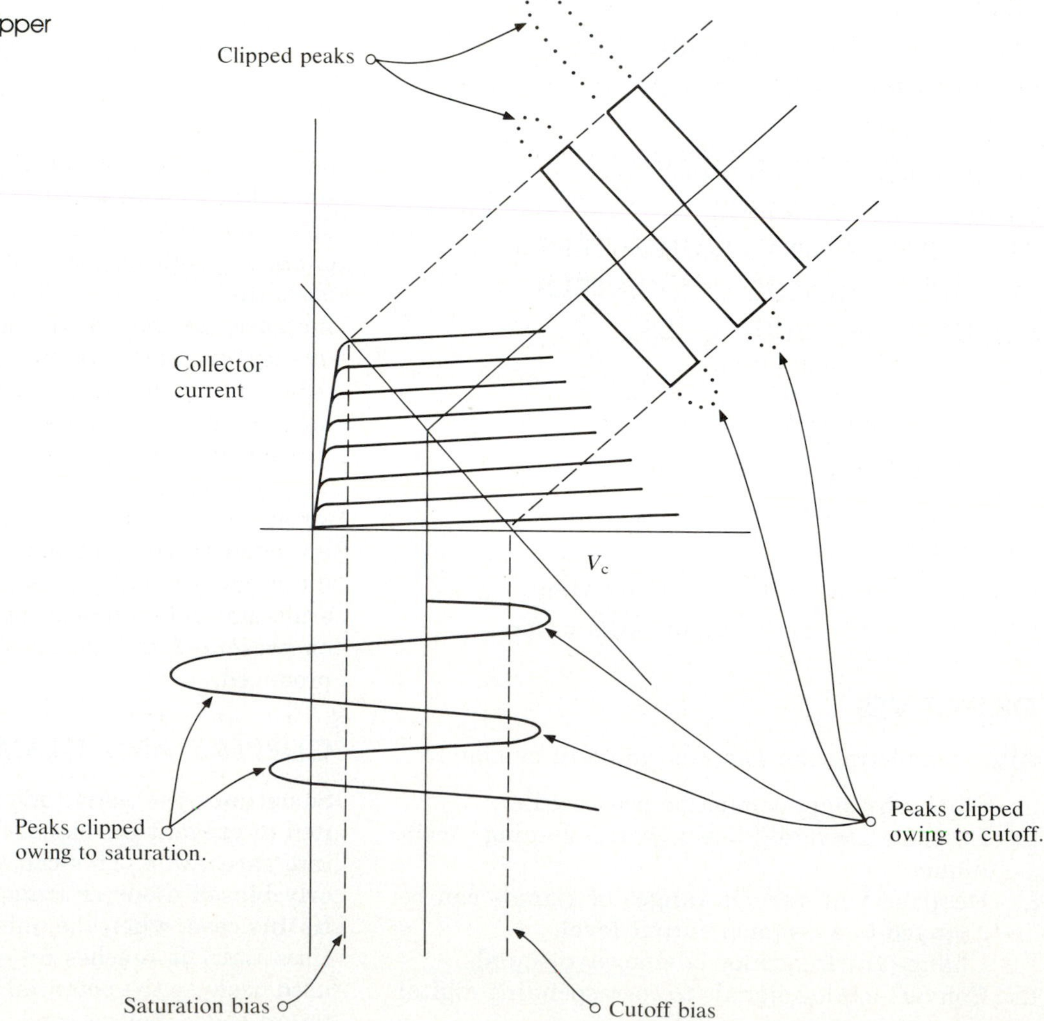

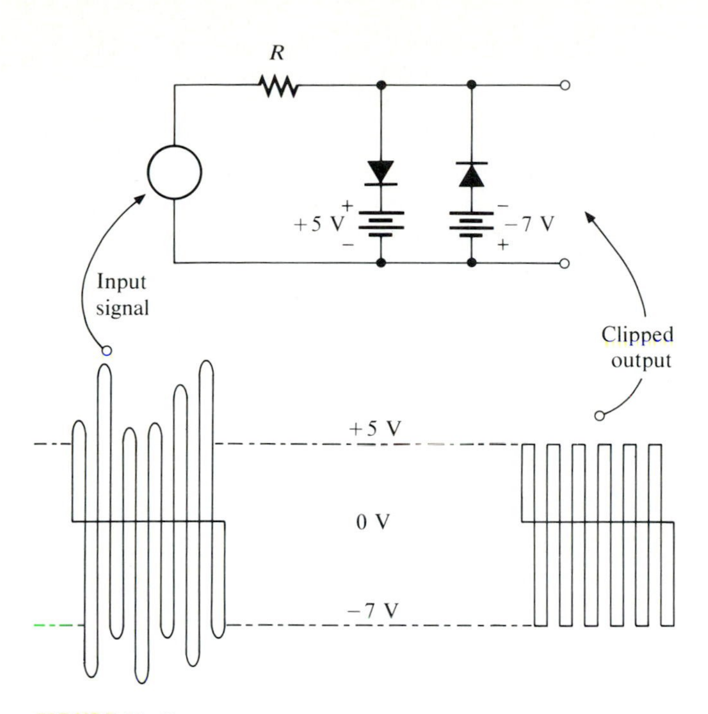

FIGURE 11–3
Circuit configuration of a double clipper

them at a chosen level (see Figure 11–3). The input signal can be clipped at any desired positive and negative level by changing the bias battery.

A signal can be clamped at any dc level by simply adding a dc source in series with the signal source (Figure 11–4). A blocking capacitor placed in series with the source can be used to clamp a signal at 0 V (Figure 11–5). Because dc cannot pass through a capacitor, the only output is ac. Other

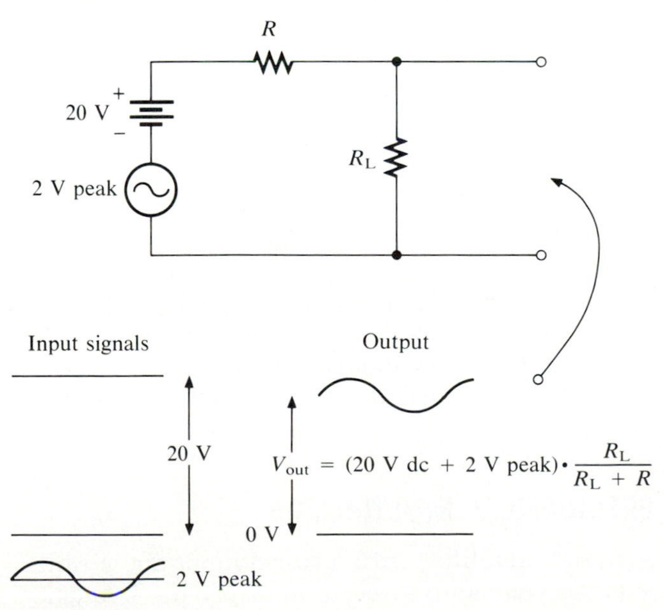

FIGURE 11–4
Signals in series add algebraically.

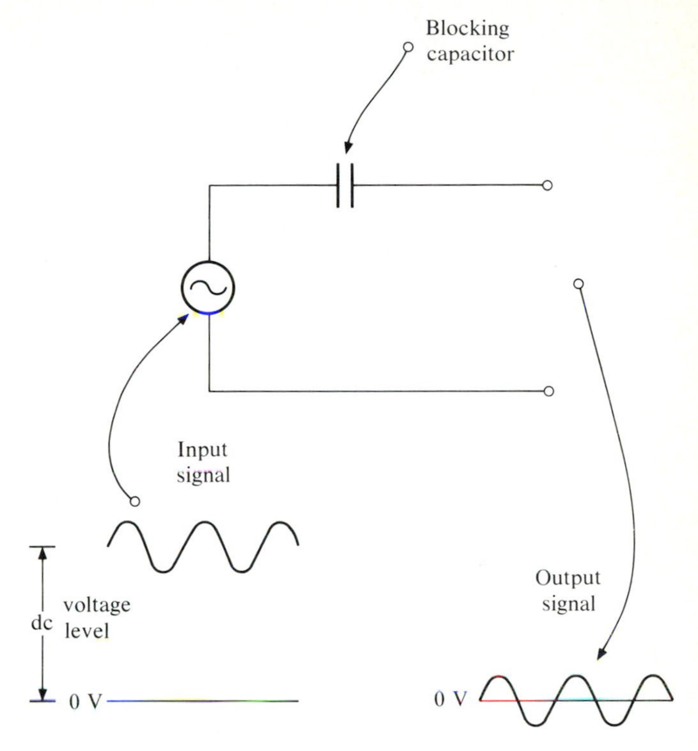

FIGURE 11–5
Removing a dc voltage

circuits using biased diodes to provide the required dc level are available.

AMPLIFIERS

Transducers and detectors typically produce voltages and signals that do not have enough amplitude or power to cause output devices to respond. If a signal is too small, an amplifier can be used to increase the voltage, current, or power to a level that will be useful for measurement or display. Other signals may have power levels so large that they can overdrive output devices or damage them. If a signal is too large, an attenuator can be used to reduce the amplitude of the signal to an appropriate level.

ATTENUATORS

Attenuators reduce the amplitude or power of a signal or voltage level. This can usually be accomplished by a voltage divider. For example, if a 25-V signal is to be measured using a device with a 0.5-V operational voltage range and a 500-kΩ resistance is used for coupling purposes, a voltage divider can be used to reduce the 25-V level to the 0.5-V level (Figure 11–6). Using voltage division in Figure 11–6,

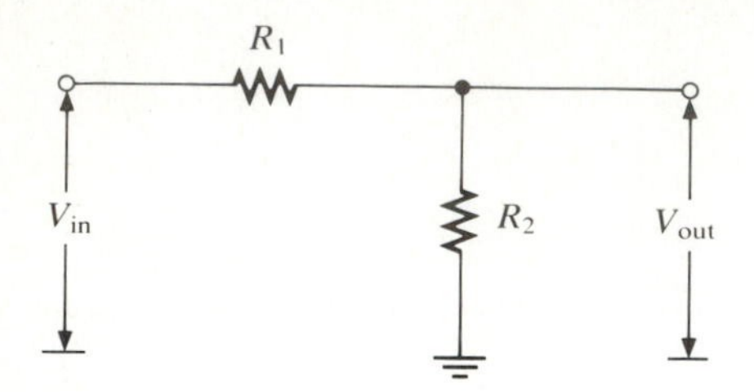

FIGURE 11–6
A voltage divider as an attenuator

$$V_{out} = V_{in}\frac{R_2}{R_t} \qquad (11\text{–}1)$$

Rearranging gives

$$\begin{aligned} R_2 &= \frac{V_{out}}{V_{in}}R_t \\ &= \frac{0.5\text{V}}{25\text{V}}(500{,}000\ \Omega) \qquad (11\text{–}2) \\ &= 10\ \text{k}\Omega \end{aligned}$$

Therefore,

$$\begin{aligned} R_1 &= 500{,}000\ \Omega - 10{,}000\ \Omega \qquad (11\text{–}3) \\ &= 490\ \text{k}\Omega \end{aligned}$$

It is sometimes necessary to design an attenuator that will allow various input levels while providing a constant output level. If an attenuator with 10 MΩ of input resistance must provide a constant output of 500 mV, a rotary switch may be used to allow input voltages of 1, 3, 10, 30, and 100 V (Figure 11–7).

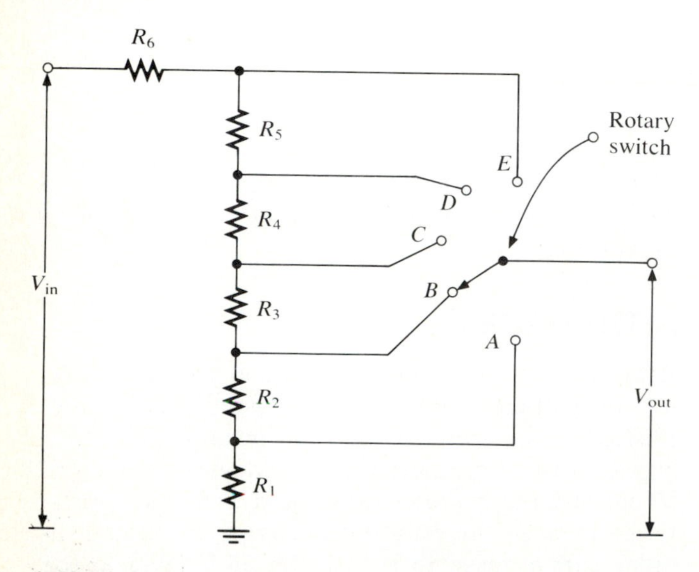

FIGURE 11–7
An attenuator as a range selector switch

When the rotary switch is in position A, the voltage divider consists of R_1, across which the voltage will be detected, and $R_2, R_3, \ldots, R_6$, across which the rest of the input voltage will be dropped. This switch position corresponds to the largest input voltage level since the voltage detected will be the smallest proportion of the input voltage. As the switch is moved to positions $B, C, \ldots$, a larger proportion of the input voltage will be selected, and so these positions must represent lower input voltage values.

In the present example, $R_t = 10$ MΩ and $V_{out} = 0.5$ V. When the switch is in position A, $V_{in} = 100$ V (maximum). From the voltage divider formula,

$$\begin{aligned} R_1 &= \frac{V_{out}}{V_{in}}R_t \\ &= \frac{0.5\ \text{V}}{100\ \text{V}}(10{,}000{,}000\ \Omega) \qquad (11\text{–}4) \\ &= 50\ \text{k}\Omega \end{aligned}$$

When the switch is in position B, $V_{in} = 30$ V, and

$$\begin{aligned} R_1 + R_2 &= \frac{0.5\ \text{V}}{30\ \text{V}}(10{,}000{,}000\ \Omega) \qquad (11\text{–}5) \\ &= 166{,}667\ \Omega \end{aligned}$$

Therefore,

$$\begin{aligned} R_2 &= 166{,}667\ \Omega - 50{,}000\ \Omega \qquad (11\text{–}6) \\ &= 116{,}667\ \Omega \end{aligned}$$

Similarly,

$$R_3 = 333\ \text{k}\Omega \qquad (11\text{–}7)$$
$$R_4 = 1{,}166{,}667\ \Omega \text{ or } 1.17\ \text{M}\Omega \qquad (11\text{–}8)$$
$$R_5 = 3.33\ \text{M}\Omega \qquad (11\text{–}9)$$

and

$$R_6 = R_t - 5\ \text{M}\Omega = 5\ \text{M}\Omega \qquad (11\text{–}10)$$

This type of attenuator resembles the input attenuator (range selector) used in the input section of multimeters and oscilloscopes. It can also be used between a measuring device and an input that is too large.

FREQUENCY MULTIPLIERS

A class-C amplifier with a tuned input and a tuned ouput is commonly used to multiply the frequency of a sinewave signal (Figure 11–8). The input parallel resonant circuit at point A is tuned to the in-

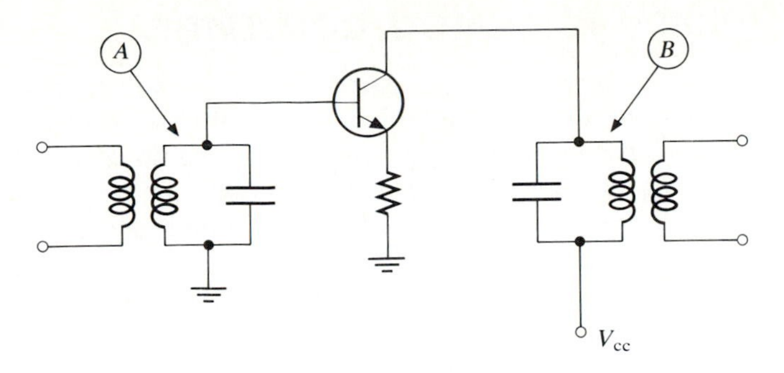

FIGURE 11–8
A class-C amplifier as a frequency multiplier

put frequency. The output parallel resonant circuit at point B can be tuned to two, three, or even four times the input frequency. Higher products are not practical since the flywheel effect of the parallel resonant circuit on the output will not respond to higher multiplying requirements. The combination of the exciting pulse from the amplifier at the fundamental frequency and the diminishing amplitude of the signal from the resonant circuit will not be sufficient to sustain the oscillation of the resonant circuit at higher harmonic values (Figure 11–9).

FREQUENCY DIVIDERS

The coupling of flip-flops and class-C amplifiers can be used to divide the frequency of sinewave signals since the output of the tuned parallel frequency of a class-C amplifier is a sinewave. An R-S (reset-set) flip-flop uses one pulse to convert the output to a high logic state and another pulse to return it to a low logic level. Because two input pulses are required to produce one output pulse, the pulse repetition rate of the input pulse train is effectively divided by 2 (Figure 11–10).

The frequency of a sinusoidal waveform can be reduced by a method called *heterodyning*. This technique involves adding the original signal and a signal of a different frequency in series across a nonlinear device. The difference between the two frequencies will be the needed product frequency. This method is also used in radio reception to con-

FIGURE 11–9
Multiplying a frequency

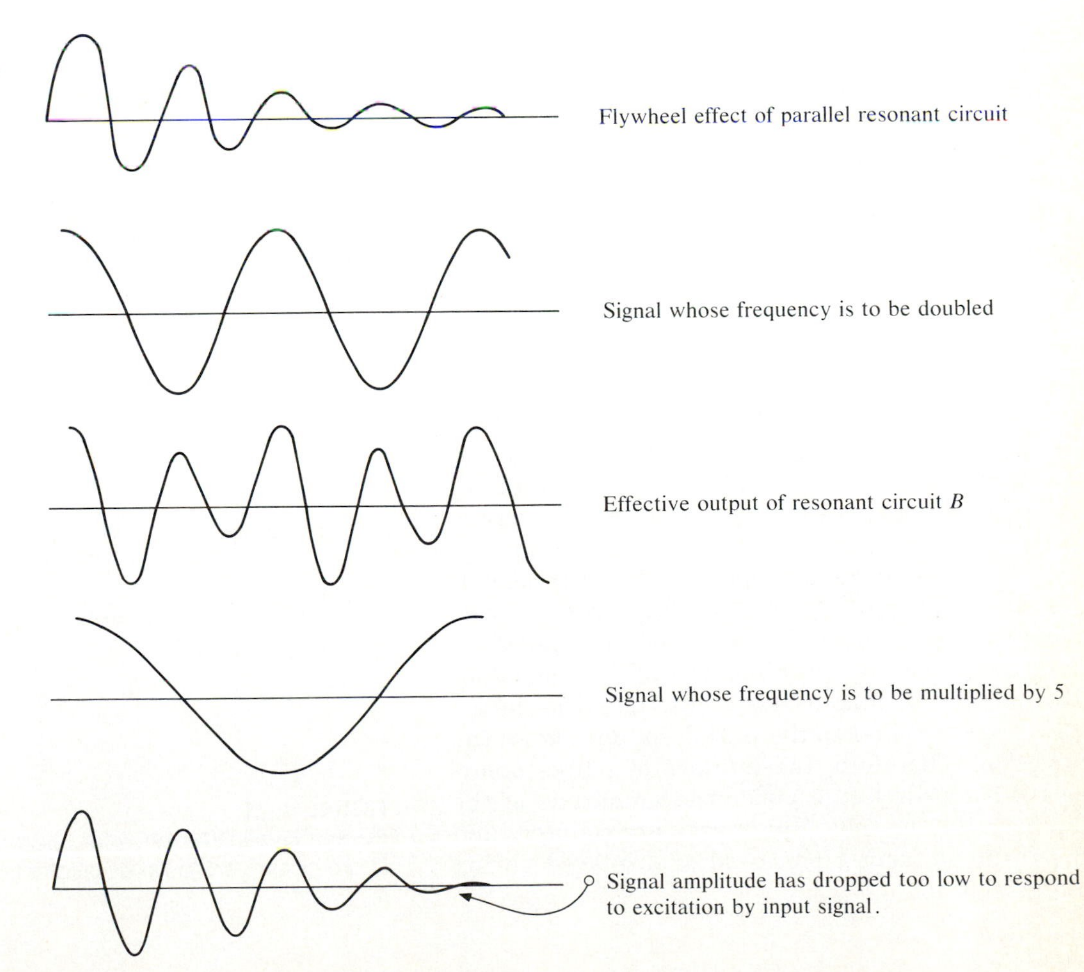

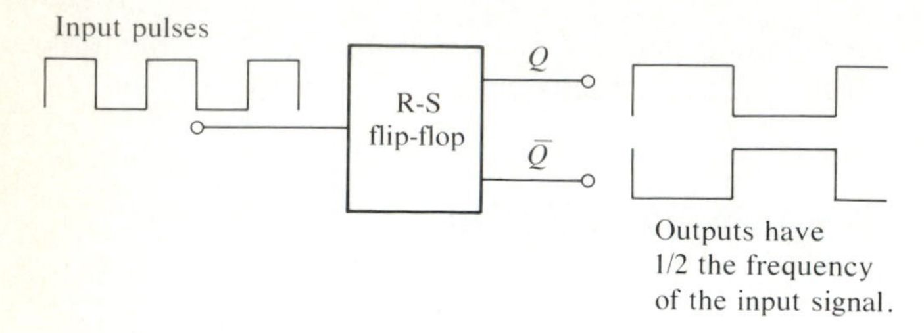

FIGURE 11–10
Dividing a pulse repetition rate

vert the incoming carrier frequency to an intermediate frequency so that the radio amplifiers (IF amplifiers) have to respond to a very small frequency range of signals.

Since signals are often nonsinusoidal and valuable information is included in the frequency of a signal, it would be improper to multiply or divide the frequency of these signals since doing so would change the information they contain. The preceding examples illustrate methods used to make the most common types of frequency multiplications and divisions.

ANALOG-TO-DIGITAL CONVERTERS

Analog-to-digital (A/D) conversion involves changing an analog (continuously changing) voltage to a digital (incremental) format. These conversions are necessary because the outputs of transducers that convert nonelectrical power levels to electrical voltage levels are analog. A/D converters are usually sampling circuits that use a series of pulses from a clock to time changes in voltage levels. The number of pulses is counted and the sum displayed using a digital output. Most analog-to-digital changes are accomplished by integrated circuits containing all of the circuitry necessary to perform the conversions.

The most common type of A/D conversion is the *integrator* type. The input voltage is allowed to charge a capacitor, and a series of clock pulses is then used to time the discharge of the capacitor. The amount of time needed to discharge the capacitor is proportional to the initial voltage across the capacitor. Therefore, the number of pulses occurring during discharge yields the amplitude of the input analog voltage. The pulses are counted, and the result is further processed or displayed during a digital output.

DIGITAL-TO-ANALOG CONVERTERS

Digital signals must occasionally be converted to analog signals for control or other purposes. Two passive networks will be presented to demonstrate how digital-to-analog (D/A) conversion can be performed using simple circuits and fundamental circuit analysis.

The input to the D/A converter in Figure 11–11 is a 4-bit binary number. If the input is a logic 1, the voltage at the input will be the indicated voltage level. If the input is a logic 0, the input will be at circuit ground. Resistance R_L is selected so that $R_L \gg R$. In the example shown, the voltage corresponding to logic 1 is chosen as +15 V. The circuit shows a binary input of 0110 (base 2), which corresponds to the decimal 6. The circuit is redrawn in Figure 11–12.

To find the analog output, we first calculate $R/4 \parallel R/2$:

$$\frac{1}{R/4} + \frac{1}{R/2} = \frac{4}{R} + \frac{2}{R} = \frac{6}{R} = \frac{1}{R'} \quad \textbf{(11–11)}$$

Therefore,

$$R' = \frac{R}{6} \quad \textbf{(11–12)}$$

Similarly,

$$\frac{1}{R} + \frac{8}{R} + \frac{1}{R_L} = \frac{9}{R} = \frac{1}{R''} \quad \textbf{(11–13)}$$

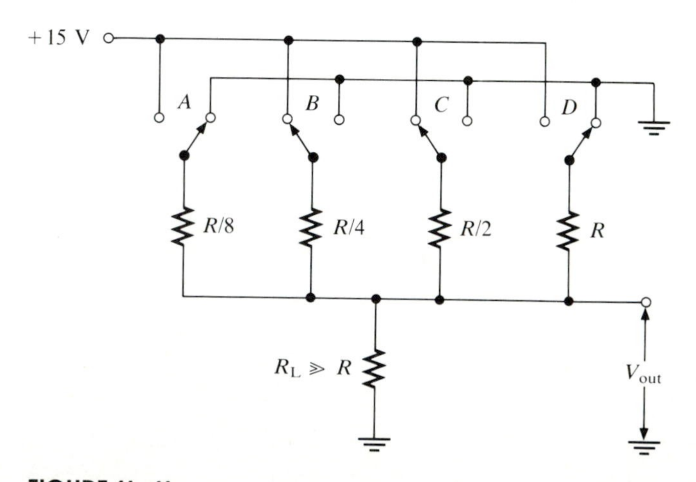

FIGURE 11–11
Binary-based D/A converter. Digital inputs a *A, B, C,* and *D* are converted to an equivalent analog output across R_L.

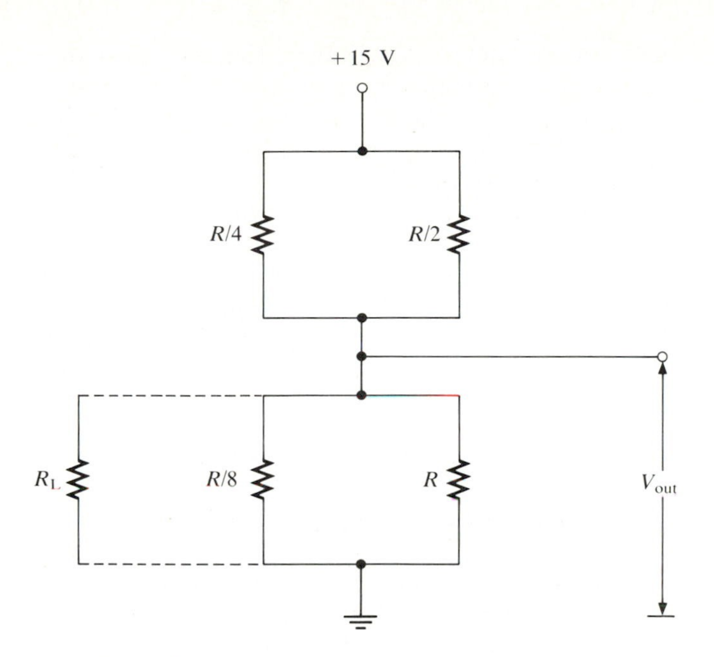

FIGURE 11–12
Developing the analog output.

($R_L \gg R$ and $1/R_L$ can therefore be ignored.) Thus,

$$R'' = \frac{R}{9} \tag{11–14}$$

The resulting circuit is shown in Figure 11–13. Voltage division is used to determine the output voltage:

$$V_{out} = 15\text{ V}\left(\frac{\frac{R}{9}}{\frac{R}{6} + \frac{R}{9}}\right)$$

$$= 15\text{ V}\left(\frac{\frac{R}{9}}{\frac{9R + 6R}{54}}\right)$$

$$= 15\text{ V}\left(\frac{\frac{R}{9}}{\frac{15R}{54}}\right) \tag{11–15}$$

$$= 15\text{ V}\left(\frac{R}{9} \times \frac{54}{15R}\right)$$

$$= 6\text{ V}$$

Hence, the binary input 0110 (base 2) has been changed to a 6-V output. The choice of 15 V for the power supply was made to simplify the mathematics for illustration purposes. No matter what logic voltage level is used, the outputs will be proportional to the value of the binary input. A larger binary input requires more branches ($R/16$, $R/32$, . . .).

It becomes difficult to acquire the necessary resistance values for larger binary inputs. A *ladder* network is used to solve this problem.

The ladder network in Figure 11–14 provides for a 3-bit binary input through terminals *A*, *B*, and *C*. This configuration shows that 110 (base 2) = 6 (base 10). Using Thevenin's theorem, the circuit can be reduced as shown in Figure 11–15. Figures 11–16A and B demonstrate that the output voltage equals 6 V. The 8-V logic level was chosen to simplify the mathematics. Notice that with resistance values of R and $2R$, the manufacturing process is greatly simplified.

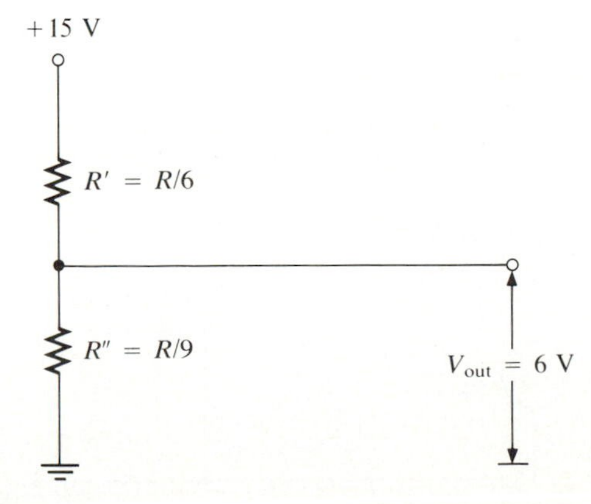

FIGURE 11–13
D/A circuit reduced to a simple voltage divider network

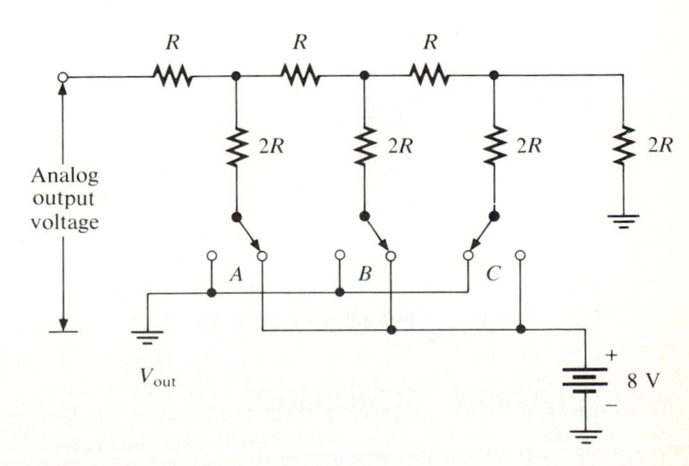

FIGURE 11–14
Ladder-type D/A converter

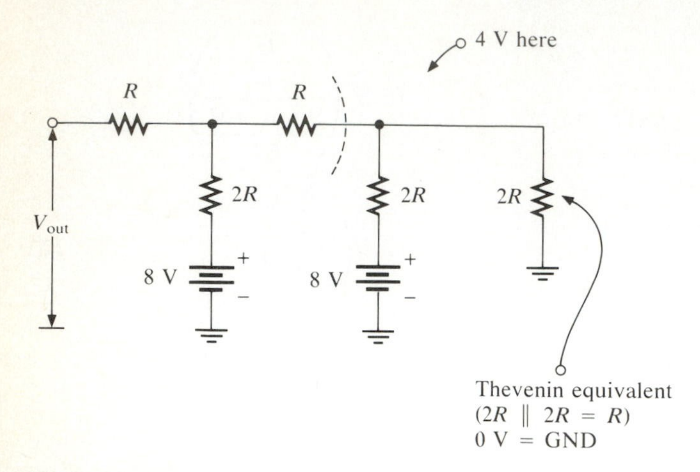

FIGURE 11–15
Developing the Thevenin equivalent

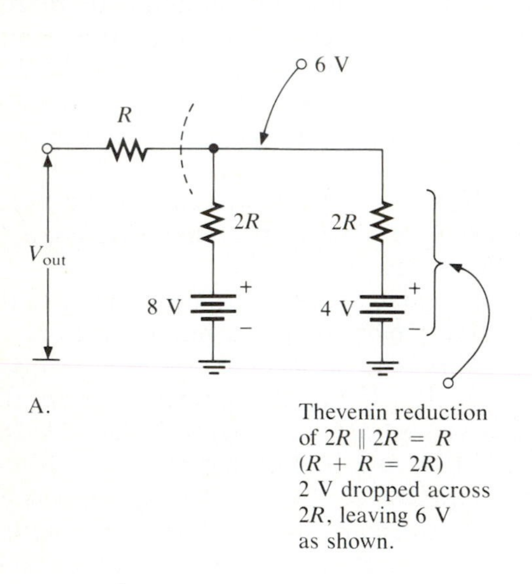

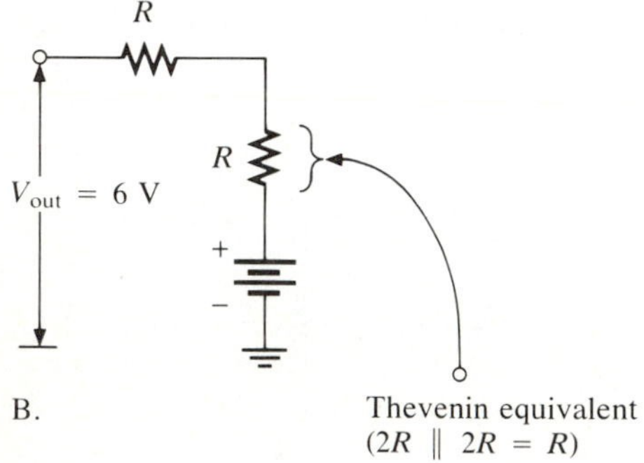

FIGURE 11–16
Continued development using Thevenin's theorem

MODULATION TECHNIQUES

Because measurements cannot always be taken at the measurement site, it is important to understand how information can be transmitted from one site to another. The information is measured, placed on a carrier wave, and transmitted to a more convenient site, where it is demodulated for use. *Demodulation* is the separation of the information from the carrier waveform.

Several types of modulation are available. Mentioned in this section are amplitude modulation (AM), frequency modulation (FM), phase modulation, and pulse amplitude modulation (PAM). Each of these techniques is popular for different applications of electromagnetic communications. They can be used in radio, television, microwave, and fiber optic systems.

Amplitude Modulation

Amplitude modulation is the use of one signal to change the amplitude of another signal. The following mathematical expression applies:

$$v = E \sin(\theta + \phi) \qquad \textbf{(11–16)}$$

where v is the instantaneous voltage, E is the maximum voltage (amplitude), ϕ is the phase angle, and θ is the radian measure of the conduction angle as a function of time ($\theta = 2\pi ft = \omega t$). As an illustration, the waveform to be changed will have the expression

$$v_c = E_c \sin \omega_c t \qquad \textbf{(11–17)}$$

This is called the *carrier* wave since it "carries" information. The waveform that will do the modulating or changing will have the expression

$$v_m = E_m \sin \omega_m t \qquad \textbf{(11–18)}$$

This is called the *modulating* wave since it contains the information that the carrier wave will carry. The composite waveform consists of the carrier wave, which has been changed in accordance with

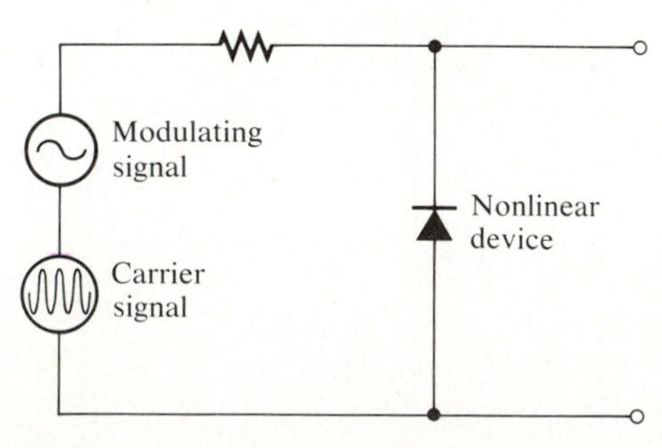

FIGURE 11–17
Mixing two signals in series with a nonlinear device

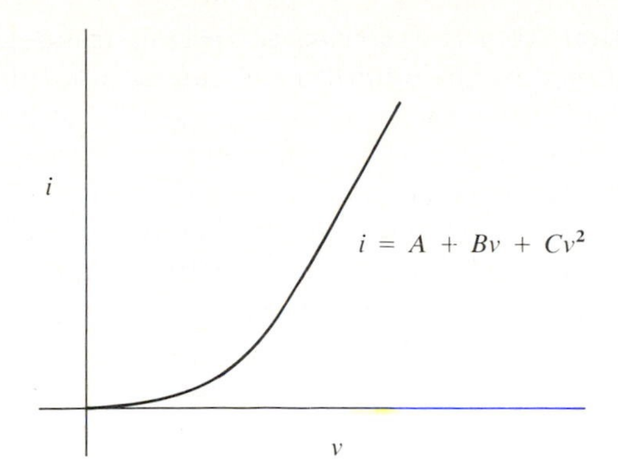

FIGURE 11–18
Approximation of the characteristic curve of a diode

the information contained in the modulating wave. These waves can be "mixed" by adding them in series with a nonlinear element (Figure 11–17). If they are added in series with a linear element such as a resistor, they will both appear at the output but there will be no new waveform. A nonlinear device is required. In practice, a class-C amplifier is usually used for this purpose.

Assume that the characteristic curve of the diode in the circuit of Figure 11–17 is approximated by the parabola in Figure 11–18. The expression

$$i = A + Bv + Cv^2 \qquad \textbf{(11–19)}$$

can be used to describe the characteristic curve. By substituting $v = E_m \sin \omega_m t + E_c \sin \omega_c t$ into the equation, it can be shown by expansion that the resulting relationship contains several interesting terms:*

a dc term

$$A$$

the modulating signal

$$BE_m \sin \omega_m t$$

the second harmonic of the modulating signal

$$\frac{CE_m^2}{2}(1 + \sin 2\omega_m t)$$

the second harmonic of the carrier signal

$$\frac{CE_c^2}{2}(1 + \sin 2\omega_c t)$$

and the term representing the amplitude-modulated signal

$$(BE_c + 2CE_m E_c \sin \omega_m t)\sin \omega_c t$$

*See Appendix C for the mathematical development.

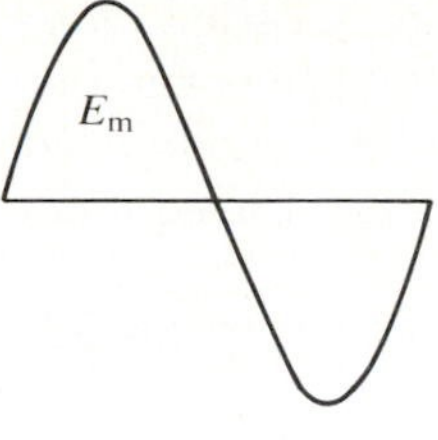

$v_m = E_m \sin \omega_m t$
Modulating waveform

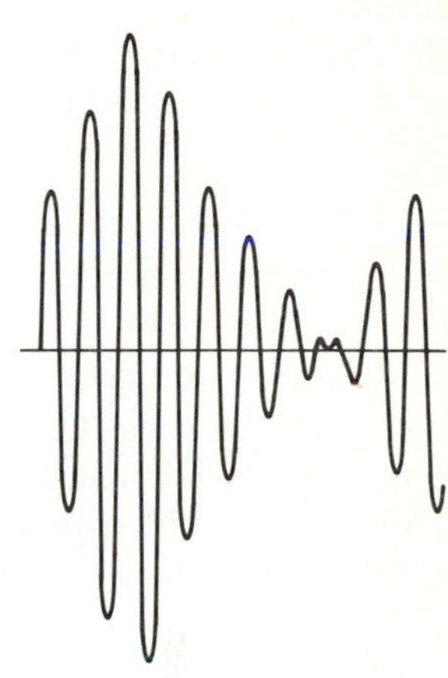

Amplitude-modulated output wave

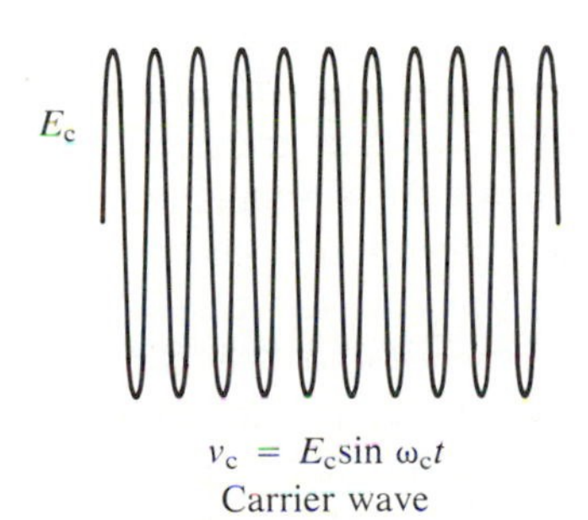

$v_c = E_c \sin \omega_c t$
Carrier wave

FIGURE 11–19
Amplitude modulation

Notice that the term representing the amplitude $(BE_c + 2CE_m E_c \sin \omega_m t)$ changes as a function of the modulating signal (Figure 11–19). This derivation is intended to show that one signal can be used to change the amplitude of another signal.

Frequency and Phase Modulation

Another look at the expression

$$v = E \sin(\omega t + \phi) \qquad \textbf{(11–20)}$$

shows that the terms ωt and ϕ can also be changed. Changing the ωt term changes the frequency of the carrier according to the amplitude and frequency of the modulating signal. The amount of this *frequency modulation* depends on the amplitude, and the rate of change is determined by the frequency. Nearly the same results can be achieved by changing the phase angle (ϕ). This change is called *phase modulation*. It is impossible to tell whether a signal has been frequency modulated or phase modulated. Figure 11–20 illustrates the effects of frequency modulation.

There are two methods of generating an FM waveform. Assume that

$$v = E_m \sin(\omega t + \phi) \tag{11–21}$$

If either the frequency term (ωt) or the phase angle (ϕ) is changed, the frequency of the carrier wave will change. The first method of FM waveform generation involves direct variation of the output frequency of an RF oscillator in accordance with a modulating signal (the ωt term). The other method involves generating an unmodulated carrier wave and then phase modulating it using a phase-shifting circuit (changing the ϕ term). In both methods, the carrier wave to be modulated and the deviation frequency are generated at a frequency lower than the output frequency of the transmitter. Frequency multipliers are then used to increase these lower frequencies to the higher output frequencies.

Appendix C contains a derivation of the development of an FM waveform. Varying the phase angle of the carrier signal, ϕ, results in the carrier and an infinite number of upper and lower side frequencies, which contain the information to be transmitted. The information imparted to the carrier wave is determined as follows. The amount of deviation from the carrier or resting frequency is determined by the amplitude of the modulating signal. The louder the modulating signal, the greater the deviation of the carrier from the resting frequency. The rate of change from the carrier signal is determined by the frequency of the modulating signal. Low frequency-modulating signals—low pitch signals—deviate the resting frequency at a low rate, and high modulating signals—high pitch signals—deviate the carrier signal at a high rate (Figure 11–21).

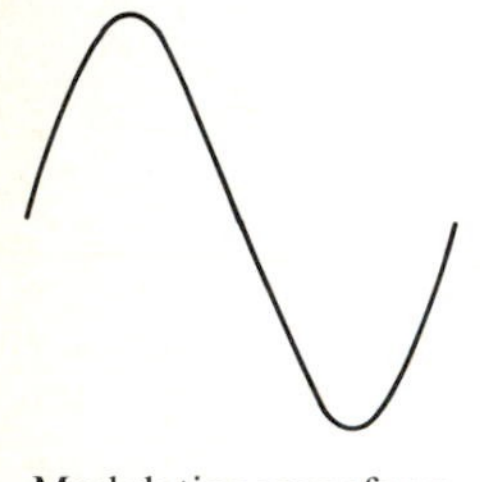

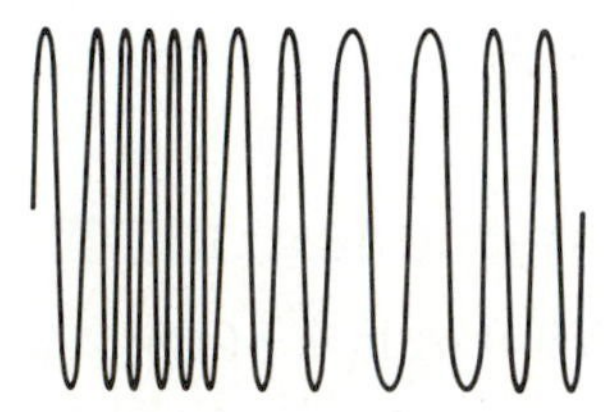

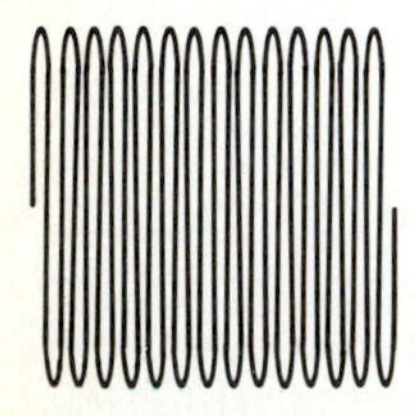

FIGURE 11–20
Frequency modulation

Another common method of altering the frequency of a signal involves heterodyning two signals of different frequencies. When two signals are mixed, the results are the sum and difference frequencies. For example, mixing a 1,030,000-Hz signal and a 1,485,000-Hz signal yields

$$2{,}515{,}000 \text{ Hz} \quad \text{and} \quad 455{,}000 \text{ Hz} \tag{11–22}$$

This technique is used in commercial radio reception to produce 455 kHz in AM radio and 10.7 MHz in FM radio. These frequencies are called *intermediate* frequencies (IF). Within the radio receiver, all incoming frequencies are converted to an IF (plus its sidebands, which contain the information). The amplifiers that boost the signal level for use by the loudspeaker can now have very large gain since they need only a narrow bandpass.

Pulse Modulation

Another method of placing information on a carrier is to use the information to change a parameter of a pulse train. This can be done by changing the amplitude (*pulse amplitude modulation,* PAM), the frequency (*pulse frequency modulation,* PFM), the position of the pulse relative to a timing mark (*pulse position modulation,* PPM), or the width of the pulses (*pulse width modulation,* PWM). Figure 11–22 illustrates these modulation techniques.

SEPARATION OF TWO SIGNALS

The device that performs the separation of the two signals involved in amplitude modulation is called a *detector*. It consists of a diode, which rectifies the AM signal, and a capacitor, which blocks the lower frequency, sending it on to the next stage, but passes the higher frequency to ground. The carrier wave has served its purpose and is no longer required. Figure 11–23 illustrates this process.

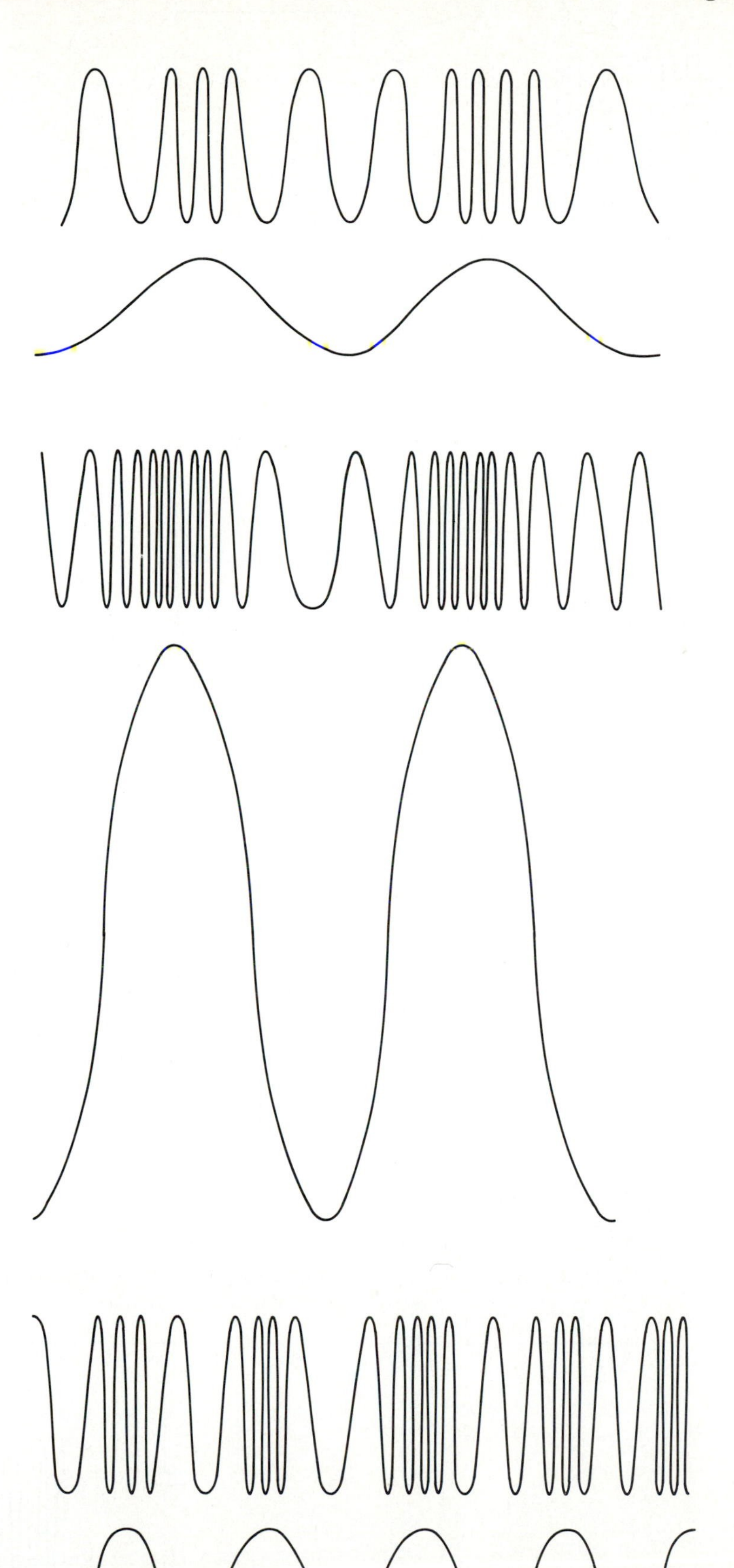

FIGURE 11–21
Volume and pitch in frequency modulation

FIGURE 11–22
Types of pulse modulation

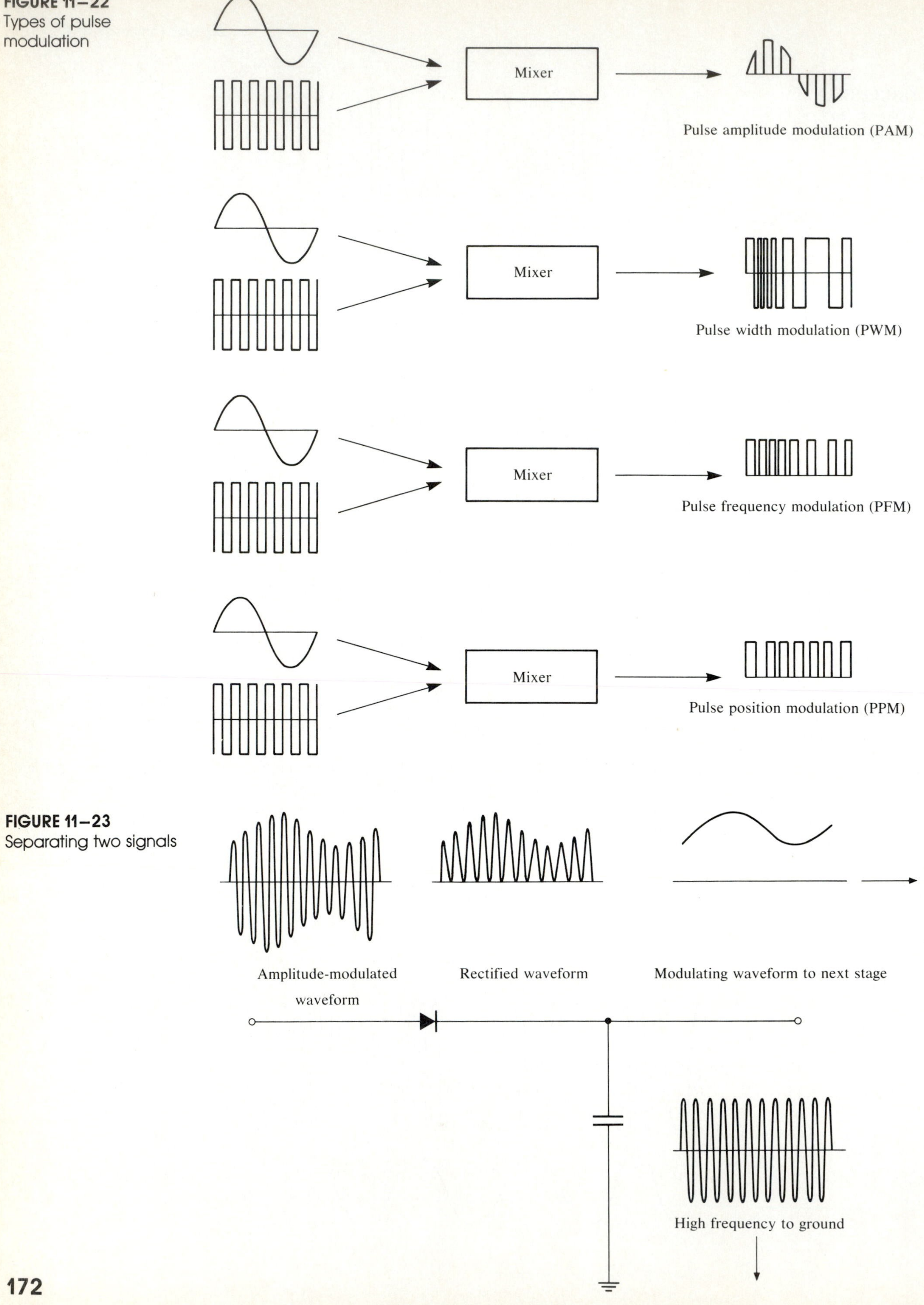

FIGURE 11–23
Separating two signals

SUMMARY

Most waveforms and electrical information that is generated by the inputs of measurement systems are not in the most useful form. The changes discussed in this chapter can be made to provide a useful transition from the input of a measurement system to the output of that system.

EXERCISES

1. The voltage developed during the muscle contraction of a frog's leg is to be recorded. A chart recorder requiring 25 mV for maximum deflection is used. A frog's muscle can develop 0.5 V upon contraction. Design a voltage divider with a total resistance of 100 kΩ that will make the input signal compatible with the recording device.
2. Design an input attenuator that will allow for input voltages of 5, 10, 50, and 100 V. The output voltage will be a maximum of 1.5 V, and the total resistance will be 1 MΩ.
3. Frequency multipliers that can double or triple the frequency of an input signal are to be used to increase a 100-kHz signal to 14.4 MHz. How many of each must be used?
4. How many R-S flip-flops are required to reduce a pulse repetition rate of 76,000 pulses per second (pps) to 9500 pps?
5. Design a circuit whose input is a 20-V peak-peak sinewave and whose output has a positive peak of 8 V and a negative peak of −2 V.
6. Describe the operation of an integrator-type A/D converter.
7. Analog information can be added to a digital pulse train in different ways. List the ways and describe what changes take place.
8. Design a 3-bit D/A converter using an R, $R/2$, $R/4$ network and prove that if a 7-V logic level is used, an input of 101 (base 2) will yield an output of 5 (base 10).
9. Design a 4-bit D/A converter using a ladder network. Show that with a 16-V logic level, an input of 1101 (base 2) will yield an output of 13 (base 10).
10. How are amplitude and frequency information contained in an AM waveform?
11. How are amplitude and frequency information contained in an FM waveform?
12. How are amplitude and frequency information separated from the carrier frequency in an AM radio receiver?
13. Why should the technician be familiar with modulating techniques?
14. Why are signal-processing networks important in measuring systems?

LAB PROBLEM 11–1
D/A Converters

Equipment

One low-voltage power supply; one DMM; various ¼-W resistors as required by the design.

Object 1

To design and test a resistor divider network D/A converter (see Figure 11–11).

Procedure

The input to the D/A converter will be a 4-bit binary number. The resistors will have values R, $R/2$, $R/4$, ... Use a 1-MΩ resistor for the load. If the power supply is 9 V, the outputs will be scaled voltages based on the supply. Design the circuit and enter the measured output voltages in Table 11–1. Calculate the output voltage for each combined input and enter the results in Table 11–1. Graph the measured output voltage and the ideal voltage on Figure 11–24. You will plot two curves, one representing the decimal value of the input versus the measured output, and the other representing the decimal value of the input versus the ideal output. Comment on the results of the D/A converter.

Object 2

To design a ladder network D/A converter.

Procedure

Design a 3-bit digital input D/A converter using a ladder network (see Figure 11–14). Use a 7-V source. Repeat the procedure using an 8-V source. Verify your re-

TABLE 11–1
Complete the table.

Decimal Input	Binary Input	Measured Output Voltage

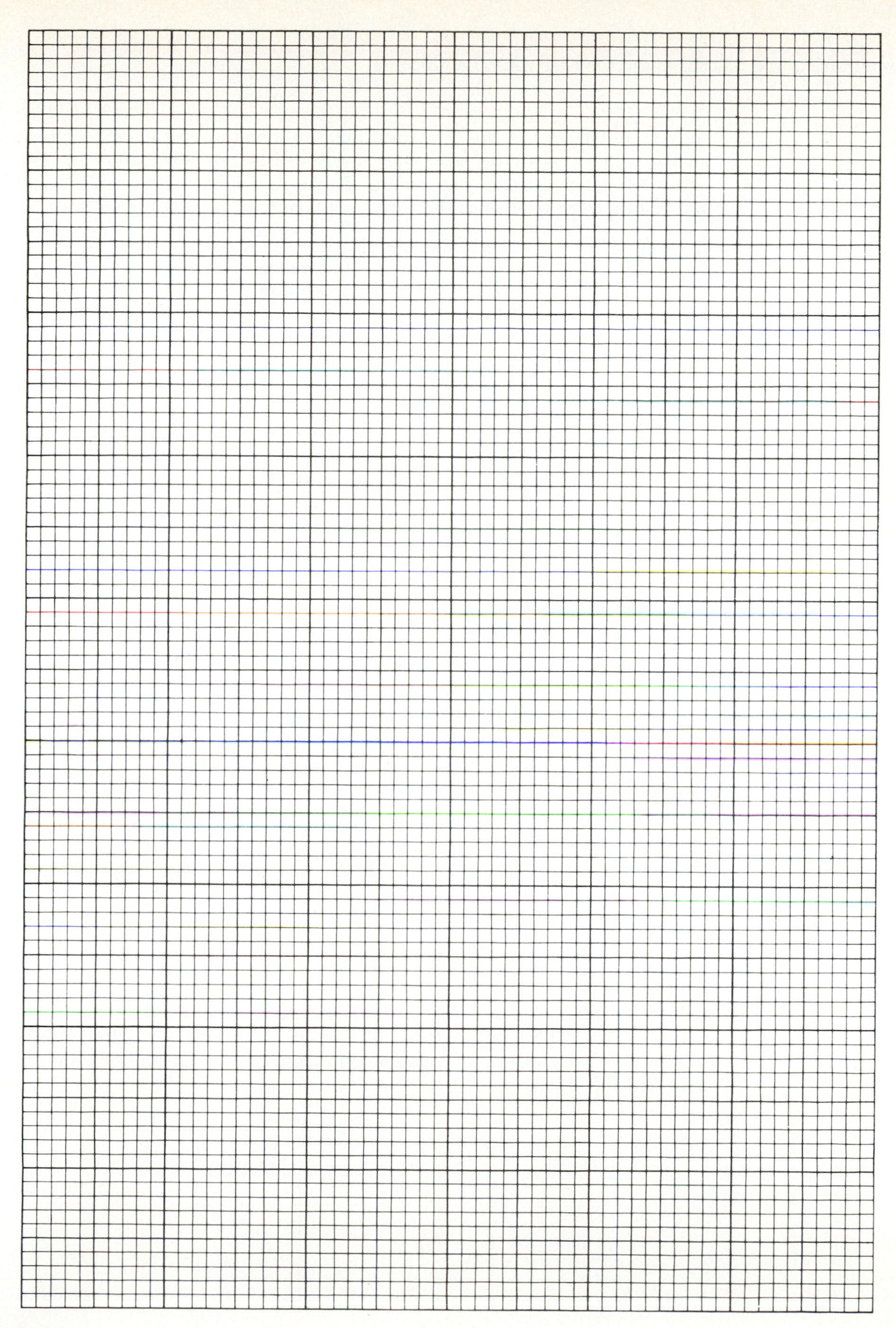

FIGURE 11–24
Graph the results.

sults as in Object 1 of this lab. Use Table 11–2 and Figure 11–25. Make sure you measure your resistances since they rarely will be the coded value and this factor will cause a deviation in your results. It is a good idea to measure the input voltage before every trial since the supply voltage may vary. If it does vary, reset it.

TABLE 11–2
Complete the table.

	Analog Output	
Binary Input	*7 V*	*8 V*

LAB PROBLEM 11–2
Input Attenuator

Equipment

One low-voltage power supply; one DMM; an assortment of ¼-W resistors as required.

Object

To design and test an input attenuator.

Procedure

Design an input resistive attenuator (see Figure 11–7). The total input resistance will be 1 MΩ. The input voltages will be 30, 10, 3, and 1 V. The output voltage will be 0.5 V for all input voltages. Show the design calculations and enter the data in Table 11–3.

TABLE 11–3
Complete the table.

R_x	V_{in} $V_{measured}$	$V_{calculated}$
	30 V	
	10 V	
	3 V	
	1 V	

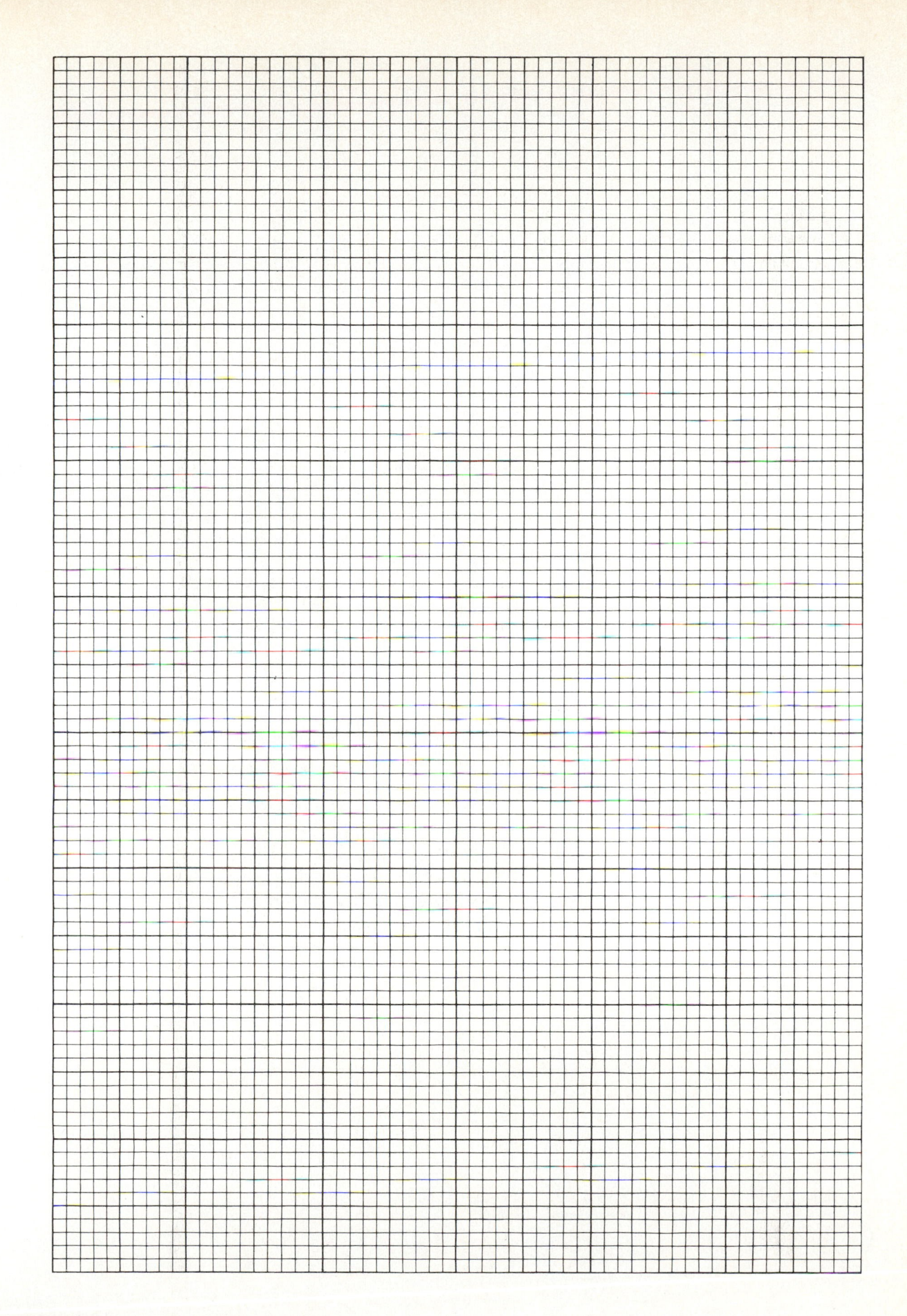

FIGURE 11–25
Graph the results.

12 Measuring Systems and Analyzers

OUTLINE

OBJECTIVES
IEEE AND GPIB BUSSES
SIGNAL SOURCES
FREQUENCY SYNTHESIZERS
LOGIC PROBES AND PULSERS
FREQUENCY COUNTERS
MICROPROCESSOR CONTROL
SWEEP GENERATORS
MICROWAVE
ANALYZERS
- Logic Analyzers
- Network Analyzers
- Network/Sprectrum Analyzers
- Waveform Analyzers
- System Analyzers
- Integrated System

SUMMARY
EXERCISES
LAB PROBLEM 12–1–Calibrating an Audio Oscillator
LAB PROBLEM 12–2—Test System Simulation

OBJECTIVES

After completing this chapter, you will be able to:

- List the building blocks of measuring systems and analyzers.
- List the various types of signal sources.
- Describe a function generator.
- Discuss other types of signal generators.
- Describe a frequency synthesizer and list some of its uses.
- Explain the operation and use of logic probes and pulsers.
- Explain the operation of frequency counters and list some of their uses.
- Describe microprocessor control as it relates to frequency sources and counters.
- Describe sweep generators and give some of their applications.
- Explain why microwave communications are important and discuss some of the problems that require different kinds of measuring devices.
- Define an analyzer.
- Explain the operation of logic analyzers, network analyzers, network/spectrum analyzers, waveform analyzers, and system analyzers.
- Appreciate how all of the individual measuring devices and techniques discussed earlier are integrated into systems that perform complex, detailed tests and measurements.

Measuring systems and analyzers are combinations of devices and circuits used to perform specific groups of tasks, some of which have been described in previous chapters. In general, these systems and analyzers consist of a signal input, processing circuits, some kind of measuring device, and the necessary interface medium. Many types of analyzers are available. This chapter presents a representative group along with a brief description of their applications. The examination begins with a brief look at interface busses. Signal sources are then discussed, followed by devices that process and analyze the input information.

IEEE AND GPIB BUSSES

Common to most analyzers is a bus that provides a path between devices for power, control signals, address information, and data. Two widely encountered terms are *IEEE-488 bus* and *GPIB* (general purpose interface bus). In 1975, the IEEE (Institute of Electrical and Electronic Engineers) published the 488 standard defining an interface for programmable instruments. It was an attempt to establish a standard for compatibility between devices. The 488 bus is usually called the GPIB. Before the GPIB was available, connecting programmable instruments to a computer or to a desktop calculator was a major job because each instrument's interface was different. Now, the IEEE standard 488 defines an interface that makes it much easier to put together computer-controlled instrument systems.

The IEEE-488 standard defines three aspects of an instrument's interface:

1. *Mechanical*—the connector and the cable.
2. *Electrical*—the electrical levels for logic signals and the manner in which the signals are sent and received.
3. *Functional*—the task an instrument's interface is to perform, such as sending data, receiving data, triggering the instrument, and so on.

Although the IEEE-488 standard defines the bus and the interface, it does not treat many other system aspects. Additional information required to use the IEEE-488 standard as a general purpose interface bus involves answers to the following questions:

1. What is the environment (repetitious testing? speed of operations?)?
2. Must the system generate stimuli? Are signal sources required for this purpose?
3. What are the data acquisition requirements (number of channels? acquisition speed? bandwidth?)?
4. Will the data require processing?
5. Are peripherals required to log results?

Although the IEEE-488 standard provides some degree of compatibility between parts of a measuring system, many factors might have to be considered in order to make the system perform as intended.

SIGNAL SOURCES

Some signals to be measured and detected are generated naturally. Such signals occur at all frequencies in the electromagnetic spectrum. Transducers are used to convert these energy sources to measurable electrical quantities.

Signal generators that produce low-frequency signals contain oscillators and amplifiers to produce electromagnetic radiations. Crystals are commonly used to produce a stable fundamental signal. Some generators contain wave-shaping circuitry to produce signals with different characteristics from the originally produced signals. These signals can then

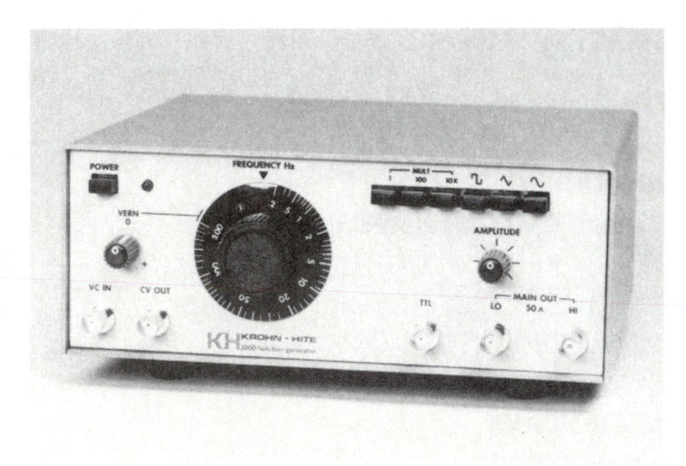

A.

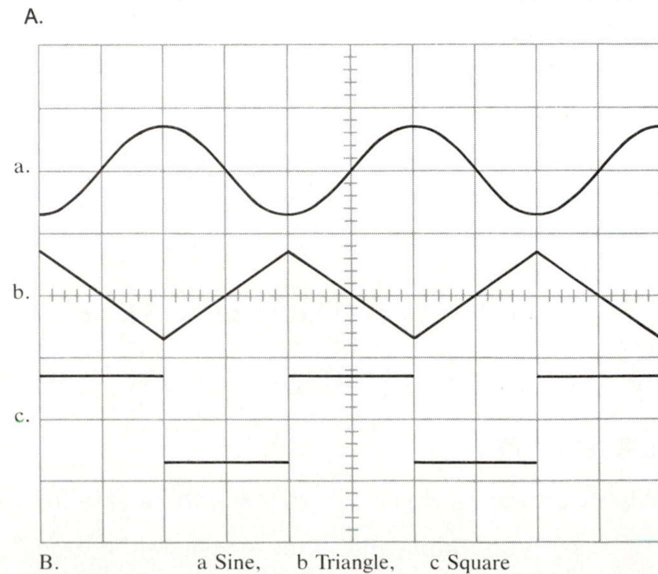

B. a Sine, b Triangle, c Square

FIGURE 12–1
A. 0.2- to 3-MHz function generator. B. Output waveforms.

be injected into the equipment being tested at critical points that will indicate whether various circuits are functioning properly. This process is called *bench-testing*. Other generators produce complex waveforms that can be applied to the input of a device to allow tracing. In some instances, a sweep generator is used to analyze the operation of a piece of equipment over a range of frequencies.

Figure 12–1A shows a 0.2- to 3-MHz function generator capable of generating a sine, square, or triangular wave with maximum output voltage of 20 V peak-peak. Its frequency range makes it useful for testing audio, low-frequency, and low-radio-frequency equipment. Figure 12–1B is an oscilloscope display of the three signal shapes.

Special types of oscillators/generators are used to produce signals in the radio-frequency (RF) range. Many of these RF generators are equipped with modulating circuitry that can impose low-frequency signals (which simulate intelligence) on the fundamental high-frequency signals. Figure 12–2A shows such an oscillator. This is an AM/FM/phase lock generator. It has a frequency range of 0.003 Hz to 30 MHz, and it can output sine, square, and triangular waves, as well as ramps and pulses. Figure 12–2B shows a display of some of its outputs. This oscillator can be used to test transmitting and receiving communications equipment.

Oscillators are available that generate signals in the microwave range (Figure 12–3). Of increasing importance are generators and frequency extenders that produce waveforms in the millimeter range, including signals in the 40- to 60-GHz range. Figure 12–4 illustrates such a generator.

FIGURE 12–2
A. AM/FM/phase lock generator. B. Output displays.

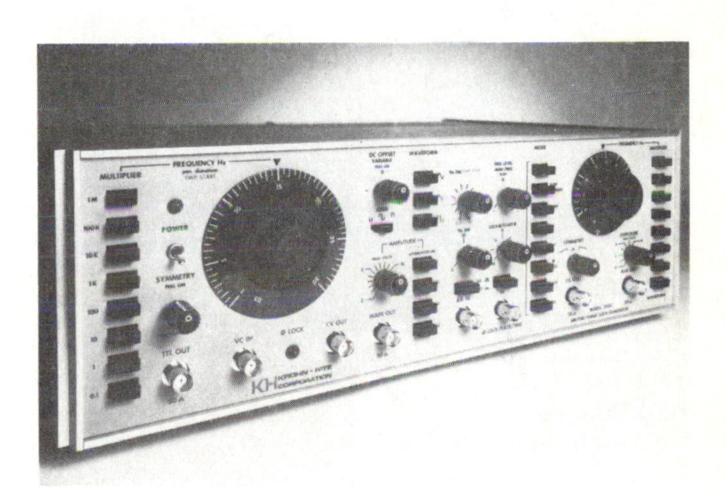

A.

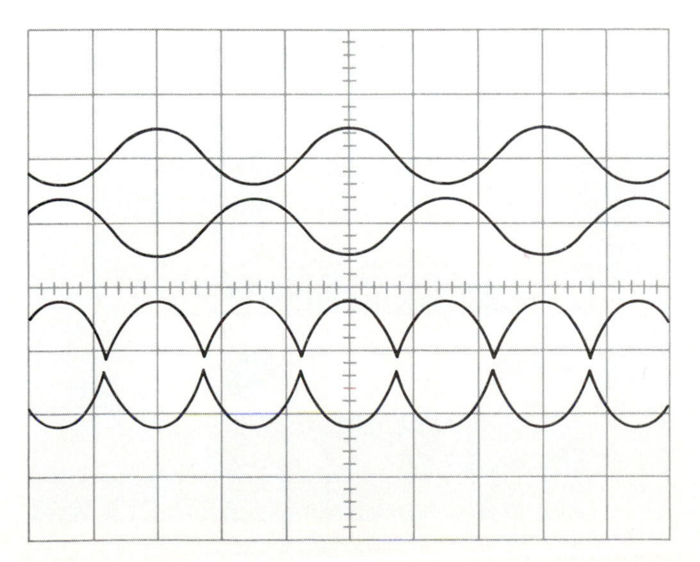

B. Internal AM (upper trace) and suppressed carrier AM (lower trace)

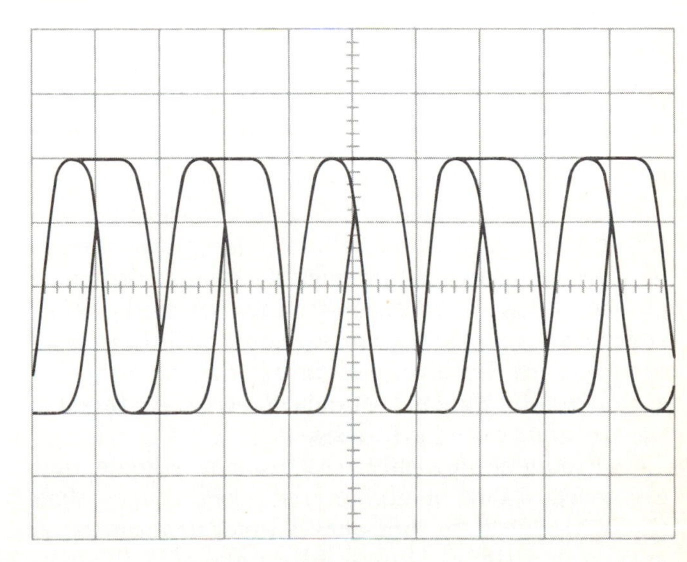

Internal FM

FIGURE 12–3
Pulse/microwave/sweep generator

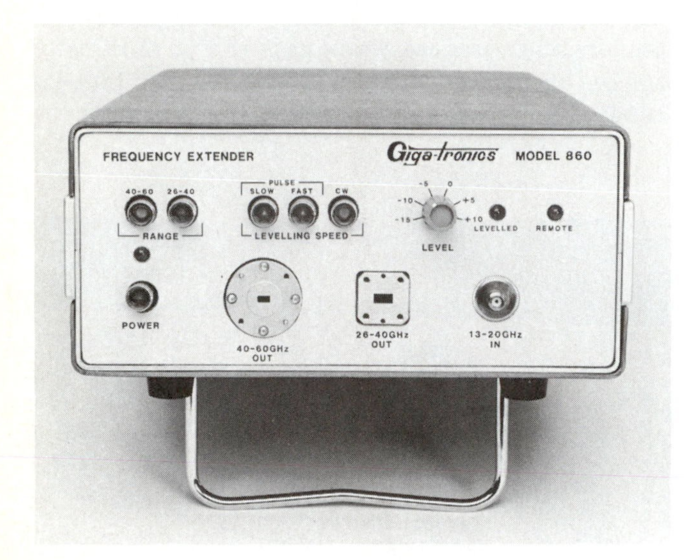

FIGURE 12–4
Frequency extender

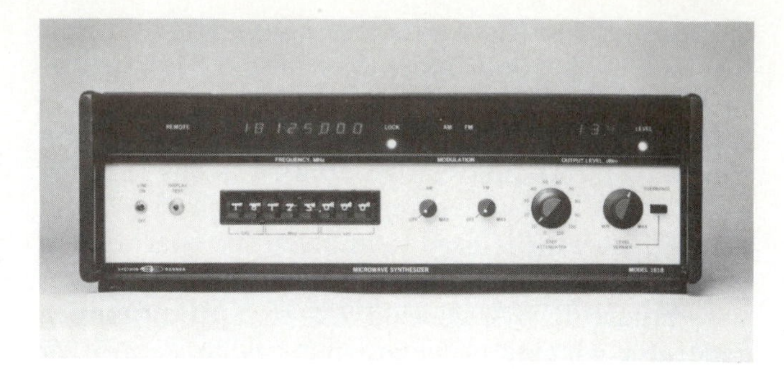

FIGURE 12–5
Microwave synthesizer

FREQUENCY SYNTHESIZERS

Frequency synthesizers are basically highly accurate, very stable variable RF generators (Figure 12–5). Within a specified frequency range, they can be programmed to produce almost any output frequency. Typical uses of these synthesizers include satellite earth station oscillators, magnetic storage media testing, crystal-production testing, and mode locking of lasers. Precision timing standards, radar simulators, and Doppler systems use frequency synthesizers. Their output signals depend either on an internal frequency standard or on a standard signal input to the synthesizer.

Frequency synthesizers produce signals that are more stable in higher frequency ranges than those produced by ordinary signal generators. A very stable signal is especially important in high-frequency applications because a 0.1% variation at 1 GHz is much greater than a 0.1% variation at 1 kHz.

In lower frequency ranges, the crystal oscillator produces the most stable signals available. However, a crystal must be cut thinner to produce higher frequencies. In higher frequency ranges, the crystal needed would be nearly impossible to produce and impractical to work with. Frequency synthesizers use processing circuits to produce high-frequency signals from stable signals with much lower frequencies. Figure 12–6 illustrates a synthesized-signal generator.

This generator is a fully programmable signal source. It can output signals from 100 Hz to 1 GHz in 100-Hz steps. The output voltage range is from 0.1 mV to 1 V rms. The signal output can be kept well within ±1 dB over this range. From 100 Hz to 1 GHz, the harmonics are better than −25 dB and the nonharmonics (spurious noise) better than −50 dB relative to the main signal. The output signal

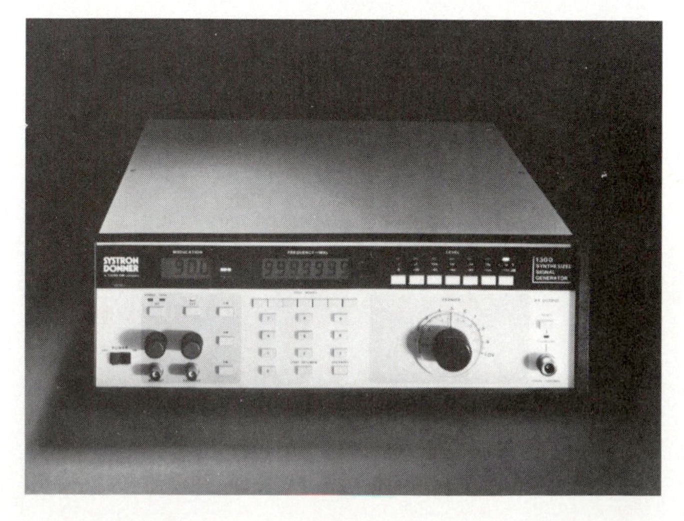

FIGURE 12–6
Synthesized-signal generator

can be amplitude or frequency modulated. The generator can be programmed through an IEEE-488 interface.

Microwave synthesizers can be used to generate 50 MHz to 18 GHz with 1-kHz resolution. They are used for communications receiver alignment, radar system test and alignment, microwave frequency measurement, and as automatic test equipment building blocks.

LOGIC PROBES AND PULSERS

Digital integrated circuits often fail totally and instantaneously. Troubleshooting with an oscilloscope requires extra work. Because pulse activity is generally present in functioning digital ICs, logic probes and pulsers can be used to easily determine if an IC is working. Logic pulsers and probes are designed to analyze and troubleshoot static and dynamic conditions of logic circuitry. Used together, they make it possible to test sequential circuits such as flip-flops, counters, and microprocessors.

A digital logic probe provides a visual display of logic states such as high-level, low-level, pulsing, or open (bad) circuits. (Figure 12–7 shows a block diagram of a digital pulse logic probe.) The input impedance of the probe tip is such that the output of a gate causes the instrument to react to whatever signal is present. Open inputs of TTL gates cause both LEDs to extinguish. The digital logic probe has built-in current-limiting circuitry for intermittent or low duty cycle overvoltages, and built-in high-speed circuitry that enables a memory when either positive or negative pulses are present. It has an automatic 50-ms reset, which allows adequate time to observe the presence of a pulse. The logic probe detects a discrete level and allows the technician to determine whether that level is appropriate. The typical logic probe can "catch" a very narrow pulse (e.g., 5 ns), which would be too fast for most other instruments to observe.

The digital pulser provides a digital logic level or pulse train to drive digital logic (see Figure 12–8). With this type of signal-inserting device, the technician can trace the behavior of digital circuitry.

FREQUENCY COUNTERS

Frequency counters are useful for applications requiring quick, accurate frequency or pulse repetition readings. The signal to be measured is processed by input amplifier/attenuator and shaping circuits to produce pulses that will trigger a counter more efficiently. The counter also receives control pulses from an oscillator which serve as a reference time base. The counter circuit (usually an IC) compares the pulses counted during each control pulse to the reference, multiplies or divides the count as required by the time base selected, and

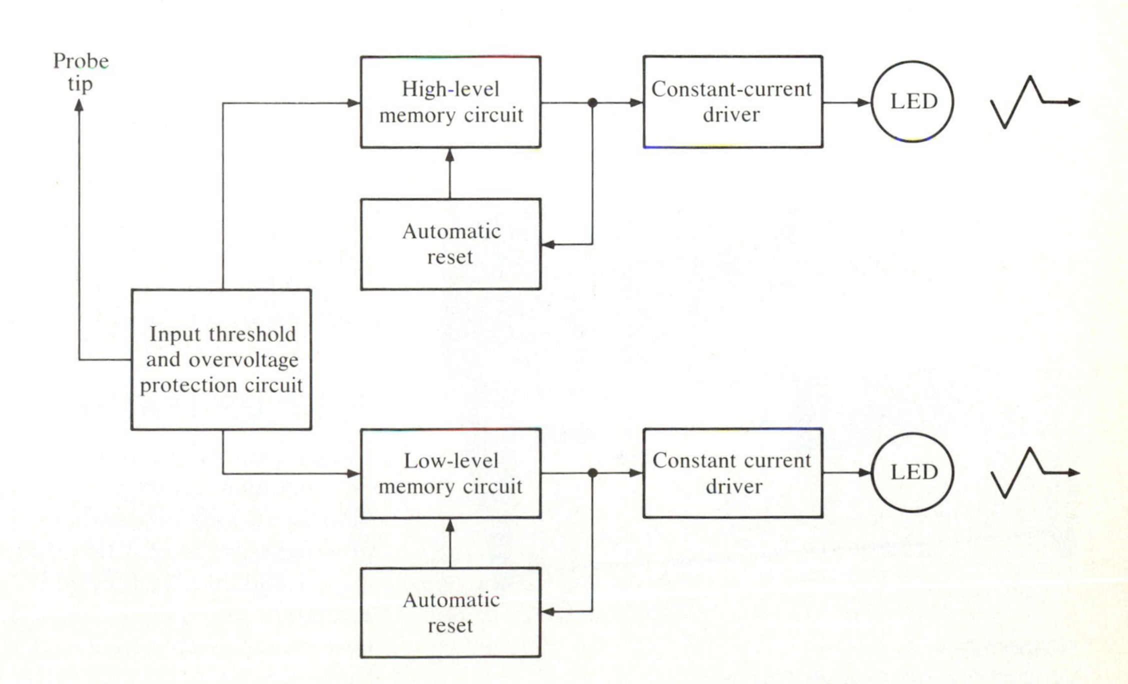

FIGURE 12–7
Block diagram of a digital logic probe

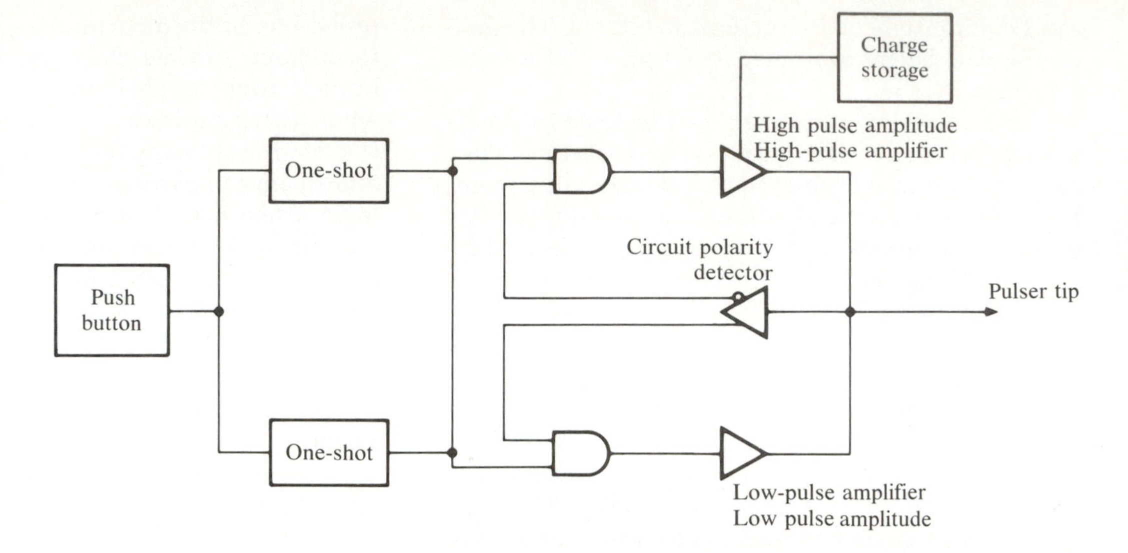

FIGURE 12–8
Block diagram of a digital pulser

yields the frequency. This information is then decoded and displayed in digital form. The frequency range of the signal source determines the frequency range requirement of the signal counter. Some frequency counters have impressive capabilities. Many can be controlled through IEEE-488 or GPIB interfaces.

The multifunction universal microwave counter shown in Figure 12–9 can count from 1 Hz to 26.5 GHz. Below 100 MHz, the counter will give eight-place resolution. This is 1-MHz resolution. The device has a built-in diagnostic package that sequences all RAM/ROM (random access memory/read only memory) locations, count chains, keyboard controls, and displays. Sequencing occurs every time the unit is turned on. Malfunctions will be detected, and a displayed code indicates what the malfunction is and where it is occurring.

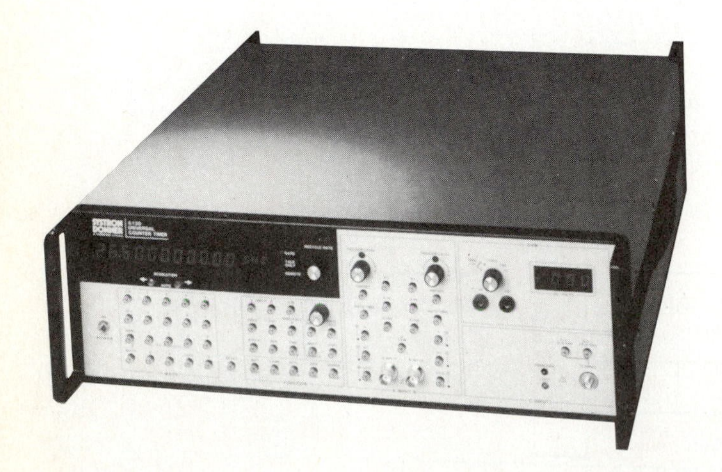

FIGURE 12–9
Multifunction microwave counter

MICROPROCESSOR CONTROL

Most signal sources and counters are microprocessor controlled. They can be interfaced with computers which can be programmed to control them. Such parameters as wave shape, frequency, pulse repetition rate, and pulse rise time are entered into the computer, which instructs the microprocessor to control the generation of signals. Many signal generators can also be programmed as stand-alone devices.

The microprocessor controllers of many counters allow high-resolution readouts over a wide range of signal inputs. These devices can be interfaced with other units through the GPIB. They can be interfaced with computers which store and analyze the data obtained.

Most microprocessor-controlled units have built-in diagnostic programs that exercise memory, counters, output devices, and any arithmetic logic operations. The appropriate readouts pinpoint malfunctions so that the technician can replace the faulty module. Maintenance and repair are much easier becuase the technician does not have to troubleshoot to the chip or pin level.

SWEEP GENERATORS

Sweep generators output a range of frequencies. For example, they may sweep from 1 MHz to 1.5 MHz in 10-kHz intervals. This means that they will produce signals of 1,000,000, 1,010,000, 1,020,000, . . ., 1,490,000, 1,500,000 Hz and then start over at 1,000,000 Hz. The purpose of this procedure is to heterodyne with other signals or to test resonant frequency points in circuits, resulting in the gen-

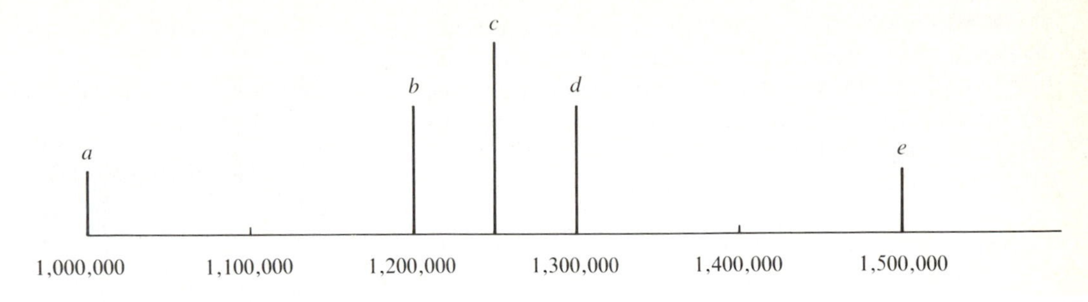

FIGURE 12–10
Pattern produced by a sweep generator

eration of markers indicating frequencies or signals that may be desirable (or undesirable). The resulting pattern may look like Figure 12–10.

The readout can be analyzed to see if the resulting resonant points (*a, b, c, d, e*) are at the desired frequency or if they are desired at all. Unwanted side frequencies such as harmonics can be detected and "tuned out" or otherwise compensated for using this detection capability of sweep generators. Figure 12–11 is an example of a sweep generator.

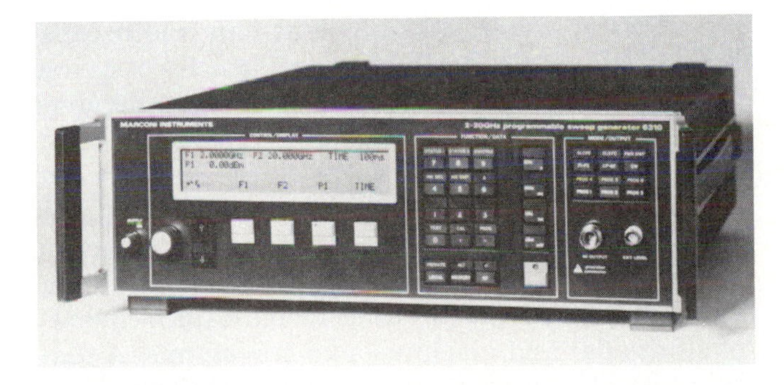

FIGURE 12–11
Sweep generator

MICROWAVE

Microwave communications require the use of higher frequencies, which allow for wider channel widths. These signals carry higher-order harmonics, and therefore the desired shape of the transmitted signals can be more accurately maintained. For example, the leading edge of a square pulse is generated by the addition of high-order harmonics. The more harmonics transmitted, the more well defined the leading edge of the pulse will be. This large number of harmonics requires great channel width to allow for the higher frequencies involved. Because of the technology required, different kinds of equipment must be used. Microwave generators, counters, and sweep generators are designed to handle the higher frequencies of the microwave range.

ANALYZERS

Four types of analyzers will be discussed here: logic analyzers, network analyzers, wave analyzers, and system analyzers. They are grouped together because they all perform similar tasks, use a similar approach, and generate similar results. In each case, the analyzer consists of many parts that work together to test the overall operation of an electronic device.

Logic Analyzers

Logic analyzers can be interfaced with a system designed to perform logic operations. Through a program, the analyzer injects appropriate logic signals at critical points in the system and monitors the outputs. The results are stored in memory or displayed on a printout. The output can then be checked or analyzed to determine whether the equipment under test is performing according to specification.

Network Analyzers

Network analyzers can be used to determine critical parameters in the testing of communications systems. By inserting the appropriate signals, the technician can automatically determine if the outputs and operation of the network are appropriate. Transmission characteristics, reflection coefficients, losses, and standing wave ratio measurements are critical in the design, manufacture, and testing of communications systems.

The first network analyzer was developed in 1930 by Bell Laboratories to measure the microbel characteristics of feedback amplifiers. In 1953, a 50-MHz band network analyzer was developed by Anritsu to perform the same task. The network analyzer has become one of the most important instruments used in the analysis of analog circuitry. Network analyzers can measure impedance and transmission characteristics (both of which are vector values) in the linear area of an analog circuit. Since analog circuit characteristics can be completely expressed by impedance and transmission characteristics, network analyzers can be used to analyze a wide variety of parts, materials, circuits, and equipment. This capability makes the network analyzer indispensable for optimizing circuit design in research and development. It also proves itself in a wide range of production and maintenance applications.

Network/Spectrum Analyzers

Network/spectrum analyzers such as the one in Figure 12–12 are no-image response network analyzers with dual input. No-image response means that the analyzer input does not respond to the image frequency signal of a mixer; the analyzer detects only the object signal, even if it exists as part of a composite signal. For example, in a single side band (SSB) microwave radio system, DADEing (differential absolute delay equalizing) is used to reduce the fading interference. For this procedure, the phase difference of pilot signals between service and protection channels (guard bands) must be measured while the radio system is in service. The no-image response network analyzer is extremely effective for performing this measurement, plus a wide range of other applications.

FIGURE 12–12
Network analyzer

Waveform Analyzers

Waveform analyzers are less automatic than the other systems discussed in this section. They perform analysis on waveforms introduced into them

FIGURE 12–13
Waveform analyzer

usually through a probe handled by a technician. The waveform analyzer in Figure 12–13 is such a device.

This analyzer has sufficient bandwidth capabilities to handle the higher frequencies required by the logic controls of a video cassette recorder (VCR)—usually in the 60-MHz range. The automatic triggering feature saves the technician considerable time. The analyzer is dual trace so that one signal can be adjusted relative to the timing characteristics of a second signal. It also has a digital output that can be used to indicate frequency of sinusoidal and nonsinusoidal waveforms. A digital voltmeter is built in to indicate the peak-peak voltage of the signals being observed. With a basic knowledge of the operation of the system under test, the technician can observe a few waveforms, isolate faults within minutes, and make adjustments where necessary.

System Analyzers

System analyzers are the most complex of the devices discussed here. They are designed to completely analyze an operating system and are closely related to automatic test equipment. A system analyzer can be programmed to insert signals of any description at any point in an operating system and display the resulting signals at most points in the circuitry. This capacity allows the technician to analyze and interpret the data, thus isolating any malfunctioning parts of the sysem. Some devices can be used as a network analyzer, a spectrum analyzer, or a system analyzer. The value of the equipment is determined, in part, by the capability of the technician to use it to its fullest potential.

Integrated System

The Fluke Scopemeter™ pictured in Figure 12–14 is an example of an *integrated system.* In one hand-held package there is a 50-MHz oscilloscope with waveform processing and memory, a multimeter, a sine/square generator, and a component tester. This system also has an RS-232 interface for full operation by remote control and a serial printer interface port. This integrated system represents an electronics "shop" in one package.

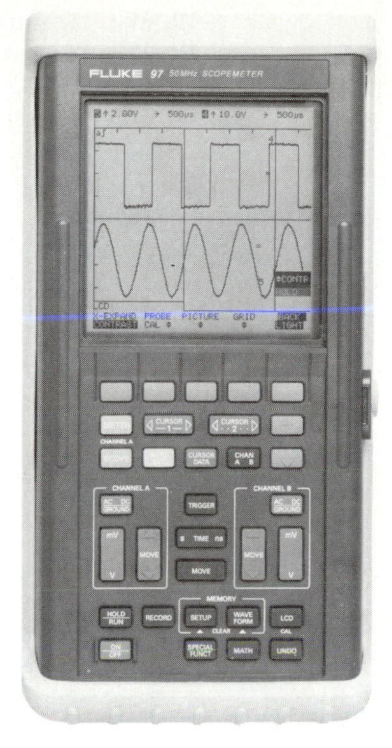

FIGURE 12–14
Fluke 97 Scopemeter

SUMMARY

This chapter considered a few examples of measuring systems and analyzers, which utilize many of the components presented individually in previous chapters. As integrated systems, analyzers and measuring systems can be used to pinpoint faults and to aid in the fine adjustments required in highly technological equipment. Selecting instruments that satisfy the needs of a measuring or testing situation requires that the technician be knowledgeable about the system under test, the measurement characteristics involved, and the test equipment.

While several systems have been presented here, technicians should become acquainted with many different types of test equipment and systems. Manufacturers are usually willing to provide information about their equipment, and many have toll-free phone numbers for this purpose. A simple phone call can often generate the answers to questions, information about products, and even applications and troubleshooting information.

EXERCISES

1. List the building blocks of any measuring system.
2. List and discuss three signal sources that occur in nature.
3. What are the most common waveforms generated by a function generator?
4. What are some of the uses for each waveform listed in Exercise 3?
5. Why must signal generators and counters be capable of handling different frequency bands?
6. What is a logic probe? Describe its use.
7. What is a pulser? Describe its use.
8. What does a frequency synthesizer do?
9. Why is microprocessor control necessary for the instruments covered in this chapter?
10. What kinds of control can a microprocessor provide to a signal generator?
11. What are sweep generators and how do they differ from common signal generators?

12. Why are high-frequency communications important?
13. Why do microwave and millimeter-wave measurements require specialized equipment?
14. How do logic analyzers, network analyzers, waveform analyzers, and system analyzers differ?
15. What is the difference between a system analyzer and automatic test equipment?
16. Analysis of ac circuits includes how the circuits respond to sinewaves. Fourier's theorm states that any periodic signal can be built up from sinewaves. The expression for a square wave can be shown to be

$$V(t) = V(\sin \omega t + \tfrac{1}{3} \sin 3\omega t + \tfrac{1}{5} \sin 5\omega t + \tfrac{1}{7} \sin 7\,\omega t + \ldots)$$

where $\omega = 2\pi f$. Using a BASIC program, show how the addition of higher-frequency components to the expansion affects the leading edge of the square wave.

LAB PROBLEM 12–1
Calibrating an Audio Oscillator

Equipment

One radio receiver capable of receiving up to 20 MHz; one function generator (audio-frequency range); one dual-trace oscilloscope.

Object

To calibrate an audio oscillator using the U.S. Bureau of Standards radio station WWV, which transmits electronic signals and geophysical information as well as the time at 2.5, 5, 10, 15, and 20 MHz. In some cases, CHU (Canada) is more easily received. Part of the transmission of these stations is a set of audio frequencies that can be used to form Lissajous figures for calibrating an audio oscillator. Signals in the audio range include 440 Hz, 500 Hz, and 600 Hz. WWV transmits on a carrier frequency of 2.5, 5, 10, 15, and 20 MHz. CHU transmits on 3.33, 7.335, and 14.670 MHz. These stations transmit 24 hours each day.

Procedure

Tune in WWV at 2.5, 5, 10, 15, or 20 MHz. If atmospheric conditions do not allow this, try tuning in CHU at 3.33, 7.335, or 14.670 MHz. Connect the output of the radio receiver to the horizontal input of an oscilloscope, and connect the output of the function generator (audio oscillator) to the vertical input. WWV transmits audio signals as part of its programming. It transmits several signals, but you should concentrate on one signal at a time (e.g., 500 Hz). Use Lissajous patterns to choose a frequency setting that corresponds to the harmonics and subharmonics of the transmitted audio signal (i.e., 500 Hz, 1 Hz, 1.5 Hz, 250 Hz, 125 Hz, etc.). Determine as many points as possible. Then concentrate on the second audio signal and repeat the procedure. Enter the data in Table 12–1 and graph the calibration curve on Figure 12–15.

TABLE 12–1
Complete the table.

Standard Frequency	Dial Setting	Standard Frequency	Dial Setting

LAB PROBLEM 12–2
Test System Simulation

Equipment

One signal generator; one dc power supply; one oscilloscope; one breadboard; Six IC chips: two 7476 J-K flip-flops, two 741 op-amps, one 7486 quad exclusive-OR, one 7432 quad OR. The pinouts for the chips appear in Figure 12–16.

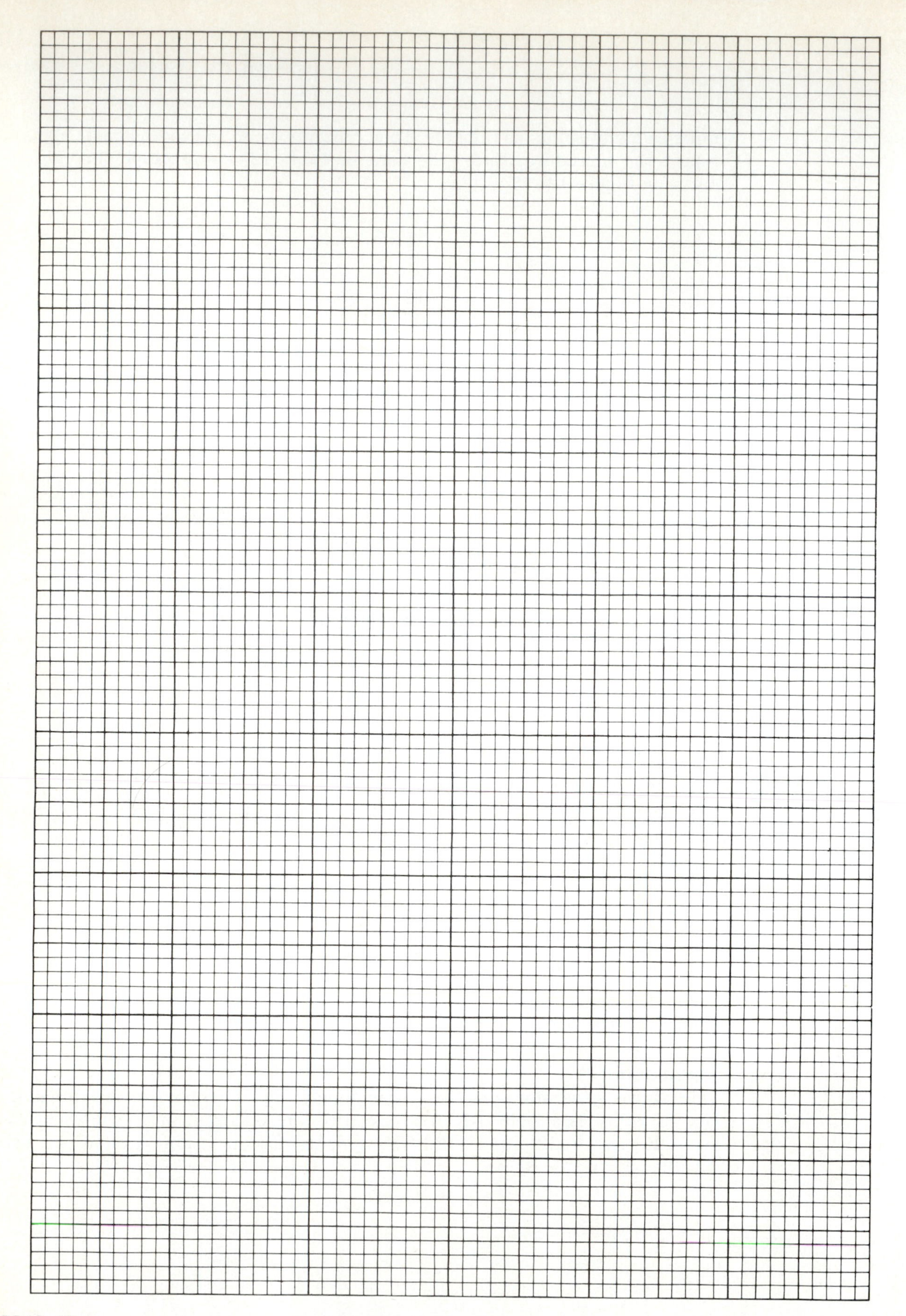

FIGURE 12–15
Graph the calibration curve.

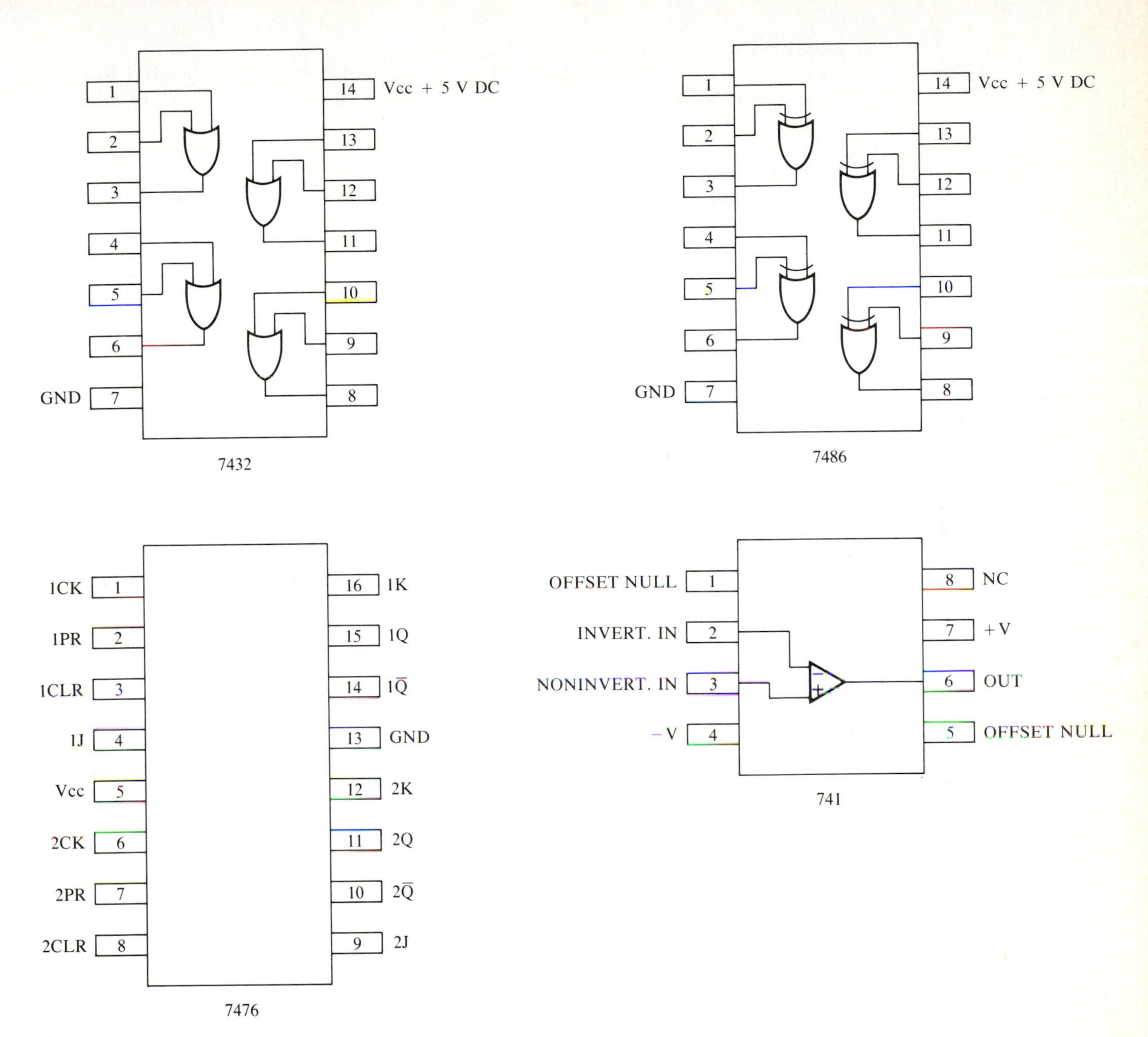

FIGURE 12–16
Chip pinouts

Object

To give a better understanding of how test systems operate. Generally, test systems consist of three major components: an input device, processing circuits, and an output device.

Procedure

Set the function generator to produce a 1-kHz square wave (5 V peak-peak). This signal will serve as a clock pulse. Wire the *J-K* flip-flops as shown in Figure 12–17 to form two identical asynchronous counters with a common clock pulse. Take care when wiring the second circuit to allow access through the connecting wires to the chip. Build the comparator circuit as shown. Connect points A_1, B_1, A_2, and B_2 to the comparator as indicated. Because the two *J-K* circuits are identical, the comparator will produce a low output. (The exclusive-OR produces a high output only if its inputs are unequal.)

For a visual output, connect channel A of your CRO to point A_1 and channel B to B_1. With all settings identical, your waveforms will be identical. Switch to A_2 and B_2, and the same results will be observed.

Now, carefully remove the second 7476 chip and replace it with another 7476. If the new chip is good, the LED will not light up.

Design a similar circuit to test the output of a 741 op-amp.

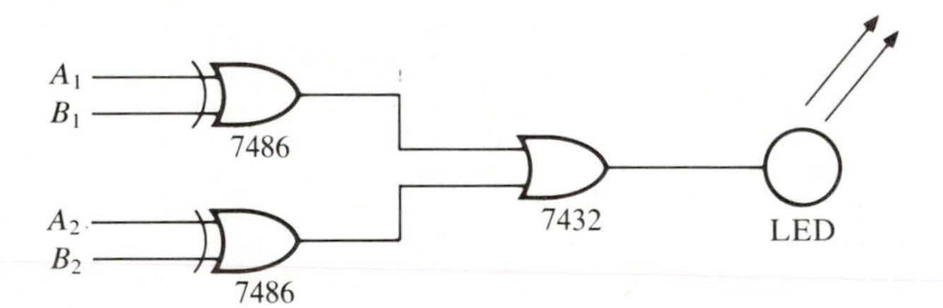

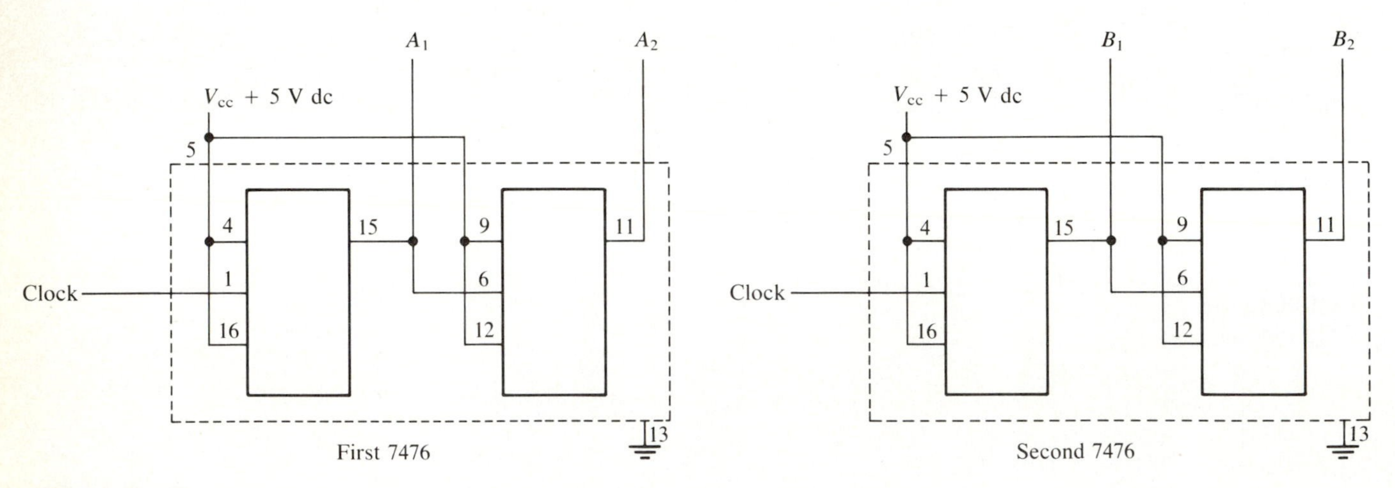

FIGURE 12–17

13 Transducers

OUTLINE

OBJECTIVES
MECHANICAL TRANSDUCERS
- ☐ Strain Gages
- ☐ Pressure Transducers
- ☐ Accelerometers
- ☐ Flow Meters

THERMAL TRANSDUCERS
- ☐ Thermocouples
- ☐ Resistance Thermometers
- ☐ Thermistors
- ☐ Infrared Detectors

OPTICAL TRANSDUCERS
- ☐ Photoconductive Cells
- ☐ Photovoltaic Cells

ACOUSTICAL TRANSDUCERS
- ☐ Dynamic Microphones
- ☐ Condenser (Capacitor) Microphones
- ☐ Ribbon Microphones
- ☐ Carbon Button Microphones
- ☐ Microphone Ratings

ULTRASONIC RANGING SYSTEMS
MAGNETIC TRANSDUCERS
- ☐ Differential-Pressure Transducer
- ☐ Linear-Variable Differential Transformer
- ☐ Hall-Effect Transducer
- ☐ Selsyn Generator

CHEMICAL TRANSDUCERS
BIOLOGICAL TRANSDUCERS
NUCLEAR TRANSDUCERS (RADIATION DETECTORS)
- ☐ Germanium Gamma Ray Detector
- ☐ Photomultiplier Tube

SUMMARY
EXERCISES

OBJECTIVES

After completing this chapter, you will be able to:

- ☐ Define transducer.
- ☐ List the kinds of transducers available.
- ☐ Choose the appropriate transducer when designing a measurement system.
- ☐ List the parameters that constitute the inputs to transducers.
- ☐ List the parameters generated by the outputs of transducers.

A *transducer* is a device that transforms energy from one system to another. Specifically, it transforms any form of energy to electrical energy. Once the energy is electrical, it is easily processed and measured.

Many forms of energy are measured in science, manufacturing, engineering, and medicine. They include light, heat, force, motion, pressure, torque, radiation, chemical reactions, and biological forces. Although most of these forms cannot be measured directly, virtually all can be converted to electrical energy by some type of transducer.

Transducers are either active or passive devices. *Active* transducers produce a voltage or current that can be measured with a voltmeter or ammeter. *Passive* devices change the electrical characteristics of an electronic circuit. In this case, the resistance, inductance, capacitance, or conductivity of the circuit is affected and therefore measured.

This chapter introduces many common types of transducers. Transducers are presented according to the type of energy they transform and according to the devices appropriate for each type of transformation.

MECHANICAL TRANSDUCERS

Mechanical transducers are used to measure force, weight, torque, pressure, acceleraton, rate of flow, and motion. Several popular types will be presented, including strain gages, pressure transducers, accelerometers, and flow meters.

Strain Gages

The resistance of a conductor can be calculated from the formula

$$R = \rho\frac{l}{A} \tag{13-1}$$

where R is the resistance of the conductor, ρ is the specific resistance (resistivity) of the conductor material, l is the length of the conductor, and A is the conductor's cross-sectional area. Quantity ρ is a constant. If a conductor is stretched, the length increases and the cross-sectional area decreases (Figure 13–1). Both of these changes increase the magnitude of the resistance of the conductor. If the length of the conductor is changed considerably, the resulting change in resistance can be measured. This transducer is called a *strain gage* (Figure 13–2).

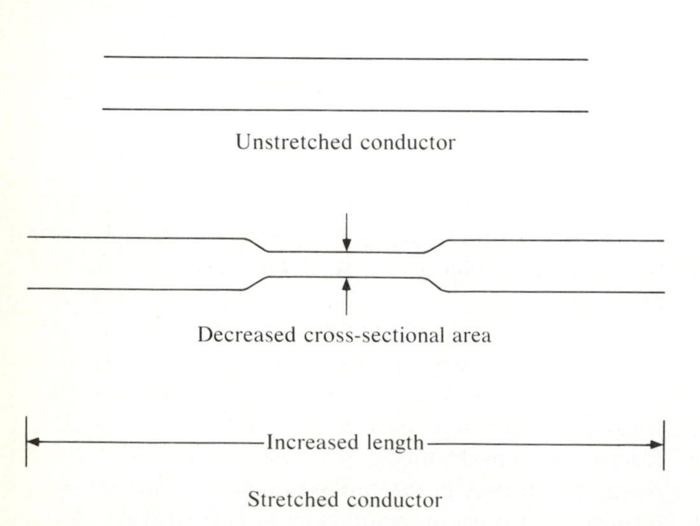

FIGURE 13–1
The changing resistance of a conductor

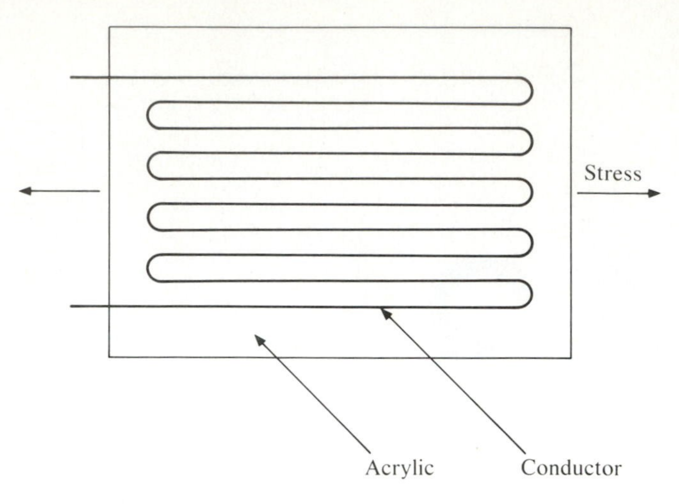

FIGURE 13–2
Strain gage

The conductor of a strain gage is sealed in a block of acrylic or some other material for support. When stress is applied, the conductor distends, changing its resistance. This resulting strain can be measured. According to Hooke's law, if the amount the conductor is stretched is within the limits of perfect elasticity, the stress will be proportional to the strain and the conductor will return to its unstretched state. To measure stress or strain in different directions, several layers of conducting material can be oriented to produce the desired measurements. This kind of strain gage is a passive transducer. Energy that the transducer can affect must be provided to the measuring circuit. A Wheatstone bridge is often used together with a strain gage (Figure 13–3).

If the original resistance of the strain gage is selected as R ohms, a balanced bridge results. Any change in the resistance R, producing $R + \Delta R$ ohms, will unbalance the bridge. This imbalance can be measured (see Chapter 10).

The *bonded resistance* strain gage is usually used for stress measurement. This type of gage is inexpensive, can be small in size, has a low mass, is minimally affected by temperature, and has a fairly high sensitivity to applied stress. When the gage is used in a Wheatstone bridge, its resistance is the only resistance that changes. The nominal resistance of a strain gage is usually 120, 350, or

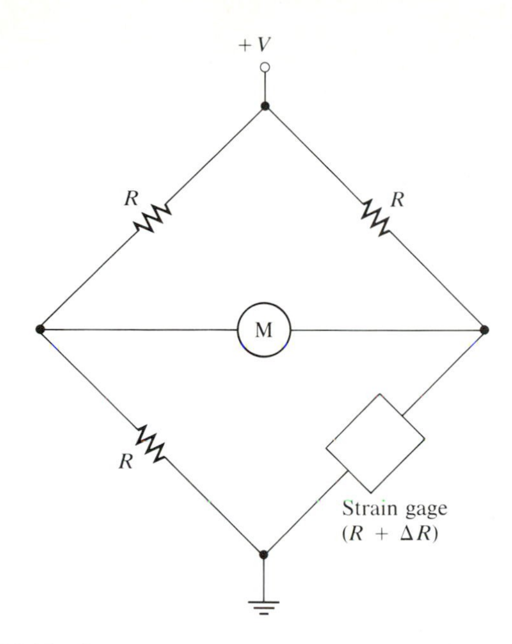

FIGURE 13–3
Strain gage in a Wheatstone bridge circuit

700 Ω depending on the type of gage used. The maximum length change of the conductor is usually <5% (typically 50,000 μm/m). If the application stays within this limit, the transducer can be used an almost infinite number of times.

The strain gage arm of the Wheatstone bridge is usually a remote sensor. Therefore, a temperature difference may exist which can affect the resistance of the gage. To compensate, a second gage (called a *dummy* gage) can be wired as an arm of the bridge while being located at the measurement site. The temperature causing the change in the measuring gage will cause an equivalent change in the dummy gage, nullifying the effect of temperature on the balance of the bridge (Figure 13–4).

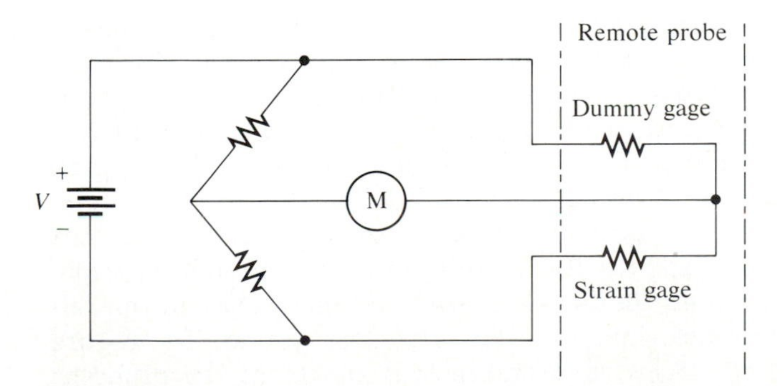

FIGURE 13–4
Use of a dummy gage

The gage selected for a particular application is determined by cost, sensitivity, size, and any special mounting hardware and adhesives required.

Pressure Transducers

One type of pressure-measuring device is the *pressure transducer,* which consists of a pressure sensor (diaphragm and strain gage) coupled with compensation networks. Pressure sensors in the form of integrated circuits with etched pressure-sensitive diaphragms are available. The ideal pressure transducer should be:

1. Accurate over a wide temperature range
2. Operable at temperature extremes
3. Durable
4. Compatible with measured gases, fluids, and slurries
5. Interchangeable
6. Readily available

One popular pressure transducer has a piezoelectric (silicon) diaphragm (Figure 13–5). The piezoresistance of a semiconductor can be described as the change in resistance resulting from a strain applied to the diaphragm. An input pressure of 0 to 100 pounds per square inch (psi) can produce an imbalance in a Wheatstone bridge, resulting in a 100-mV output. The cost of pressure transucers varies according to the design and application requirements.

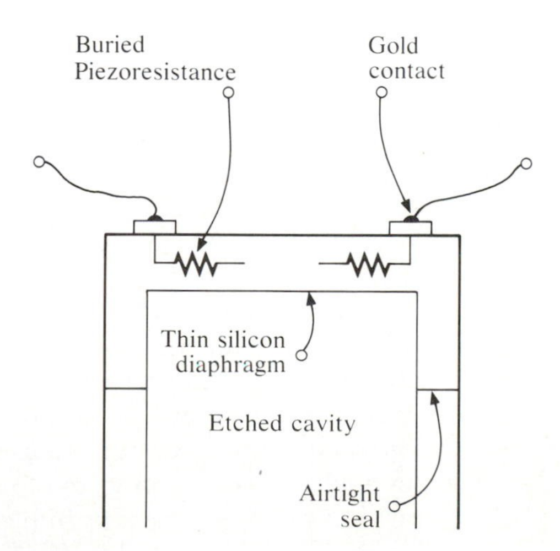

FIGURE 13–5
Solid-state pressure transducer

Accelerometers

Accelerometers provide accurate measurement of acceleration forces through the use of four strain gage elements. Acceleration is measured in g's (1 g = 15 psi). The sensitivity of accelerometers varies from 0.2 mV to 1000 mV per g. These devices are used in safety and automotive testing as well as in measuring elevator shock, aircraft takeoff and landing shock, and physiological acceleration effects.

Accelerometers are a special application of pressure sensors. Acceleration produces a force, defined by

$$F = ma \qquad (13\text{–}2)$$

which causes the displacement of a mass. The displacement can be measured by converting the mechanical displacement to a variable electrical parameter such as resistance, inductance, or capacitance, or simply to an input to a microprocessor that calculates the acceleration of the body in question.

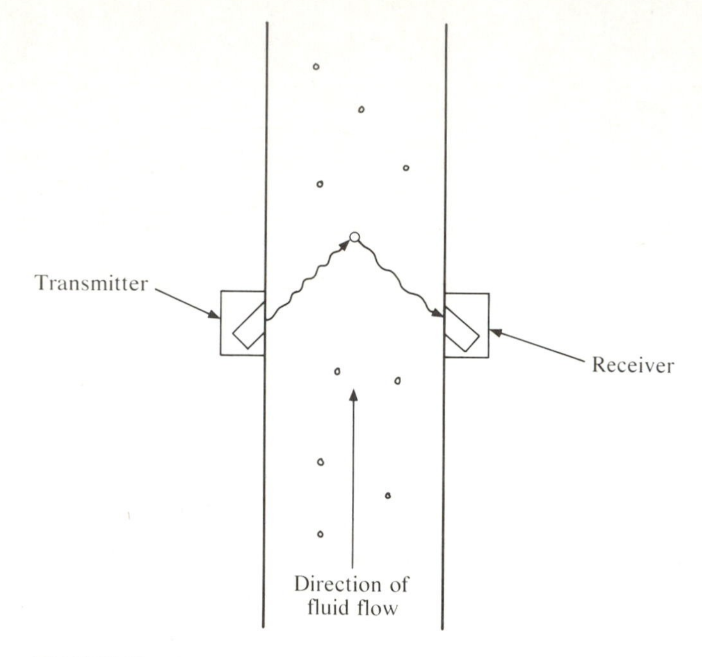

FIGURE 13–6
Ultrasonic flow transducer

Flow Meters

Flow meters monitor the motion of a fluid through a pipe. Their operation is similar to that of radar used to measure the speed of a moving object. As energy waves bounce off a moving object, the reflected frequency changes in relation to the speed of the object. Similarly, a flow meter projects an ultrasonic beam, produced by piezoelectric transducers, through the walls of the pipe at an angle to the direction of flow. The beam is deflected by particles (including bubbles) contained in the fluid (Figure 13–6). The frequency of the reflected sonic signals changes as a function of the velocity of the liquid. If the frequency of the energy waves is known, the resultant sum or difference of frequencies of the transmitted wave and the fluid can be calculated. The change in frequency due to relative motion is known as the *Doppler effect.* A receiver measures the reflected signal and sends the result to a microprocessor, where it is filtered, compensated for temperature, and analyzed to determine the flow rate. The particles in the moving fluid produce the frequency change. This change in the electrical property of the transmitted signal can be measured. These flow-sensing devices can be connected to LED or LCD outputs or to alarm systems that indicate a change in the flow by flashing a light. These devices must be used with processing units. They are passive transducers since they do not produce electric power.

THERMAL TRANSDUCERS

Thermal transducers, including thermocouples, resistance thermometers, thermistors, and infrared detectors, measure changes or differences in temperature.

Thermocouples

In 1826, Thomas Seebeck discovered that an emf can be produced in a circuit composed of two different metallic conductors whose junctions are maintained at different temperatures (Figure 13–7). This device is known as a *thermocouple.* With the reference junction at a fixed temperature (e.g., temperature at which water freezes = 0°C) and the test probe placed in an environment where the temperature is other than that of the reference junction, current will flow through the circuit. The current depends on the metals forming the thermocouple and on the temperature difference between the reference junction and the test probe. In Figure 13–7A, wire A and wire B represent the different metal conductors listed in Table 13–1. The leads of the galvanometer (G) are copper. A galvanometer

TABLE 13–1
Thermocouple Composition

Thermocouple Type	Composition
J	Iron/constantan
K	Chromel/alumel
T	Copper/constantan
E	Chromel/constantan
R	Platinum/platinum 13% rhodium
S	Platinum/platinum 10% rhodium
B	Platinum 6% rhodium/ platinum 30% rhodium

is used because the direction of current flow changes according to which junction is at the higher temperature. (Figure 13–7B shows an alternate wired diagram for the circuit in Figure 13–7A.) The service temperature of a type K thermocouple is from −330°F to 2300°F, and the response time is 50 ms (the response time is the time required to reach 63% of final voltage, for an instantaneous temperature change).

Thermocouples are active devices. They produce a voltage in the tens of millivolt range. They can be used over a much wider temperature range than conventional thermometers and to measure the temperature of corrosive materials or materials under dangerously high or low pressure.

Thermocouples as small as AWG 33 (0.008 in. in diameter) are available for use in hypodermic probes. They have uses in biophysics, in medical research, and in space vehicles where size is critical. Thermocouples can be designed for virtually any temperature range. They cost from a few dollars to hundreds of dollars depending on construction, materials, and packaging.

A. Test junction, Wire A, Wire B, G

B. Probe, Cu, Cu, G, Junction at reference temperature

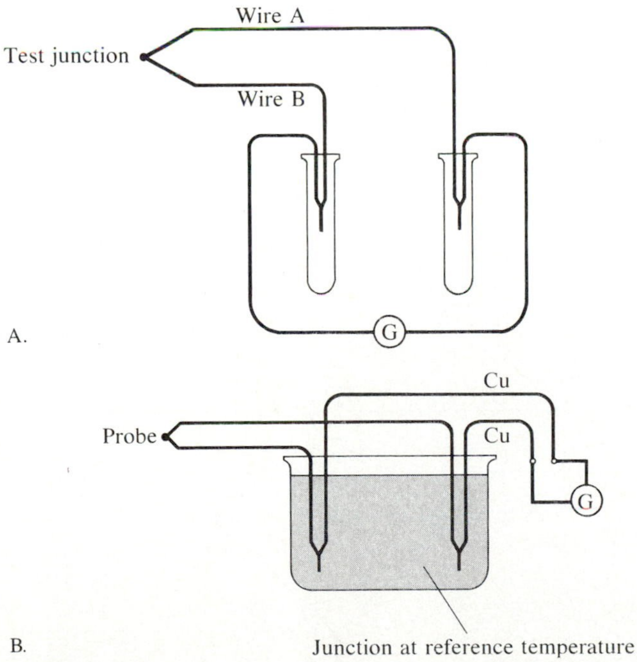

FIGURE 13–7
Thermocouples

Resistance Thermometers

The *resistance thermometer,* also known as a *resistance temperature detector* (RTD), operates on the principle that a change in temperature results in a change in the resistance of a conductor. In general, the temperature coefficient of conducting metals is a positive number, which means that as temperature increases, resistance increases. Carbon, for example, has a negative temperature coefficient, which means that its resistance decreases as the temperature increases.

The resistance thermometer shown in Figure 13–8 contains a platinum conductor wound around

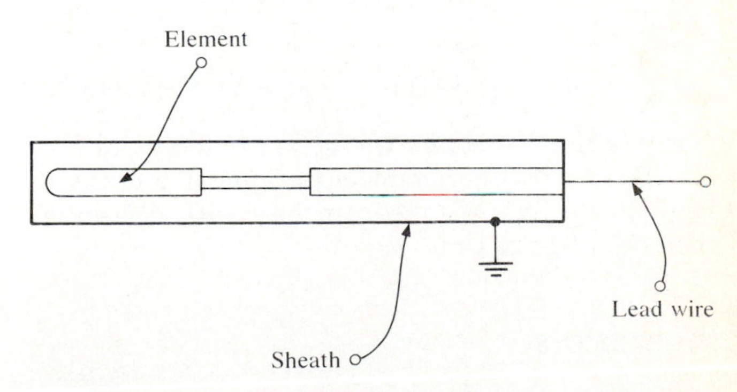

FIGURE 13–8
Resistance temperature detector

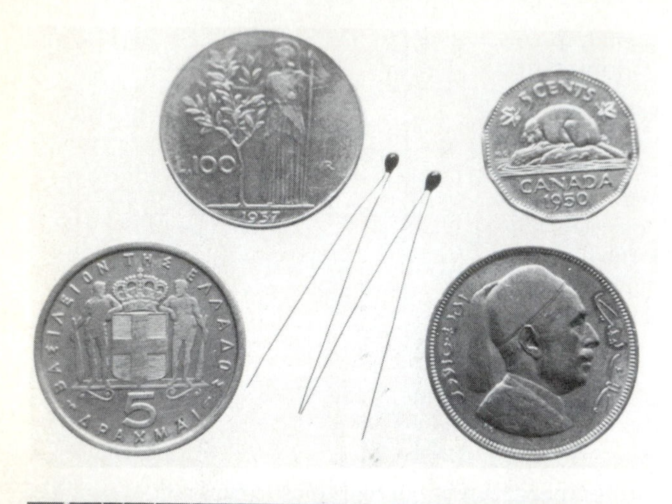

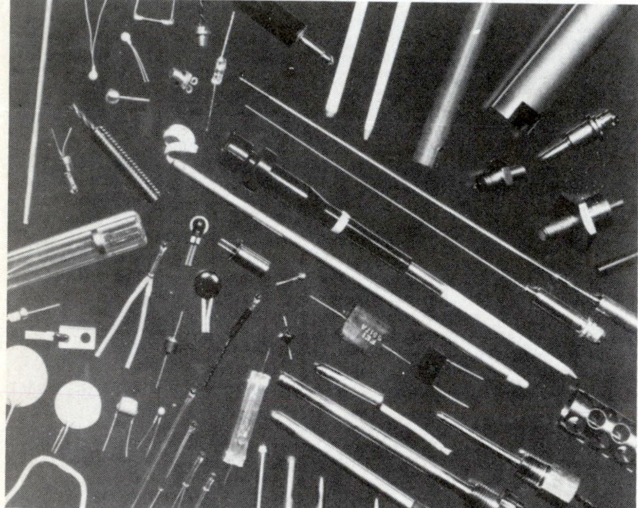

FIGURE 13–9
Thermistor types

a ceramic core. This sealed unit is the element. The resistance of the platinum is usually controlled to be 100 Ω at 0°C. It changes at 0.00392 Ω per ohm per degree Celsius. For example, the resistance range of a platinum detector (100 Ω) at 200°C would be

$$\Delta R = 0.00392\ \Omega \times 100 \times 200 = 78.6\ \Omega \qquad \textbf{(13–3)}$$

The resulting resistance would be 178.6 Ω at 200°C.

Resistance thermometers are passive devices. The change in resistance can be easily measured using a bridge circuit.

Thermistors

An assortment of thermistor types appears in Figure 13–9. *Thermistors* are passive devices whose conductivity (reciprocal of resistance) depends on

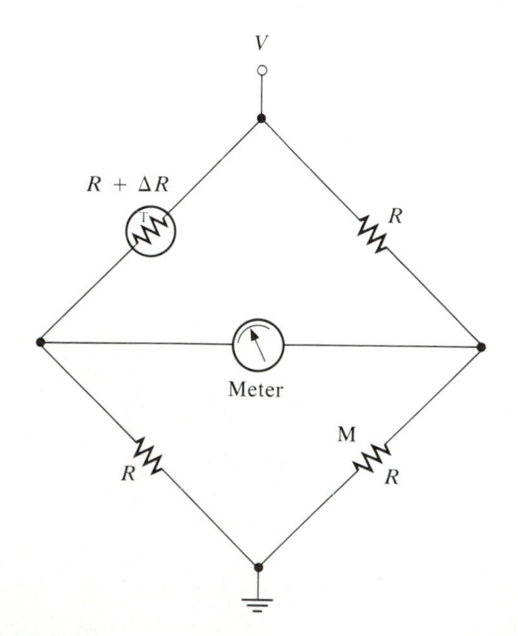

FIGURE 13–10
Thermal transducer in a bridge circuit

temperature. Manufactured in many resistance ranges, these devices can be used to compensate for the effect of temperature changes in electric circuits. Thermistors can be used for temperature measurement by placing them in a Wheatstone bridge. As the temperature of the thermistor changes, the change in resistance (conductivity) causes a deflection of the meter, which allows for easy measurement (Figure 13–10).

Infrared Detectors

Infrared detectors use the infrared emission from heated objects to determine their temperatures. Because infrared radiation is a type of electromagnetic radiation, it behaves like visible light. It can be detected at a distance from its source, making remote measurement possible. Infrared detectors use filters to limit the radiation input to the 8- to 14-μm range, eliminating other "background" light which would interfere with the input. An infrared-sensitive device is used in a microprocessor-controlled circuit that zeros the instrument, converts degrees Celsius to degrees Fahrenheit, stores readings, finds the maximum and minimum temperatures, and calculates the average and difference of the temperatures.

OPTICAL TRANSDUCERS

Two types of *optical transducers* are photoconductive (passive) cells and photovoltaic (active) cells. These cells convert changes in light intensity to changes in electrical circuit behavior.

Photoconductive Cells

Photodiodes

The conductivity of a *photodiode* depends on the free current carriers available at the P-N semiconductor junction. Figure 13–11 illustrates a photodiode with a window situated at the P-N junction. As photons of light energy strike atoms near the junction, the collision of the photons with inner-shell (low-energy) electrons elevates these electrons to a level of free electrons. The electrons then become part of a current flowing in the external circuit connected to the P-N junction. This increase in current carriers causes the conductivity to increase according to the number of incident-light photons. The circuit in Figure 13–12 shows how a photodiode can be used to measure light intensity. The

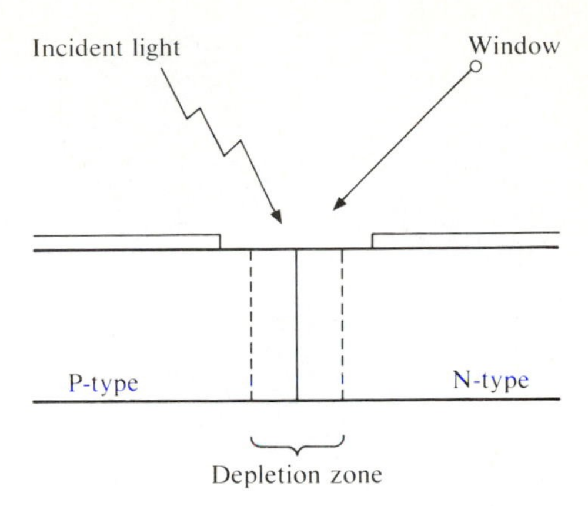

FIGURE 13–11
Solid-state light transducer (photodiode junction)

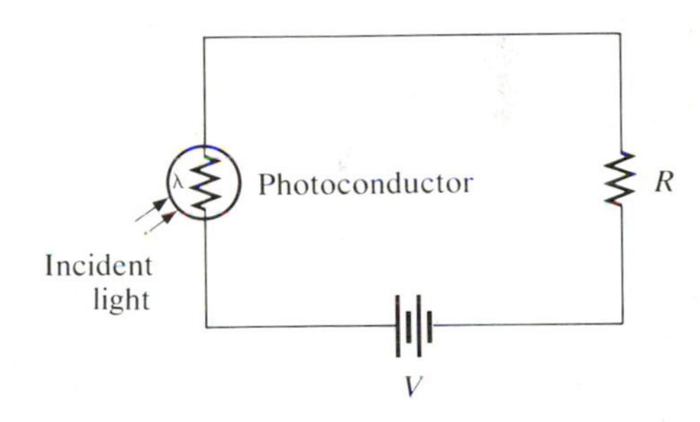

FIGURE 13–12
Phototransducer (denotes a wavelength of light)

current through resistance R varies as the intensity of the incident light changes the conductivity of the photoconductor (Figure 13–13). By measuring the voltage across R, the intensity of the incident light can be determined.

Phototransistors

The window of a *phototransistor* is situated at the base-emitter P-N junction. The incident light, which affects the conductivity, is usually enhanced

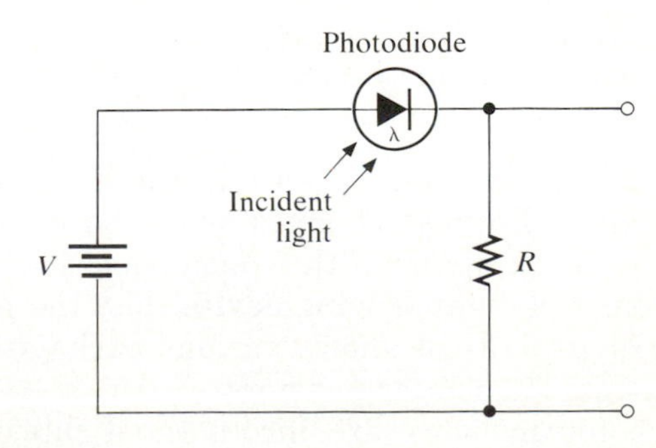

FIGURE 13–13
Photodiode circuit

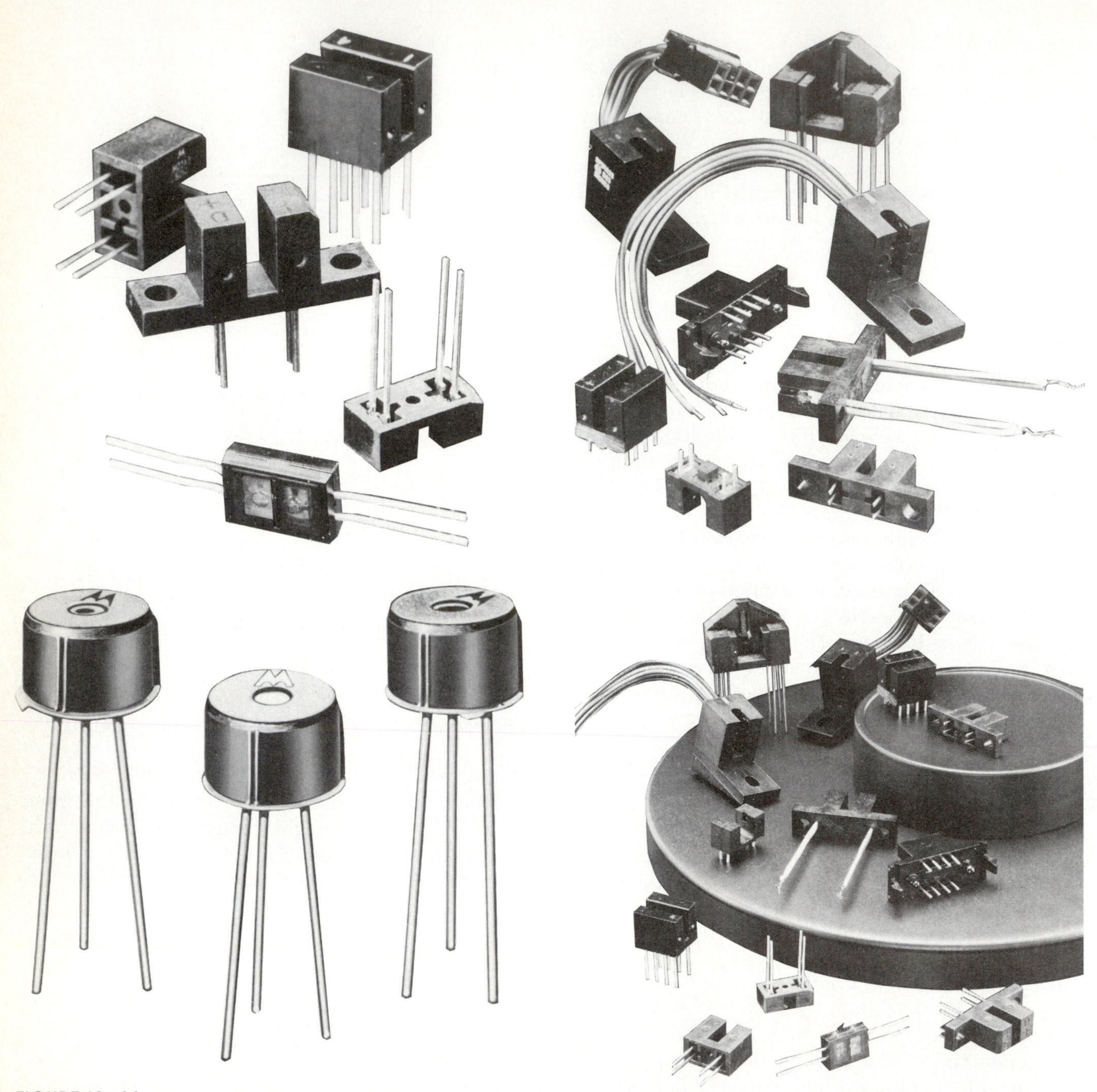

FIGURE 13–14
Types of phototransistor devices

by optical lenses that concentrate the light on the desired area. Because changes at the base-emitter junction are amplified, the phototransistor is a more sensitive light-sensing device than the photodiode. Figure 13–14 shows various packaging arrangements for phototransistors.

Phototransistors are incorporated into many analog and digital light meters and alarm systems. They are used extensively as receiving elements in fiber optic systems. Figure 13–15 shows the symbols for phototransistor elements.

Photovoltaic Cells

The *photovoltaic cell* is a semiconductor that can be energized by light (or heat) to produce voltage. Photons of light incident upon silicon crystals can knock an electron out of its orbit and create not

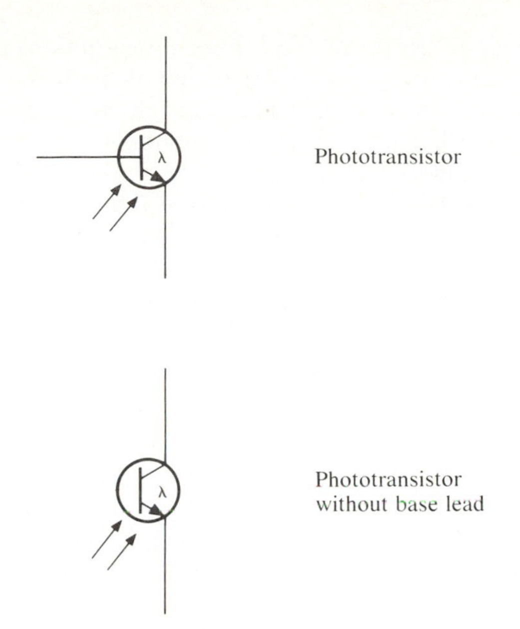

FIGURE 13–15
Phototransistor symbols

only the free electron but also an empty bond, or *hole*. If the electron is allowed to recombine with the hole, the photon's energy is converted to heat energy. To avoid this, an energy barrier is constructed using N- and P-type materials. This P-N junction is only about 0.00005 in. thick. The resulting diode junction becomes a light-activated battery. Free electons from the P-type material flow across the energy barrier, and electrons build up in the N-type material. If a load is connected, the excited electrons in the N-type material flow freely through the external circuit back to the P-type material. The voltage produced depends on the type of semiconductor material used and is usually less than 1 V. Higher voltages can be achieved by connecting several cells in electric series.

Several types of losses limit the theoretical efficiency of the silicon "solar cell" to 22%. They include infrared and ultraviolet light, which are contained in sunlight but do not help voltage generation. Infrared has too much energy, and much of the excess energy is wasted as heat within the crystal. Ultraviolet light does not have enough energy to separate the electrons from the holes in the crystal. Figure 13–16 shows the construction of a typical commercial silicon cell.

The antireflection coating is designed to pass maximum light through to the P-N junction without reflection. The contact bar and fingers are kept to minimal dimensions to allow for the maximum passage of light. Silicon photovoltaic cells are usu-

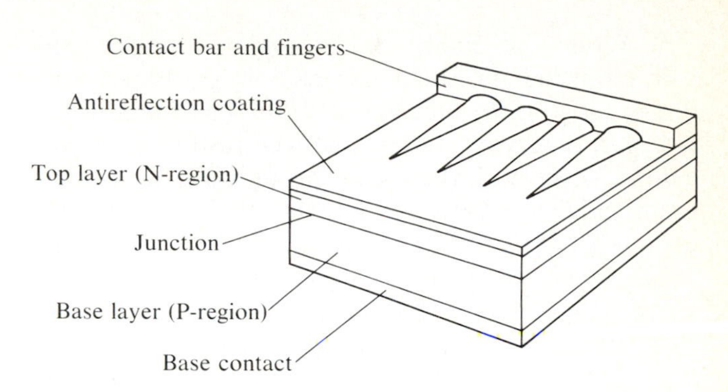

FIGURE 13–16
Schematic of a typical commercial silicon solar cell

ally 0.004 in. thick. This thickness was chosen for maximum cost efficiency.

In light-measuring systems, the signal from the photovoltaic cell is processed and used to drive measuring instruments.

ACOUSTICAL TRANSDUCERS

Acoustical transducers convert sound energy (a type of mechanical energy) to electrical energy. The conversion is performed by a device commonly called a *microphone*.

Dynamic Microphones

Dynamic microphones produce a current proportional to the volume (loudness) of the sound, which alternates at a frequency determined by the pitch (tone) of the sound. An example of a dynamic microphone (active device) is shown in Figure 13–17.

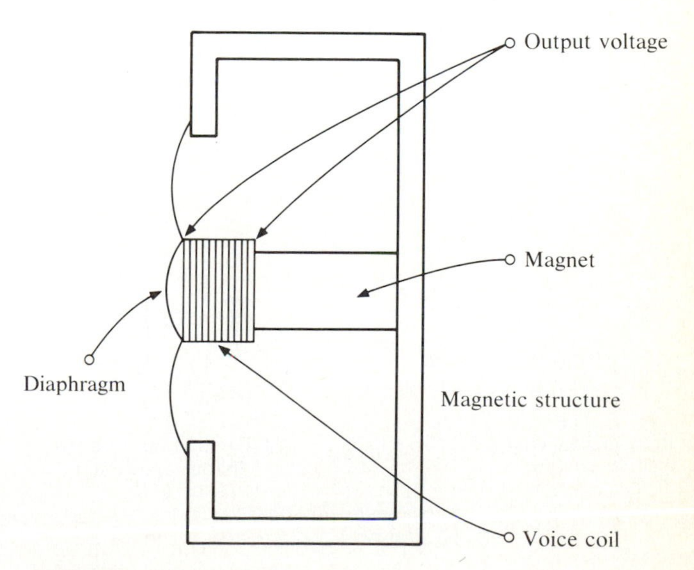

FIGURE 13–17
Dynamic microphone

When the diaphragm is made to vibrate, the coil mounted on the cylinder vibrates back and forth, causing the coil to cut the lines of flux of the magnet. A current (ac) is produced in the coil which can be detected (processed) through the leads. The amplitude of vibration is determined by the amount of diaphragm displacement, which depends on the loudness of the incident sound waves. The rate at which the coil cuts the flux lines is determined by the frequency of the incident sound waves.

Dynamic microphones are renowned for their ruggedness and reliability and are capable of smooth, extended response. The output level is high enough to work directly into most microphone inputs with an excellent signal-to-noise ratio. These microphones need little or no regular maintenance, and with reasonable care will perform well for many years.

Condenser (Capacitor) Microphones

Condenser or *capacitor* microphones use a lightweight membrane as one side of a capacitor and a perforated plate as the other side. Sound waves cause the polymer film to move, thus changing the capacitance of the circuit. The charge in capacitance in turn creates a change in the electric circuit.

Two basic types of capacitor microphones are currently available. One uses an external power supply to provide the polarizing voltage needed for the capacitive circuit. This type is used almost exclusively in professional recording studios. A more recent development is the *electret* condenser microphone. In this type, the polarizing voltage is impressed on a coating bonded to the perforated backplate during manufacture. This charge will remain, unaltered, indefinitely. No external power is needed, although an FET impedance-matching circuit is typically required, which is powered by a small low-voltage battery in the microphone itself (Figure 13–18). Electret microphones are suitable for high-fidelity and semiprofessional recording as well as major studio applicatons. The low mass of the diaphragm permits extended high-frequency performance, while the nature of the design ensures outstanding low-frequency respones.

Ribbon Microphones

Ribbon microphones are capable of excellent bidirectional response. A metallic ribbon is suspended between the poles of a magnetic circuit. This type of microphone must be protected against high acoustical pressure or wind. For this reason, ribbon microphones are rarely seen in nonstudio recording applications.

Carbon Button Microphones

The simplest microphone type is the *carbon button* microphone, shown in Figure 13–19. The conductivity of the carbon granules depends on the density of the carbon. As the diaphragm vibrates, the granules are constantly compressed and released, which changes their density. Changes in density cause the conductivity of the carbon to change as a function

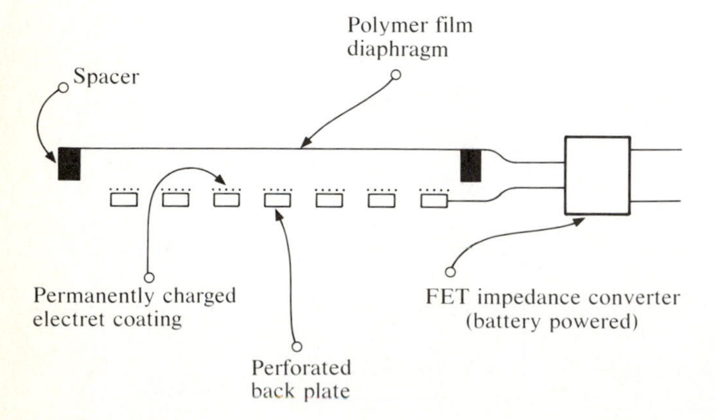

FIGURE 13–18
Electret condenser microphone

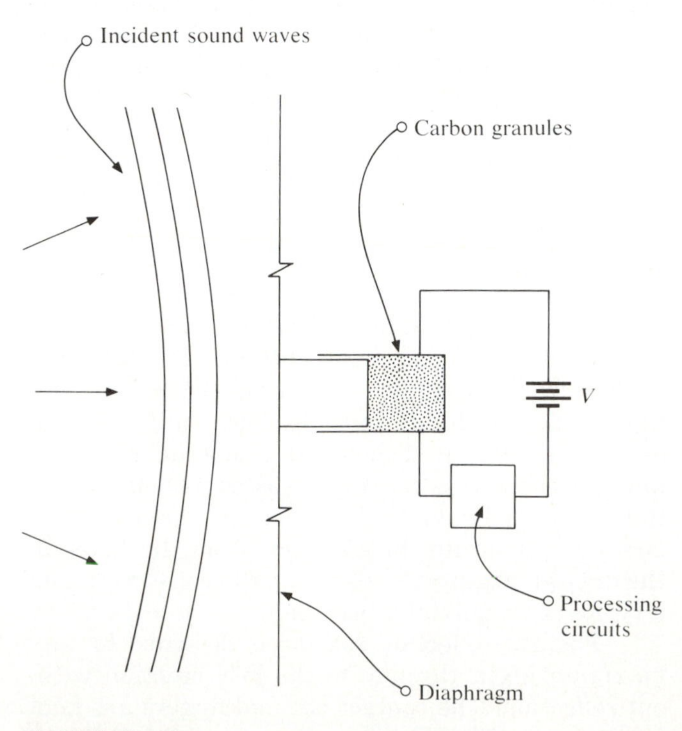

FIGURE 13–19
Carbon button microphone

of the incident sound waves. Because this is a passive transducer, an external power source is required.

Carbon button microphones are low-cost, rugged devices. They are popular in telephones and in mobile installations such as citizens's band (CB) radios. The carbon granules tend to "pack" after extended use and the bandwidth (fidelity) is limited.

The carbon button microphone is an example of a change in resistance, produced by a mechanical vibration, causing an electrical change, which can be measured or processed and used.

Microphone Ratings

Microphones are rated according to their sensitivity. Most manufacturers use a reference level well above the output level of the microphone. Thus, a microphone with a sensitivity rating of -55 dB will provide more signal output than one rated at -60 dB. A typical sensitivity reference level is 0 dB = 1 mW/10 dynes/cm^2. This means that one milliwatt of electric power will be produced if a force of 10 dynes per square centimeter is applied to the microphone.

ULTRASONIC RANGING SYSTEMS

Polaroid has developed the Polaroid Ultrasonic Ranging System, which is used to perform precise focusing of a camera lens. A transducer, mounted on the camera and acting as a loudspeaker, transmits a short ultrasonic pulse. The echo is received by the same transducer, now acting as a microphone. The time between the transmission and reception of the ultrasonic signal is proportional to the distance traveled. This information is electronically processed. With this data as a control, a motor rotates the front element of the lens, resulting in proper focus of the lens for the given distance. The sound-producing and -sensing transducer is illustrated in Figure 13–20.

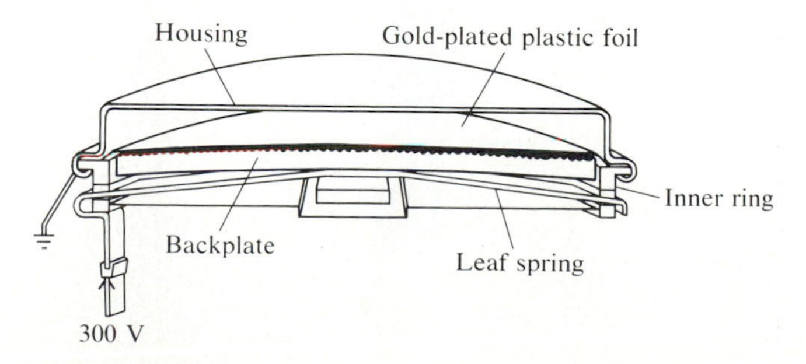

FIGURE 13–20
Range finder

MAGNETIC TRANSDUCERS

Magnetic transducers include the differential-pressure transducer, the linear-variable differential transformer, the Hall-effect semiconductor, and the Selsyn generator.

Differential-Pressure Transducer

The *differential-pressure transducer* illustrated in Figure 13–21 is an example of a variable-reluctance transducer. It uses a diaphragm of magnetically permeable stainless steel clamped between two sets of electromagnets that is deflected when a pressure difference exists between the two ports. A core and coil assembly is embedded in each block with a small gap between the diaphragm and the core. This arrangement is symmetric, resulting in a condition of equal inductance when the diaphragm is not deflected. When a difference in pressure occurs between the ports, the gap in the magnetic flux path of one core increases, and the flux path of the other core decreases. The effect is a differential change in the inductance of the two coils, which can be measured. Coil L_1 is excited by 115 V ac or 28 V dc. The output is taken from coil L_2, amplified, and chopped, resulting in a dc voltage proportional to the pressure being measured.

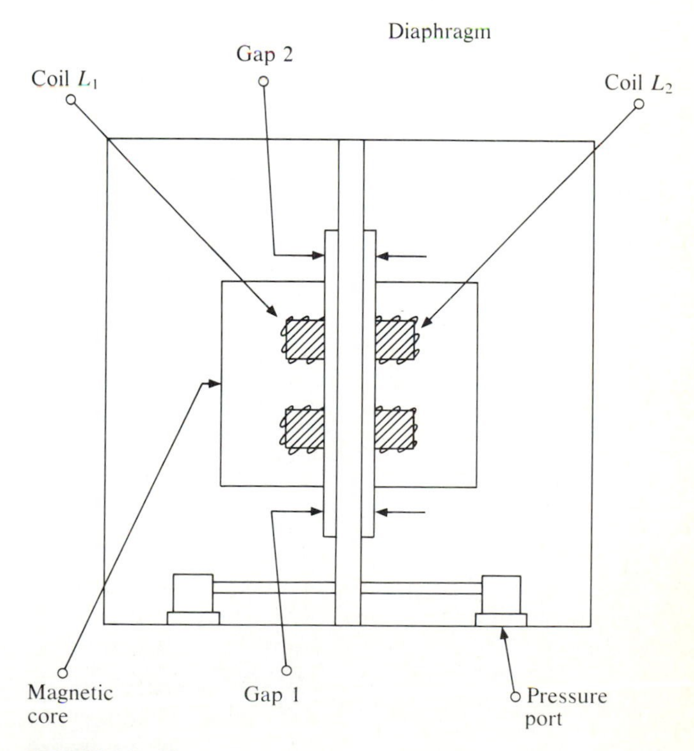

FIGURE 13–21
Differential-pressure transducer

Linear-Variable Differential Transformer

The *linear-variable differential transformer* (LVDT) winding arrangement allows the core to be moved in relation to the two transformer windings (Figure 13–22). This movement changes the coupling between the primary and secondary windings of the transformer. Because of the direction of the various windings, a differential linear voltage output results. Detection of the output signal yields a sensitive measurement of the magnitude and direction of the displacement of the movable core. The LVDT can be used to measure very small movements.

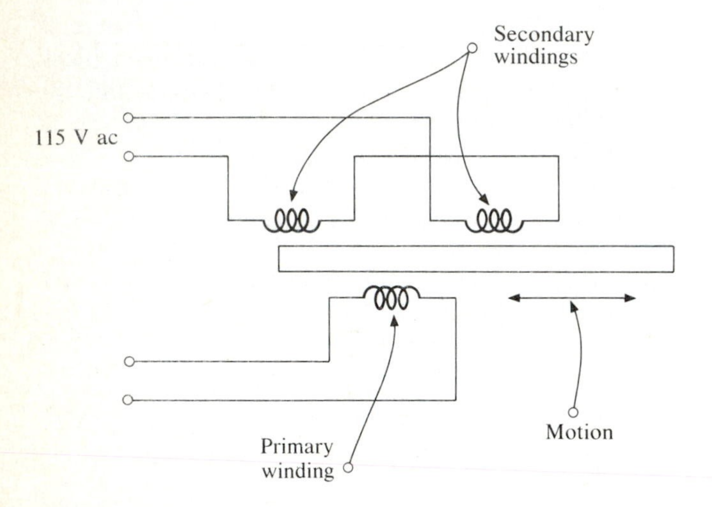

FIGURE 13–22
Linear variable differential transformer

Hall-Effect Transducer

When current passes through a conductor, a magnetic field is produced around that conductor. If this conductor is then brought near another magnetic field, additional voltage is produced. This voltage is proportional to the flux density (B) of the second magnetic field. Because this increase in voltage is measurable, it allows flux density to be measured (see Figure 13–23).

This effect, known as the Hall effect, was discovered by Dr. Edwin H. Hall in 1879. Today, *Hall-effect transducers* are used to measure magnetic field densities and currents. Figure 13–24 shows a Hall-effect transducer and Figure 13–25 shows a hand-held device that measures dc current, ac current, and ac-on-dc current. This device is clamped onto the conductor under test; thus it does not physically affect either the insulation or the conducting characteristics of that conductor. The magnetic field around the conductor is concentrated in a magnetic core within the jaws of the probe. The Hall-effect transducer (also known as a Hall-effect generator) is located in an air gap in the core and produces a voltage directly proportional to the current in the conductor. This voltage (V_H) can then be measured using an oscilloscope or a voltmeter, or it may be recorded for later analysis. The Hall-effect generator is a passive transducer.

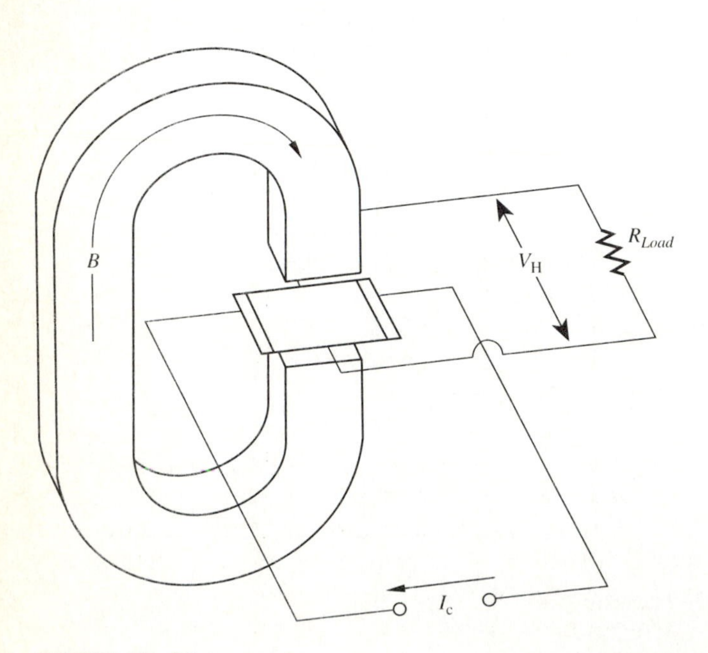

FIGURE 13–23
Measurement using the Hall-effect transducer

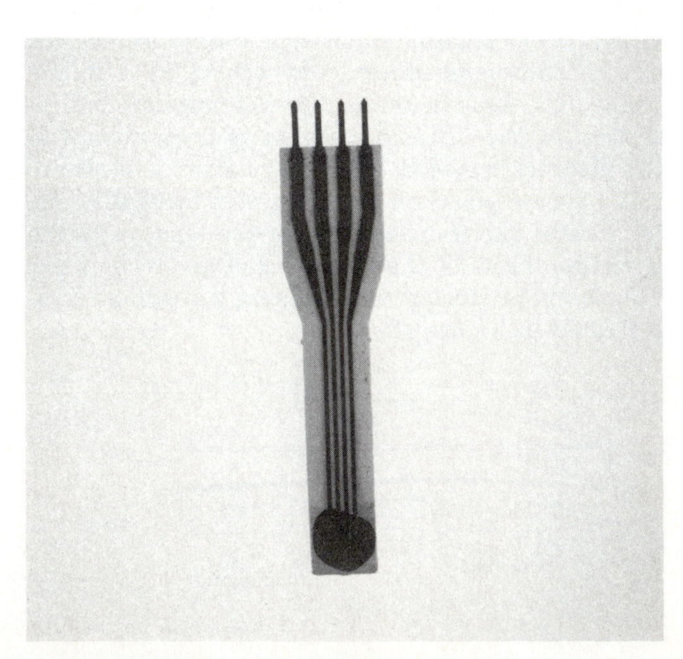

FIGURE 13–24
Hall-effect transducer

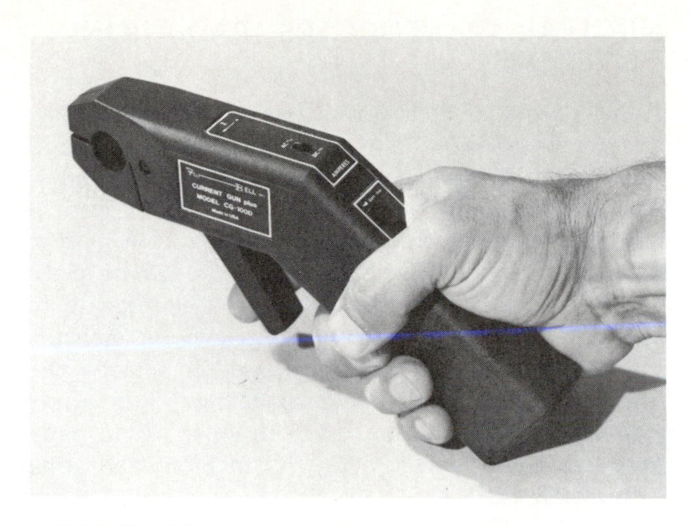

FIGURE 13–25
Clamp-on ammeter

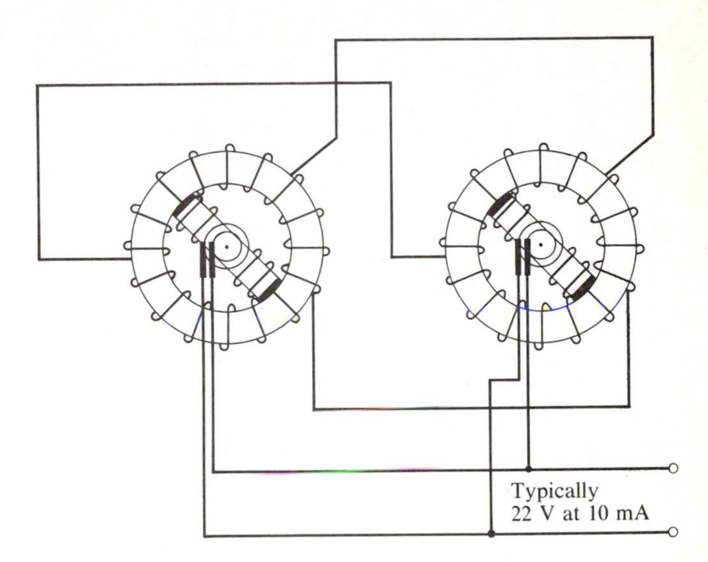

FIGURE 13–26
Selsyn position "motor"

Selsyn Generator

Another electromagnetic system is the *Selsyn position "motor."* This device converts electrical energy to position indications, and position indications to electrical energy. The orientation of a device can be measured at a distance using this system (Figure 13–26).

The transmitting and receiving units are identical. Each involves a rotor and a stator; the rotor is provided with a single-phase winding, the stator with a polyphase winding. The single-phase windings are excited from a common ac source, and the polyphase windings are interconnected. The rotors of the two units tend to assume duplicate angular positions so that when one is attached to a measuring element, the other will provide a remote indication of the position of the first. The system requires three line conductors in addition to the pair forming the power supply. The Selsyn generator is particularly adaptable to telemetering mechanical displacements. Position controls and indicators of television antenna direction use a similar arrangement.

CHEMICAL TRANSDUCERS

The most common *transducers* in the chemical industry that are electronic in nature measure the degree of ionization of a fluid. When a potential is placed across two electrodes immersed in an electrolyte, the amount of current flowing between the electrodes is a function of the number of ions or current carriers in the electrolyte. The conductivity of the electrolyte can be used as an indication of the amount of chemical that has reacted or the amount of product that has been produced. In a more basic system, two different materials produce current carriers (active transducer). This transducer is called a *battery*. Much material is available describing the types of batteries and their voltages and structures.

BIOLOGICAL TRANSDUCERS

There are more than 30 *biological* measuring applications. The transducers involved include:

1. Surface electrodes
2. Needle (hypodermic) electrodes
3. Light source and photocell combinations (piezoelectric)
4. Strain gage pressure transducers
5. Electromagnetic flow meters
6. Doppler flow meters
7. Accelerometers
8. Thermistors
9. Microphones

If a gauze pad is soaked in saline solution and attached to a piece of wire, it can be used as a surface electrode. Because of the unreliability of this method, disposable disk electrodes have been developed. The electrode detects electrical impulses associated with muscle movement such as heart beat. To measure signals from discrete portions of tissue, needle electrodes made of zinc, stainless steel, tungsten, or platinum wire are used. Recently, teflon-coated (except for the tip) electrodes have been introduced. As with surface electrodes, the pupose is to detect electrical signals given off by a part of the body.

Electroencephalograph (EEG) potentials can be as low as 1 μV. Electromyocardiograph (EMG) signals vary from 20 μV to 4 mV. The frequency response of bioelectric voltages ranges from 0.01 Hz to 10 kHz. Different kinds of processing circuits are required for recording various types of physiological activities.

A piezoelectric pressure transducer will produce an electrical signal whenever it is vibrated by the force of a peripheral pulse pressure. Like a crystal phonograph pickup, the piezoelectric transducer has an extremely high impedance. Its output must be fed into a system with an input impedance of 1 MΩ or greater.

In cardiac catheterization and chronic blood-pressure measurements, a subminiature (1-mm-diameter) catheter containing a solid-state strain gage and bridge circuit is used. The transducer is encased in pure titanium.

Flow meter probes are basically electromagnets that produce a magnetic field and two electrodes (180° apart) that pick up the flow of body fluids (Figure 13–27). The polarity of the induced voltage is determined by the direction of fluid flow, and the magnitude by the fluid's velocity. Doppler-effect flow meters depend on the change in frequency of an ultrasonic signal when it encounters particles (or gas bubbles) in a flowing fluid.

Accelerometers, thermistors, and microphones were discussed earlier in the chapter. Their applications are as indicated in those discussions.

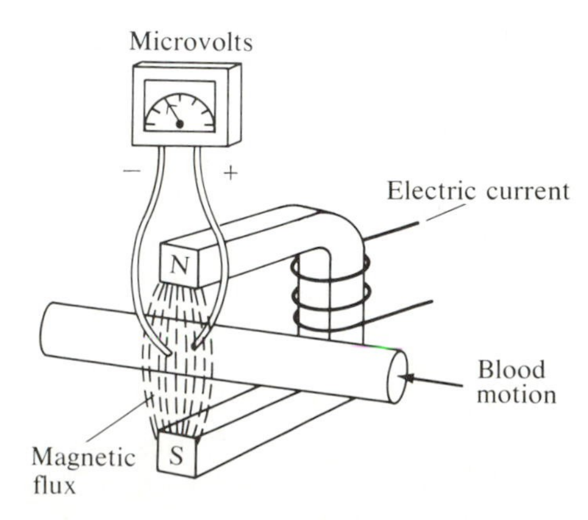

FIGURE 13–27
Magnetic flow meter

NUCLEAR TRANSDUCERS (RADIATION DETECTORS)

Henri Becquerel discovered that gases become electrical conductors as a result of exposure to radiation. When a high-speed particle passes through a gas, it causes the removal of an electron from a neutral atom, which in turn causes an ion pair to form. If two oppositely charged electrodes are present, the members of the ion pair will migrate to their respective electrodes. In one type of measuring instrument, batteries are used to charge the electrodes. The result is a current flow in an external circuit proportional to the number of ion pairs produced and, therefore, to the incident radiation (Figure 13–28).

In the CD Model V-700, when radiation penetrates the tube through the window (which is covered with paper to shield out alpha particles), a gas molecule is ionized. The resulting ion pair is accelerated toward the electrodes by the electric field. Because of the accelerating voltage, creation of additional ions is rapid, producing an avalanche, which results in a discharge in the tube. The discharge causes a pulse in the external circuit. The

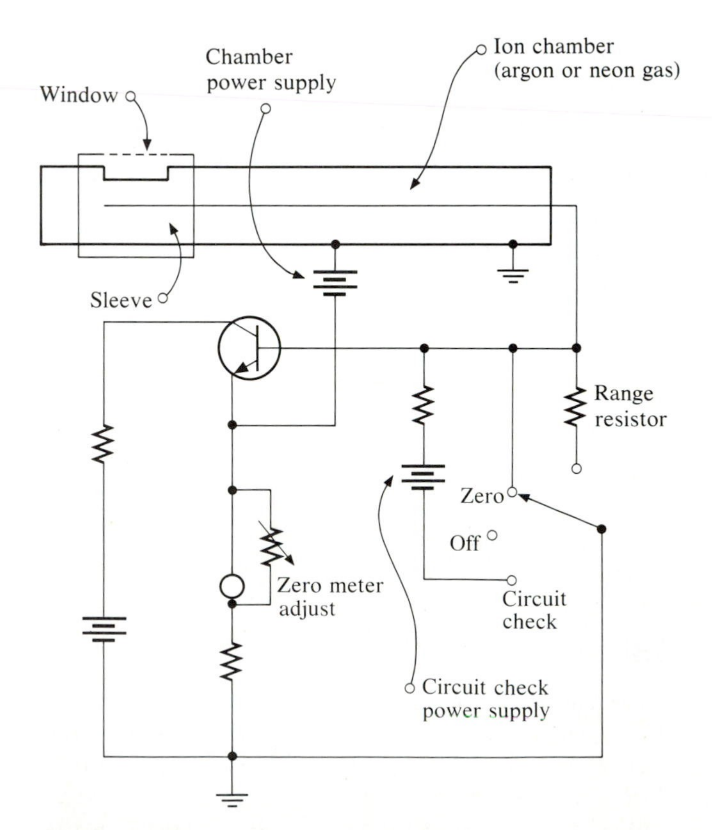

FIGURE 13–28
CD Model V-700 nuclear transducer

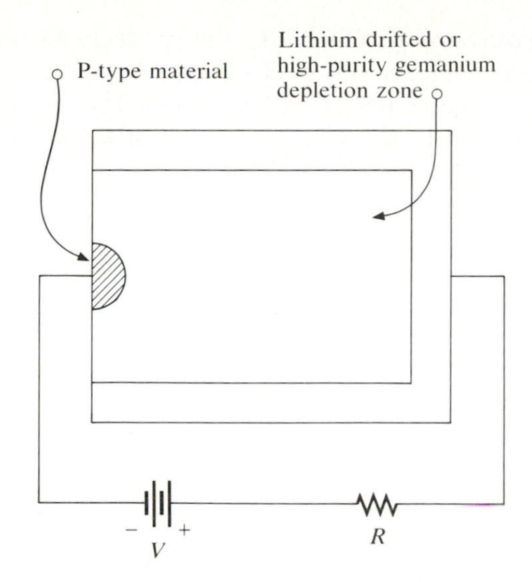

FIGURE 13–29
Germanium gamma ray detector

frequency of pulses is proportional to the strength of the radiation field. A small amount of halogen gas in the tube helps to stop each discharge and restore the tube to its initial condition. The pulse output from this *Geiger tube* is amplified and measured by a meter that sums the radiation effects in the form of a reading in either counts per minute or, in the case of gamma radiation, millirems per hour. The pulses from the Geiger tube can be made to produce audible clicks. A movable metal shield or sleeve can be positioned across the window to eliminate beta radiation from the tube and therefore allows only gamma radiation to be indicated.

The Model V-700 can measure up to 50 millirems/per hour. It detects beta and gamma radiation and is not affected by direct sunlight or electromagnetic interference.

Field service is usually limited to replacing batteries, zeroing the meter, and eliminating the effects of radiation contamination. The latter can be accomplished by cleaning the instrument with a cloth dampened in a mild soap solution.

Germanium Gamma Ray Detector

Figure 13–29 illustrates a *germanium gamma ray detector*. Gamma radiation interacts in the depletion region, where it forms electron-hole (ion) pairs. These pairs are collected by a relatively high voltage. The collected charge pulse of electrons and holes passing through the resistor produces a voltage pulse (Ohm's law). These detectors are used at liquid nitrogen temperatures. Because of their high cost and difficulty of operation, other instruments are usually used.

Photomultiplier Tube

A *photomultiplier tube* is sometimes used to amplify the results of an incoming ray (Figure 13–30). The inside surface of the window is coated with a material that releases electrons when struck by radiation. The electrons are attracted to the highly positively charged dynodes. Upon striking a dynode, each electron "knocks off" an average of three electrons (secondary emission). Each photomultiplier tube contains from 10 to 14 dynodes, which produce a gain of about 1,000,000. The anode collects the electrons and produces a pulse with an amplitude proportional to the intensity of the incident radiation. A multichannel pulse-height analyzer stores the numbers of pulses at each amplitude. A gamma ray energy spectrum can be displayed as the result.

In each of these transducers, radiation produces a change in an electric circuit. This change yields measurable data about the incident radiation. The instrument *transduces* radiant energy to electrical behavior, which can be measured.

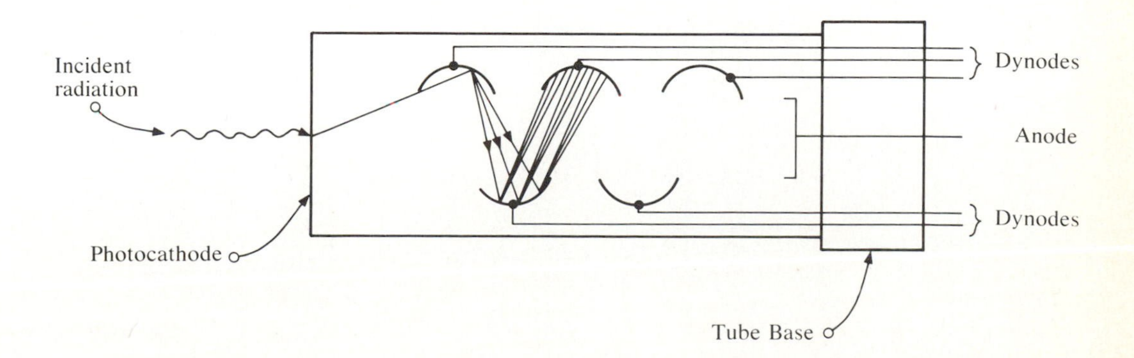

FIGURE 13–30
Photomultiplier tube

SUMMARY

The transducers discussed in this chapter are representative of the types of devices available. Using these or related devices, the technician should be able to devise a measurement system for almost any requirement. Once the energy transformation has been made, common electrical or electronic processing circuits are used to prepare the signal for presentation to the output portion of the measurement process.

EXERCISES

1. What is a transducer?
2. Explain the difference between active and passive transducers.
3. Explain how the resistance of a strain gage changes.
4. Why is a dummy gage commonly used in bridge measuring circuits?
5. Describe a use for each of the following:
   a. Pressure transducer
   b. Accelerometer
   c. Flow meter
   d. Thermocouple
   e. Thermistor
6. Calculate the new resistance of a 100-Ω (0°C) platinum resistance thermometer measuring a temperature of 300°C.
7. Describe a use for each of the following:
   a. Infrared detector
   b. Optical transducer
   c. Differential-pressure transducer
   d. Linear-variable differential transformer
   e. Hall-effect transducer
8. Explain how the following microphones convert acoustical (mechanical) energy to electrical energy:
   a. Carbon button
   b. Condenser
   c. Electret
   d. Ribbon
9. Chemical and biological transducers are usually combinations of transducers. Explain.
10. How do nuclear transducers convert nuclear radiation energy to electrical energy?

14 Fiber Optics

OUTLINE

OBJECTIVES
WHY LASERS?
LIGHT WAVE PARAMETERS
OPTICAL FIBER LINKS
A BASIC COMMUNICATIONS SYSTEM
TRANSMISSION LINE CHARACTERISTICS
ADVANTAGES OF OPTICAL FIBERS
DISADVANTAGES OF OPTICAL FIBERS
SPACE OPTICAL SYSTEMS
LIGHT SOURCES
LIGHT DETECTORS
USE OF OPTICAL FIBERS
AMPLITUDE MODULATION VERSUS PULSE AMPLITUDE MODULATION
SUMMARY
EXERCISES
LAB PROBLEM 14–1—Free-Space Transmission

OBJECTIVES

After completing this chapter, you will be able to:

- ☐ Relate some of the history of the development of optical fibers as a communications medium.
- ☐ Discuss the parameters of optical fiber transmission (frequencies and wavelengths).
- ☐ Differentiate among simplex, half-duplex, and full-duplex communications modes.
- ☐ Compare the characteristics of optical fiber, copper wire pairs, and coaxial cable.
- ☐ List the advantages of optical fibers over other transmission media.
- ☐ List the disadvantages of optical fibers as a transmission medium.
- ☐ Discuss the advantages of optical fiber over free-space optical communications systems.
- ☐ Discuss the characteristics of light sources used in fiber optic systems.
- ☐ Discuss the characteristics of light detectors.
- ☐ Describe the characteristics of optical fibers as a transmission medium.
- ☐ Describe some problems associated with splicing optical fibers
- ☐ Discuss the advantages of optical fibers in pulse amplitude modulation and pulse code modulation.

Although optical fibers have been around for a long time, transmission of information through them is a relatively new technology. Reflection of sunlight from a shiny surface was the first type of light beam communication. The first transmission of speech using a light beam was made February 19, 1880, by Alexander Graham Bell. Because of serious limitations of the technology, this means of communication was ignored until the 1960s when the development of the laser (*l*ight *a*mplification by *s*timulated *e*mission of *r*adiation) focused attention on the possibility of introducing a light beam into a "glass pipe," or optical fiber. The electromagnetic radiation output (light) from a laser is monochromatic, and the rays are parallel and in phase. In

1962, the gallium arsenide light-emitting diode (LED) was developed, which produces a light near the infrared end of the visible spectrum. The semiconductor laser was also developed in 1962.

Optical fibers became of interest as a closed-system conductor of light from a transmitter to a receiver. Low-loss silica fibers were developed in 1966; however, these still had losses rated in thousands of decibels per kilometer. Corning developed optical fibers in 1970 characterized by a loss rating of 20 dB/km. In 1979, the Japanese developed supperpure silica fibers with a 0.2-dB/km loss. Advances in outer-space manufacturing technology have led to the possibility of ultrapure silica, which could be used to produce an optical fiber transmission line with nearly zero loss. Since the fibers are manufactured in zero gravity, air bubbles and impurities which cause losses from internal reflection of light are eliminated.

WHY LASERS?

For light beams to enter and travel down an optical fiber, the incident light should be *coherent*. Coherent light has parallel light rays that are in phase and monochromatic. Light not displaying these characteristics is termed *noncoherent*.

Light is electromagnetic radiation and is similar to electricity in many respects. If two light waves reach peak values together, they are in phase and the effect is additive. This relationship is called *constructive interference*. If two light waves are 180° out of phase, they cancel each other. This effect is termed *destructive interference*. Parallel light rays follow the axis of direction of the optical fiber without reflecting from the fiber's interior walls, reducing destructive interference and loss.

Figure 14–1A shows coherent light. Figure 14–1B shows cancellation due to a phase shift resulting from internal reflection of nonparallel light rays. Figure 14–1C shows cancellation due to the phase difference between two out-of-phase light rays.

All of the light rays from a monochromatic light source have the same wavelength. If the light rays are not of the same wavelength, losses will result because of destructive interference. Figure 14–1D illustrates this cancellation.

Light emitted from a laser is characteristi-

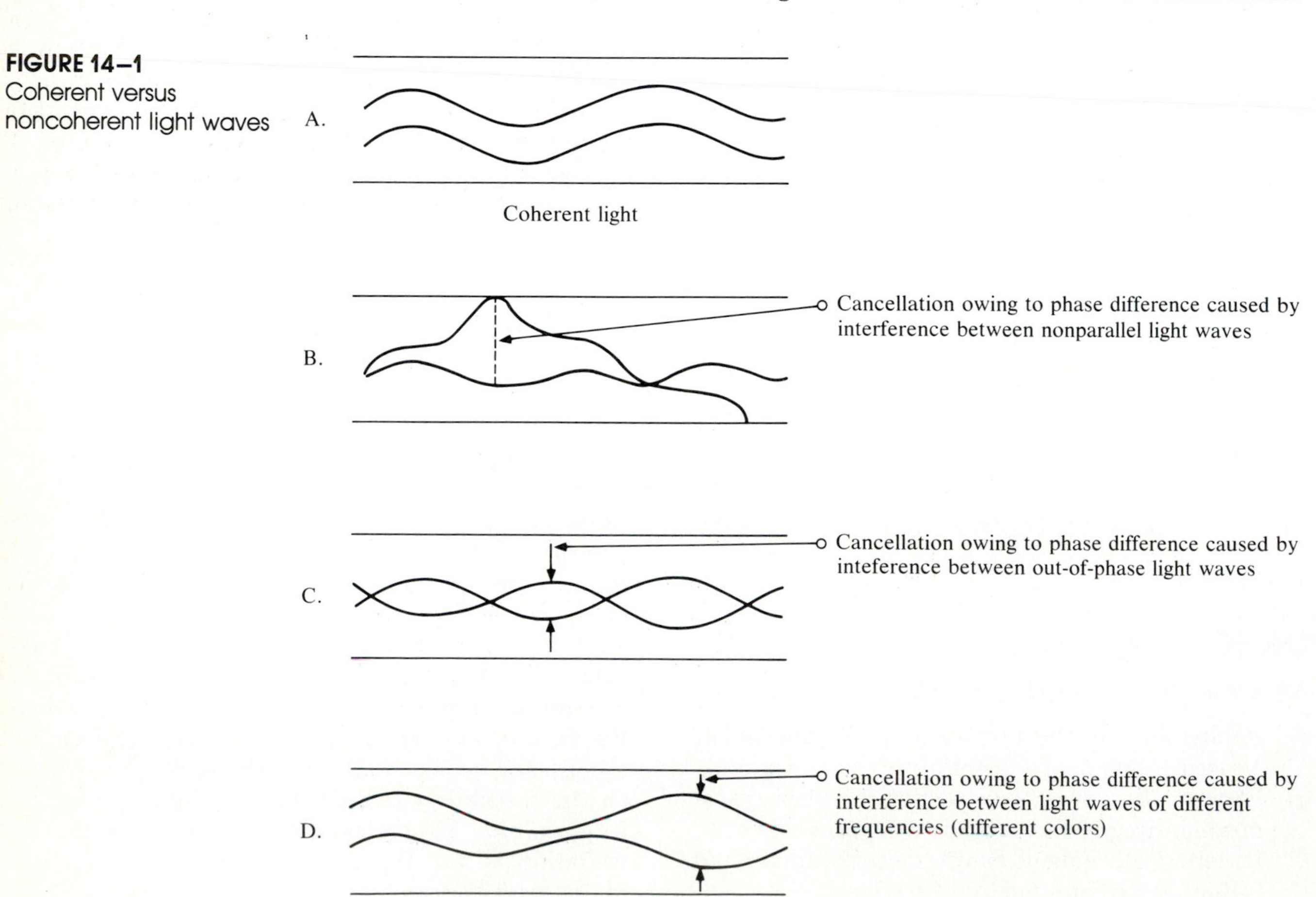

FIGURE 14–1
Coherent versus noncoherent light waves

cally coherent and thus desirable for light beam communications through optical fibers.

LIGHT WAVE PARAMETERS

The relationship between the wavelength and the frequency of light can be expressed by the formula

$$f = \frac{c}{\lambda} \quad (14\text{–}1)$$

where f is the frequency of the light waves in hertz, c is the speed of light (300,000,000 m/s), and λ is the wavelength of the light waves in meters.

Light wavelengths are very small. They were at one time measured in angstrom units (Å), where

$$1\ \text{Å} = 10^{-10}\ \text{m} \quad (14\text{–}2)$$

The modern measurement is in nanometers (nm):

$$1\ \text{nm} = 10^{-9}\ \text{m} \quad (14\text{–}3)$$

Another unit of measurement sometimes used is the micrometer (micron, μm):

$$1\ \mu\text{m} = 10^{-6}\ \text{m} \quad (14\text{–}4)$$

Ultraviolet light ranges from 10 nm to 390 nm, visible light from 391 nm to 770 nm, and infrared light from 771 nm to 1000 nm. These numbers are not absolute since many people can see light with measurement into the infrared range.

A gallium arsenide laser emits a light of 904 nm. In this case,

$$f = \frac{300 \times 10^{6}\ \text{m/s}}{904 \times 10^{-9}\ \text{m}} = 3.3 \times 10^{14}\ \text{Hz} \quad (14\text{–}5)$$

which is 330,000,000 MHz! Very small wavelengths translate into very high frequencies, which opens up a part of the electromagnetic spectrum previously not used for communications purposes. High frequencies permit the use of very broad bandwidths which are required for data communications.

OPTICAL FIBER LINKS

Many modes of communication are possible between two stations (a transmitter and a receiver). *Simplex* transmission occurs between a single transmitter and a single receiver. It is one-way communication. An example is television transmission. The receiver cannot be used to respond to transmissions from the television camera. *Half-duplex* transmission is bidirectional with communications conducted one way at a time. An example of half-duplex is citizen's band (CB) radio. In this case, a transceiver is located at each station, and either transceiver can be used to transmit or receive, but not both at the same time. *Full-duplex* transmission involves a bidirectional link that permits simultaneous two-way communication. An example of full-duplex is the telephone.

If a duplex network containing optical fibers involves more than one transmitter or receiver, the switching system is expensive compared to electrical or electronic switching since optical fiber links are mechanically connected and mechanical switches must be used.

A BASIC COMMUNICATIONS SYSTEM

A signal is usually generated by electronic means. The signal is called a *carrier* since it will carry the information to be transmitted. The information is "placed" on the signal, and a transducer converts the electrical signal to a light wave signal. The resulting signal is then transmitted through an optical fiber.

At the receiving end, the optical signal must be changed back to an electrical signal. This transformation is accomplished by another transducer (usually a photodiode or phototransistor). The electrical signal can then be amplified to drive yet another transducer, producing sound, light, or any other type of energy required. Information is placed

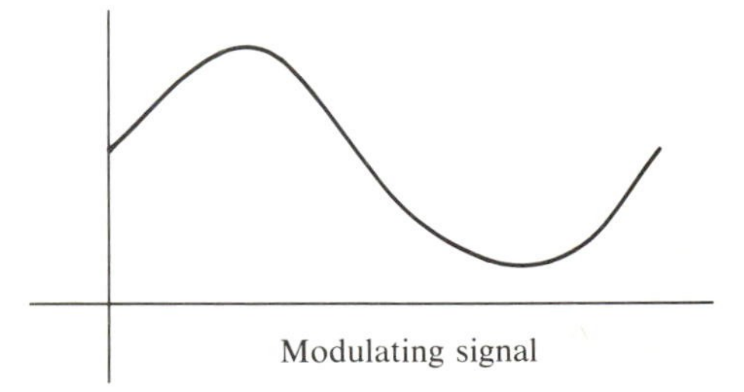

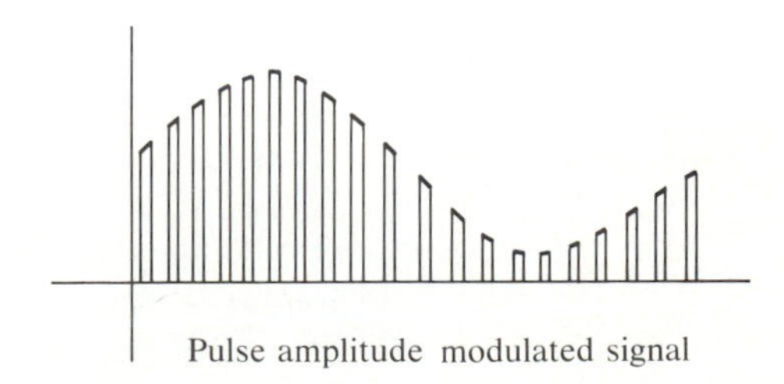

FIGURE 14–2
Pulse amplitude modulation

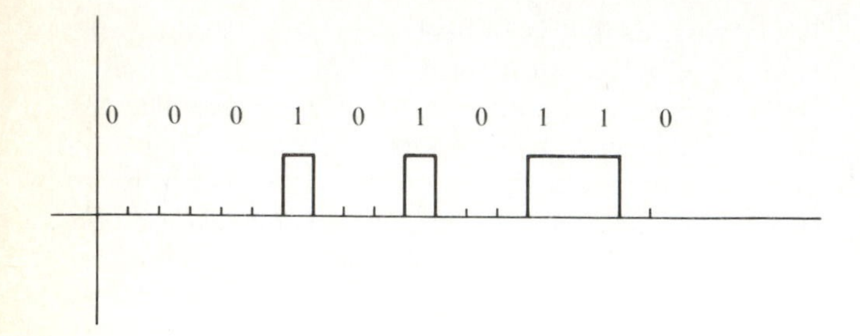

FIGURE 14–3
Pulse code modulation

on the light beam through pulse amplitude modulation or pulse code modulation.

In *pulse amplitude modulation,* the amplitude (intensity) of the light beam is changed to coincide with an information signal known as a modulating signal (Figure 14–2). In *pulse code modulation,* the information to be transmitted is coded into a pulse string, which is transmitted along the optical fiber (Figure 14–3). In both cases, the information must be extracted from the received waveform. This demodulation process is a function of the receiver circuitry.

TRANSMISSION LINE CHARACTERISTICS

Table 14–1 can be used to compare the characteristics of various types of transmission lines. Column 1 lists the type of line. Column 2 lists the bandwidths the various lines can handle and thus provides a description of the frequency range or signal-handling capabilities of each line. Column 3 indicates the channel capacity of each line when used as a transmission cable. Column 4 is the attenuation in decibels per kilometer or the maximum repeater spacing required in kilometers. (Repeaters receive a weak signal, amplify it, and reradiate it.) The data show that, in general, optical fibers are superior in most ways to other transmission media.

ADVANTAGES OF OPTICAL FIBERS

The following are some advantages of optical fibers used as transmission media.

1. As noted, optical fibers offer less attenuation of transmitted signals.

2. Optical fibers can handle data at a wide range of transmission rates. Signals from audio-frequency ranges to 140 megabytes per second can be sent, making optical fibers useful for transmission of voice, music, radio, and television carrier signals and for data transmission. Even high-speed computer data can be handled by optical fibers.

3. The frequency of the light signal used in optical fiber transmission is very high (10^{14} Hz). This range provides wideband capabilities. The commercial FM frequency range of 88–108 MHz allows for 100 channels, each 200 kHz wide. The visible light spectrum is 3.9×10^{14} Hz to 7.67×10^{14} Hz. This represents a band of 3.77×10^{14} Hz as compared to the FM band of 2×10^{7} Hz.

4. Optical fibers can transmit power. However, this capability is limited.

5. The security of communication systems involving optical fibers is much higher than for other systems. Physical connections and capacitive or inductive pickups can be used to intercept or detect a signal carried by metal conductors. Any attempt to physically intercept the light traveling down an optical fiber will change the characteristics of the signal. Thus, fibers are less susceptible

TABLE 14–1
Transmission line characteristics

Cable Type (1)	Frequency Bandwidth (2)	Channel Capacity (3)	Attenuation or Repeater Spacing (4)
Wire pairs	1–140 kHz	Up to 240 (16 pairs)	0.1–0.3 dB/km
Multipair cable	12–250 kHz	1400 on 2 pairs	3 dB/km
Coaxial cable			
Type L1	64–3,096 kHz	1,800	12.6 km
Type L3	308–8,320 kHz	16,740	6.4 km
Type L4	0.5–17 MHz	32,400	3.2 km
Type L5	3–51 MHz	90,000	1.6 km
Circular waveguide	50 GHz	Several hundred thousand	0.5–4 dB/km (30–75 km)
Optical fiber	10^{14} Hz	Several hundred thousand	0.2 db/km

to unauthorized access. Since the information is not electrical, capacitive or inductive pickup is impossible.

6. Optical fibers are small. A fiber the thickness of a human hair has the same signal-handling capability as a copper cable 3 in. in diameter containing 900 twisted pairs.

7. Optical fibers weigh very little. A 1-km length of polymer-coated fiber weighs 167 g. If the same cable is strengthened and jacketed, it will weigh 13 kg. A 1-km-long copper cable with the channel capacity of a single optical fiber would weigh tons.

8. Optical fibers are flexible. Constant vibration or bending does not have the same effect on optical fibers as on metal conductors.

9. Optical fibers have a tensile strength of about 500,000 pounds per square inch (psi). This is stronger than steel!

10. Immunity to natural and man-made electromagnetic interference (EMI) is important in communications. Since optical fibers do not carry electrical signals, essentially no EMI occurs.

11. Metal conductors carrying radio-frequency signals tend to radiate energy, which can be picked up by adjacent conductors. This phenomenon is called *crosstalk*. Optical fibers are immune to crosstalk.

12. Because optical fibers are made of glass, which is a nonconductor, lightning strikes will not damage optical fiber transmission line.

13. Since optical fibers do not carry electricity, no shock hazard exists.

14. Likewise, short circuits are impossible between optical fibers.

15. Optical fibers can be used in hazardous atmospheres since sparking will not occur.

16. With optical fibers between the transmitter and the receiver, the two devices are electrically isolated. Ground loops are not a problem.

17. Unlike metal conductors, optical fibers are highly resistant to corrosion.

18. Optical cables are much more temperature resistant than metal conductors. If a copper cable is exposed to high temperatures, the insulation between the conductors can melt, causing short circuits. If optical cables are exposed to the same conditions and the cladding on optical fibers melts, no interaction between the conductors can occur.

19. Because of the nonelectical characteristics mentioned, technicians can perform repairs on fiber optic systems without shutting the systems down.

20. Optical fiber is more economical than other conducting materials (Table 14–2).

TABLE 14–2
Cable costs

Cable Type	Attenuation at 100 MHz (dB/km)	Cost ($/m)
133 Glass	10	0.75
RG/8	66	1.14
RG/58	135	0.40
RG/174	289	0.47

DISADVANTAGES OF OPTICAL FIBERS

The following are some disadvantages of optical fibers used as transmission lines.

1. Optical fibers cannot carry electricity. This is a major drawback since virtually all equipment used in measurement, entertainment, and communications is electrical or electronic in nature.

2. Because optical fibers must be drawn from very pure materials, manufacturing techniques are more complicated than those used to produce metal conductors.

3. Optical fibers are so small that they are difficult to handle. Coupling and repair by splicing fibers is more complicated than soldering copper conductors.

4. Optical fibers are vulnerable to radiation. Ionizing radiation in some parts of the spectrum can change the index of refraction of the fibers. This alteration affects the propagation of the light and the character of the signal being transmitted. This type of ionizing radiation usually occurs immediately after a thermonuclear explosion. Plastic fibers are less vulnerable than silica fibers.

The advantages of using optical fibers far outweigh the disadvantages. Optical fibers represent a superior pathway from a measuring transducer to a processing network or to some remote output device.

SPACE OPTICAL SYSTEMS

Free-space systems are composed of a light transmitter and a receiver, with no physical path between the two. These line-of-sight systems require less power than electrical or fiber optic systems, need no antenna, and have no licensing regulations. In outer space, where fewer particles exist to scatter the light beam, these systems are less troublesome than they are on earth, where their reliability is determined by atmospheric conditions. Such common occurrences as rain, snow, fog, or dust can severely diminish system reliability. Microwave links are more reliable in fog. Free-space systems have stringent alignment requirements. The transmitter and receiver must be accurately aligned to prevent severe loss of information. Ambient sunlight and artificial light sources cause background noise in open-space systems. Finally, open-space optical links could pose a hazard since a high-energy beam might cause eye damage to a person walking into the path of radiation. Open-space optical links have many short-range applications, primarily in manufacturing processes and security systems.

LIGHT SOURCES

The primary light sources in light transmitters are *light-emitting diodes* (LEDs) and *injection laser diodes* (ILDs). ILDs provide smaller sources than LEDs and are therefore easier to couple to optical fibers. ILDs produce a more directional light beam and have a shorter response time. They can respond to a change in electrical signal amplitude with a corresponding change in light intensity more quickly than an LED. ILDs can be modulated up to a few gigahertz whereas LEDs can be modulated up to only a few hundred megahertz. LEDs with physical and optical focusing are called *domed emitters*.

LEDs have a longer projected life than ILDs. The power output of LEDs drops to one half in about 1,000,000 hours. The output of ILDs drops to one-half power in 10,000 to 100,000 hours. ILDs are more susceptible to excessive current burnouts (overloading) than are LEDs. See Table 14–3 for a cost comparison of LEDs and ILDs. Selection of an ILD or an LED is usually dictated by the specific application and economic considerations.

TABLE 14–3
Light source costs

Device	Cost
LED (5–10 mW)	\$2–\$10
Domed emitter	\$100
Low-power injection laser (1-W pulse)	\$10–\$20
High-power ILD	\$30–\$150

LIGHT DETECTORS

The most commonly used detectors are photodiodes and phototransistors. Both devices depend on the generation of current carriers in the vicinity of the P-N junction, which occurs when photons of light energy strike electrons in the lower-energy levels of the atom. This action converts low-energy electrons to higher-energy levels. The electrons can then become free electrons. Free electrons are current carriers that effect the conduction state of the semiconductor junction. The greater the intensity of the light, the greater the number of free electrons.

Phototransistors can be used in active or passive applications. When used in the active mode, these devices are photovoltaic. In the passive mode, they are photoconductive.

Other examples of photovoltaic devices are silicon solar cells and PIN photodiodes. The solar cell has a very low frequency response and is impractical for high-speed transmission requirements. The "I" in PIN denotes the special *intrinsic* layer between the P-type and N-type materials forming the P-N junction. PIN photodiodes have a rise time of 1 ns. When used in either active or passive modes, they can produce about 0.6 mA of current per milliwatt of incident radiant energy. Phototransistors, whether bipolar or FETs, can be used to tens of megahertz. All of these devices are used to convert light energy to electrical energy for use by the receiver circuit.

USE OF OPTICAL FIBERS

Optical fibers conduct light in the direction of the major axis. Any light rays not parallel to the axis are reflected when they reach an interface between

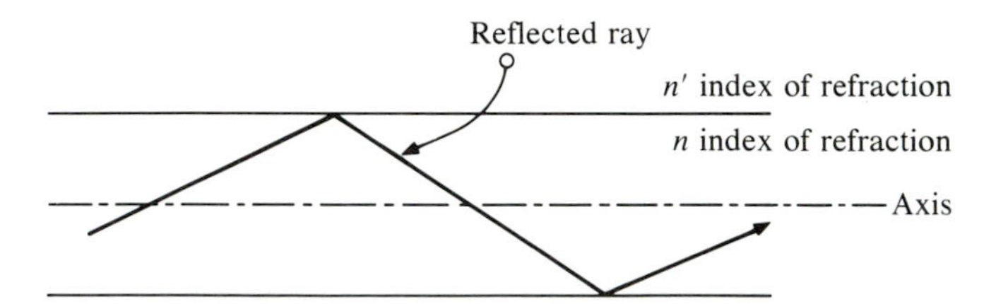

FIGURE 14–4
Reflected ray inside an optical fiber

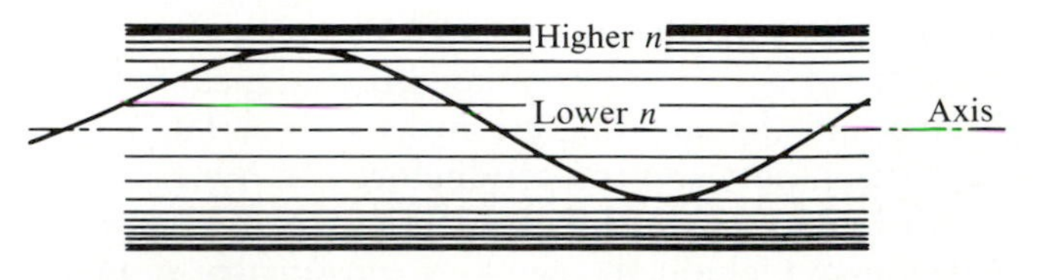

FIGURE 14–5
Graded index of refraction

the glass and any medium with a different index of refraction (Figure 14–4).

Another type of optical fiber has a graded index of refraction, with the outer edge of the fiber having a greater index of refraction than the inner part. This arrangement acts as an optical lens to bend the light wave back toward the center of the fiber (Figure 14–5).

Energy losses occurring in optical fibers include the following:

1. Fresnel losses caused by internal reflections
2. Losses from varying indices of refraction
3. Source/detector coupling losses
4. Natural losses caused by fibers
5. Splicing losses (see Figure 14–6).

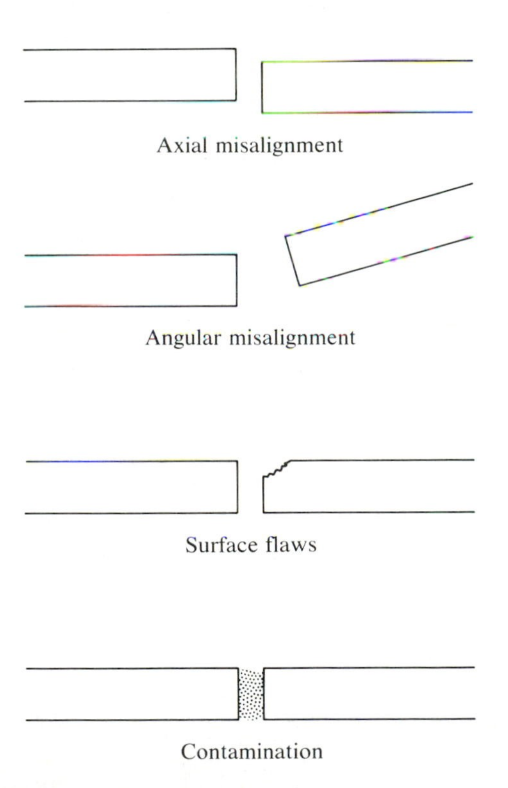

FIGURE 14–6
Splicing losses

AMPLITUDE MODULATION VERSUS PULSE AMPLITUDE MODULATION

Using fiber optic systems to transmit amplitude-modulated (AM) signals requires simple low-cost circuitry. Unfortunately, these systems are susceptible to electrical noise and have high power requirements. Much power is wasted during transmission because most of the AM power is discarded by the receiver without being used. By comparison, pulse modulation applications are relatively immune from noise. They require low-power circuitry, and more of the transmitted power is useful at the receiver. The circuits are more complex and, therefore, more expensive. Figures 14–7 and 14–8 illustrate simple circuits for the transmission of voice by amplitude modulation. Figures 14–9 and 14–10 illustrate the transmission of data by pulse amplitude modulation.

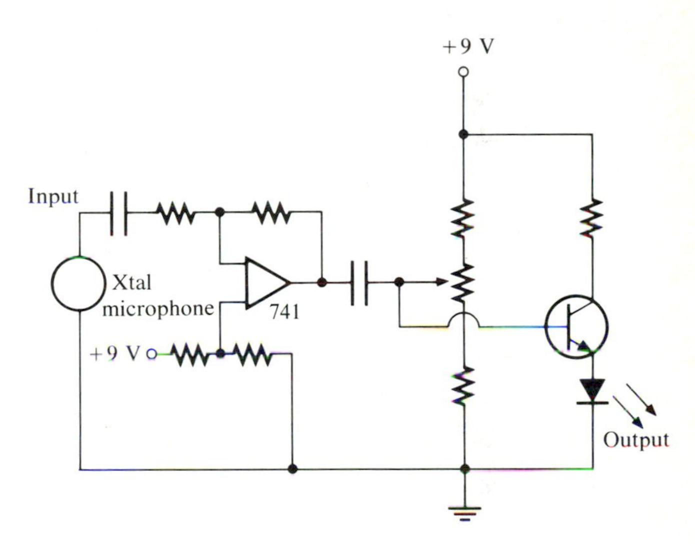

FIGURE 14–7
Voice-modulated AM LED transmitter

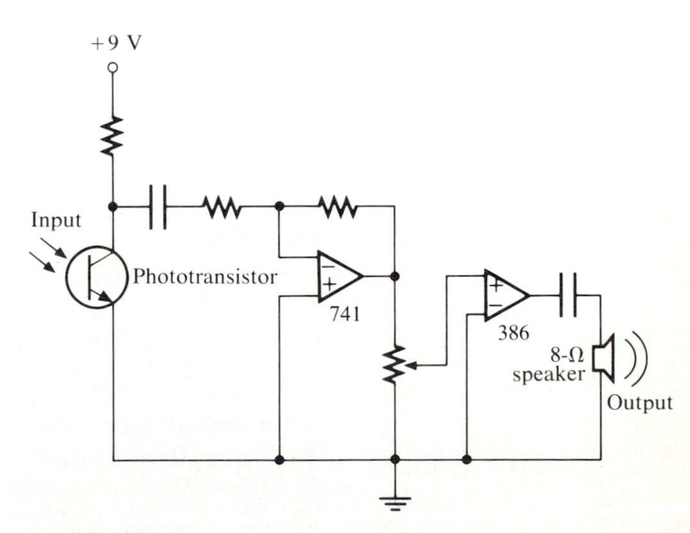

FIGURE 14–8
Voice-modulated receiver

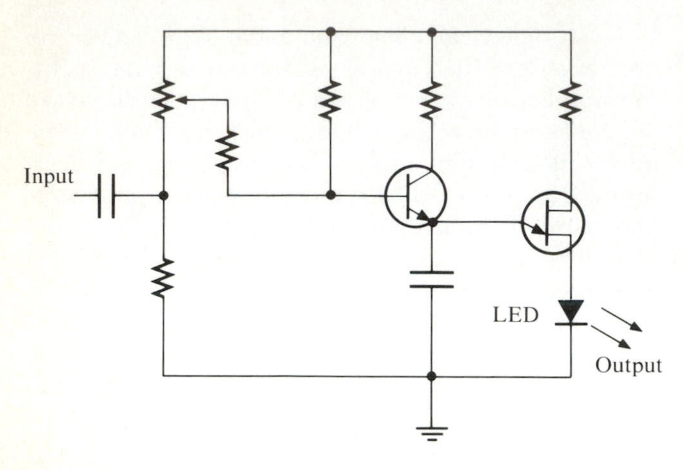

FIGURE 14–9
Pulse frequency–modulated transmitter

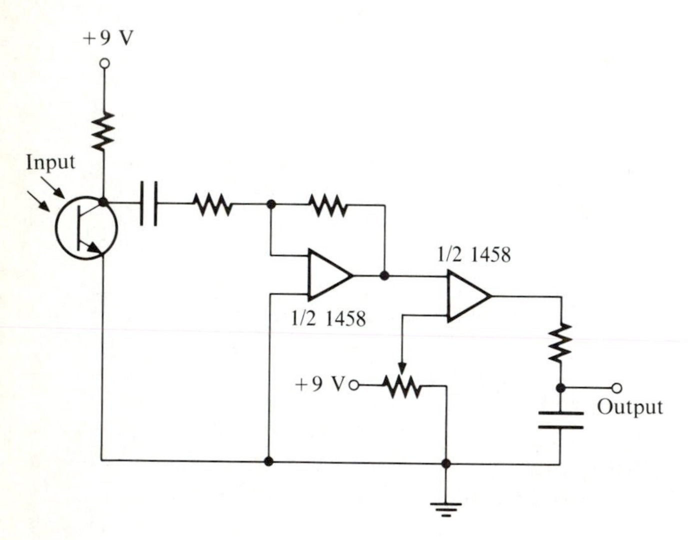

FIGURE 14–10
Pulse frequency–modulated receiver

SUMMARY

This chapter presented the general principles of fiber optics as a transmission medium. Optical systems are being used to transmit measured data to receiving stations. Bandwidth and the resulting frequency-handling capabilities of optical communications make them compatible with the high-speed data transmission rates required by modern computer systems.

Because optical fibers cannot be substituted for electronic components in existing systems, entire electronic systems must be replaced with equivalent fiber optic systems. Developing technology indicates that the transition to these systems is not only feasible but also economically advantageous. Unfortunately, although fiber optics demonstrate superior data-handling capacities, manufacture and repair pose considerable problems.

EXERCISES

1. List several advantages of optical fibers over wire conductors in communications systems.
2. List several advantages of wire conductors over optical fibers in communications systems.
3. What is the frequency of blue light with a wavelength of 400 nm?
4. What is the wavelength of an infrared light wave with a frequency of 3.9×10^{14} Hz?
5. If an optical fiber has a tensile strength of 500,000 psi, what would be the tensile strength of a single optical fiber 0.003 in. in diameter?
6. If optical fibers cost more than some metal conductors, why are they more desirable for some applications?
7. List the kinds of fiber losses that can diminish the power of a light signal.
8. What kinds of coupling losses cause a reduction in the amplitude of a light signal transmitted through an optical fiber?

LAB PROBLEM 14–1
Free-Space Transmission

Object
To illustrate a free-space light transmission system.

Equipment
AM radio receiver; light source (try a flashlight bulb and reflector); cardboard tube; solar cell; amplifier; speaker.

Procedure
Assemble the optical transmission system shown in Figure 14–11.

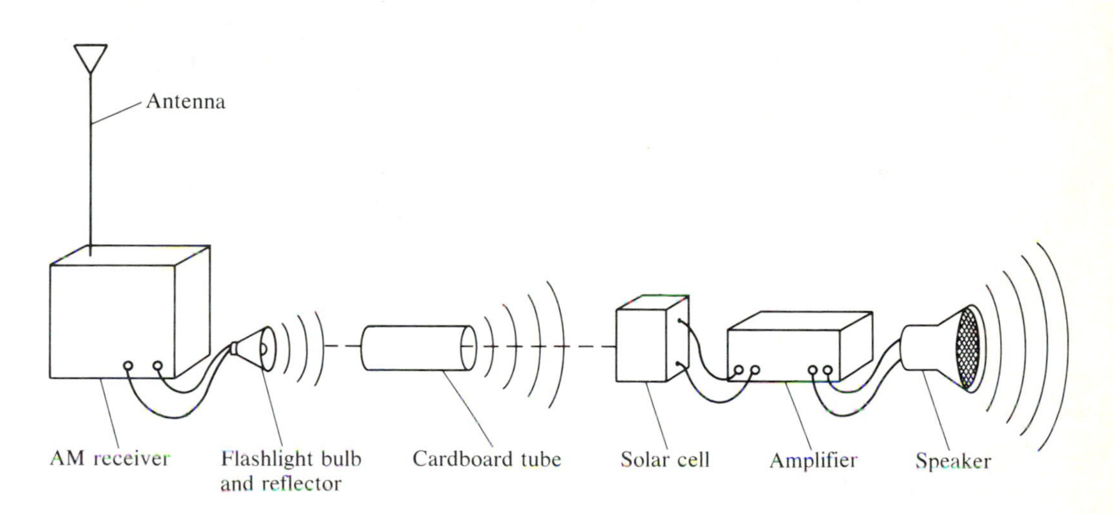

FIGURE 14–11
Free-space transmission system (amplitude modulation)

The intensity of the light varies with the amplitude of the voltage (audio) at the radio speaker. Use the cardboard tube to aim the light beam. If the system is properly aligned, the output should be what is expected from the AM station to which the receiver is tuned. Interrupt the transmission by blocking the light beam with your hand. Note that if the light in the room (artificial or natural) is reduced, the background noise is also reduced. Try reflecting the beam at right angles with a mirror. Comment on the success of this experiment. Discuss the effect of alignment problems, distortion, and background noise.

15 Automatic Test Equipment (ATE)

OUTLINE

OBJECTIVES
AUTOMATIC TEST EQUIPMENT
WHY ATE?
TYPES OF FAULTS
MANUAL TESTING
ATE IN-CIRCUIT TESTERS
TEST EQUIPMENT REQUIRED
DIGITAL TESTING
COMPONENT ISOLATION
- Analog Testing
- Digital Testing

FLIP-FLOPS
BUSSED DEVICES
SUMMARY
EXERCISES
LAB PROBLEM 15–1—ATE Isolation

OBJECTIVES

After completing this chapter, you will be able to:

- State the purpose of automatic test equipment.
- Describe the kinds of faults that can occur during the manufacture of a printed circuit board.
- Describe what is involved in manual testing of PC boards.
- Discuss the capabilities of an in-circuit tester.
- Describe the general architecture of an in-circuit tester.
- Explain the differences between analog and digital testing.
- Discuss some programming techniques used in ATE.
- Describe how components can be electrically isolated for in-circuit testing.
- Describe how backdriving is used in digital testing.
- Discuss some of the problems encountered in the testing of bussed devices and how these difficulties can be overcome.
- Describe the role of ATE and its significance in manufacturing technology.

Automatic test equipment (ATE) is today's technology. This chapter discusses some reasons for the existence of ATE. Specifics of isolating components by the use of programming (software) for the purpose of testing will be shown. A few examples of how the programs control the electrical or electronic tests are given. The intention is to provide an overview of the operation of ATE and an appreciation for the integration of analog, digital, and hybrid circuits that make ATE possible.

AUTOMATIC TEST EQUIPMENT

A major use of ATE is to test assembled printed circuit (PC) boards. ATE is computer-controlled testing. A computer program directs the machinery to perform many tests on mounted components and on entire boards to improve the product yield of a manufacturing or assembly process.

A PC board can be installed on a 'bed-of-nails' fixture associated with ATE. The operator enters a program into the ATE memory or calls up a test program from a disk. The machine performs the necessary in-circuit testing of each component and tests the entire PC board as a unit to ensure that it is operating properly. Any faulty results are printed out so that corrections or repairs can be made.

Very large scale integration (VLSI) devices such as central processing units (CPUs), memories, and programmable logic arrays (PLAs) are becoming more and more complex. Board designers are integrating more and more analog functions onto what were once mainly digital board designs. ATE test pins can drive all the inputs to a device under test and at the same time sense all device outputs. Test data can be sensed or stored in random access memory (RAM) since each test pin usually has its own dedicated RAM. Many events under different test conditions can be monitored simultaneously.

WHY ATE?

In a typical manufacturing process, components are received at the manufacturing facility from a supplier. Upon arrival, they are usually inspected and tested. The components are then mounted on PC boards, typically by automated techniques. The PC boards are assembled to form systems. The systems are installed at the customer's site. Table 15–1 illustrates the relative expense of finding and fixing a faulty component or device during each of these steps.

TABLE 15–1
Relative costs of finding and fixing a fault

Point at which fault is detected	Cost
When components are received	*C**
When boards are tested	10*C*
When systems are tested	100*C*
During service call to customer site	1000*C*

**C* is a base cost for the detection and repair of a fault.

During the inspection process, the faulty device can be simply discarded. However, every device is not always tested. A statistical sampling process is often used because of the large number of incoming components.

If the faulty component is discovered after it has been mounted on a PC board, it must be unsoldered and replaced. The cost of correction at this point may be 10 times the cost of detection at the original inspection. If a faulty component is not detected until the system has been assembled and is being tested, the tracing process and correction are much more extensive, and the cost may be multiplied by 10 again. If a system has been installed and a field service engineer is required to correct a fault, the cost is increased by another factor of 10. It is clearly more cost effective to detect a fault in a device early in the development of a system than it is to send out a highly paid employee to deal with an irate customer.

TYPES OF FAULTS

Since most PC boards are assembled by automatic processes, it is useful to consider the types of problems encountered during assembly. *Wave soldering* involves soldering an entire PC board after the components have been mounted. This process ensures uniform soldering through very carefully controlled mass soldering. Occasionally a solder bridge occurs. This is an inadvertent path of solder that jumps across the solder runs, providing a short-circuit path that will likely lead to an expensive repair later on (Figure 15–1A).

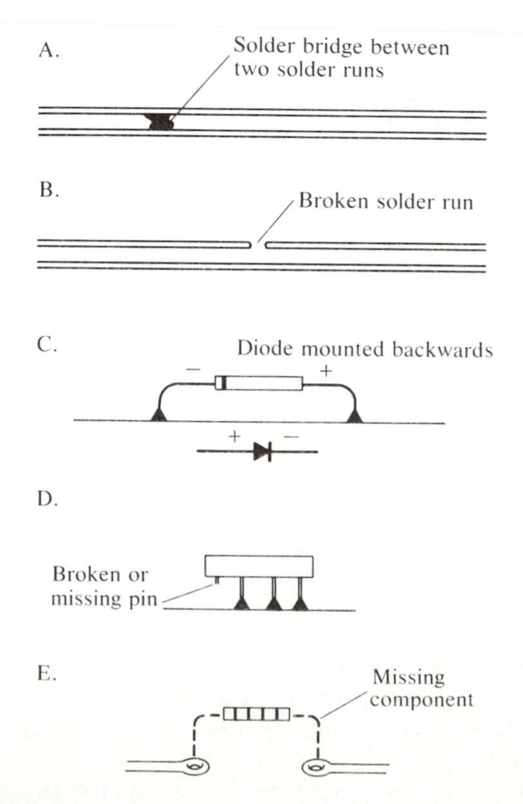

FIGURE 15–1
Types of faults

Figure 15–1B shows a broken solder run. An incomplete soldering process will cause an open circuit where one should not exist. The result will be a malfunctioning board.

Figure 15–1C illustrates a component that has been mounted backwards. Figure 15–1D shows a broken or missing pin.

In automatic component mounting, the components may jam in the magazine that feeds them to the inserting mechanism, or the magazine may simply run out of components. In either case, a necessary component would be omitted (Figure 15–1E).

All of these faults can be detected when the board has been assembled and should be capable of performing the functions for which it was designed. This is the ideal place in the manufacturing process for ATE.

MANUAL TESTING

Before a PC board is tested manually, the technician must understand how all of the circuits on the board operate. From this knowledge, appropriate tests can be developed. All circuits must be tested under normal operating conditions. Once the tests have been developed, the necessary test equipment must be wired to the board. Manual testing requires assembling and adjusting oscilloscopes, voltage sources, signal generators, and meters. When the equipment has been turned on, if all goes well the board will operate properly. If it does not, the technician must find the fault and repair it. Then every combination of operating conditions must be simulated and checked. This is a difficult, time-consuming, and expensive process.

ATE IN-CIRCUIT TESTERS

With proper programming, a microprocessor can establish all of the test procedures and perform all of the required tests in a very short time. The test equipment is electronically wired to the board under test, the tests are performed, and the results obtained.

ATE tests each component, one at a time, for unwanted open and short circuits. The tester then isolates and tests each component to see if it is performing its specific function properly. The overall performance of the board can then be tested.

To perform all of these tests, in-circuit testers must have access to each component and be able to isolate each for testing. Accessing circuit connections is accomplished by a *"bed of nails"* (Figure 15–2). The "nails" are mounted in sockets properly located on the "bed" so that each nail lines up with a circuit connection on the board. A vacuum attachment draws the board tightly to the array of nails or pins to ensure a good electrical connection. The pin array usually includes hundreds to thousands of pins appropriately spaced (usually at 0.1-in. intervals (Figure 15–3). Each pin is spring loaded to provide contact while minimizing pin wear. A beveled top provides contact if the board connections are improperly aligned. The wire-wrapped connections provide access to each pin through the input and output ports on the side of the ATE unit.

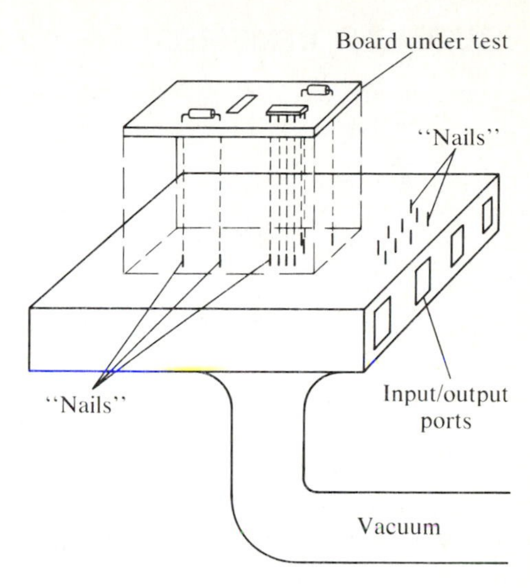

FIGURE 15–2
"Bed of nails"

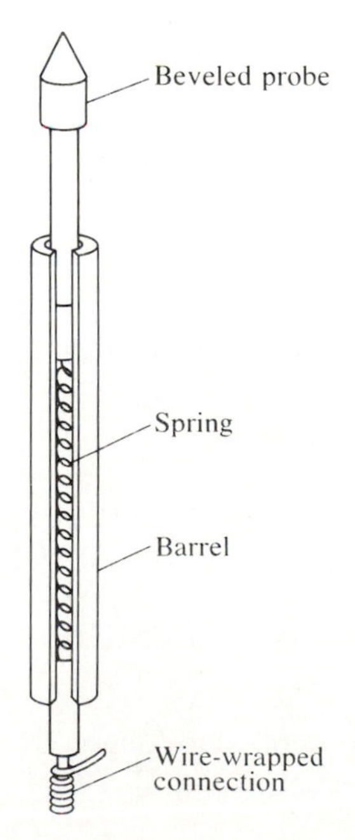

FIGURE 15–3
Test pin or nail

TEST EQUIPMENT REQUIRED

To perform analog and digital tests, the following equipment must typically be available:

1. dc current source
2. dc voltage source
3. dc voltmeter
4. dc ammeter
5. ac impedance measurement module

Figure 15–4 illustrates the connections that can occur between the nails and the test equipment. The relays or switch arrays are electronic connections that can be activated by program instructions. Figure 15–5 shows the switch layout of the relays providing paths from the instrument unit to the bus through the analog instrument multiplexer and from the pins to the bus through the scanner. The first program command

SET MUX AT (CHA = 6: CHB = 8)

closes the relay connecting pin 6 of the analog instrument unit to channel A and pin 8 to channel B. The switching unit that gives access to various analog test equipment is called the instrument *multiplexer* and is denoted in these programming examples as MUX. The unit that provides access to the pins is called the *scanner* and is denoted as SCAN. Figure 15–6 illustrates the relays activated by command 1. The second command

SET SCAN AT (CHA = 18: CHB = 49)

closes the relays connecting the same MUX channel A to test nail 18 and channel B to test nail 49 (Figure 15–7).

Assume that a resistor is mounted across nails 18 and 49 and that 0.2 V is to be applied across it. The resulting current can be measured and compared to preset data or stored in memory. In this example, the result will be stored in memory location IRESULT:

```
SET MUX AT (CHA = 6: CHB = 8)
SET SCAN AT (CHA = 18: CHB = 49)
SET DCS V = 0.2
MEAS DCI INTO IRESULT
```

The programming shown is a combination of BASIC and Assembly language. DCS is a dc voltage source to be set at 0.2 V. DCI is a dc ammeter or sensor.

If proper interfacing is available, other test equipment can be used. For example, much ATE comes equipped with an IEEE-488 controller which enables the setup and operation of IEEE-488-compatible equipment.

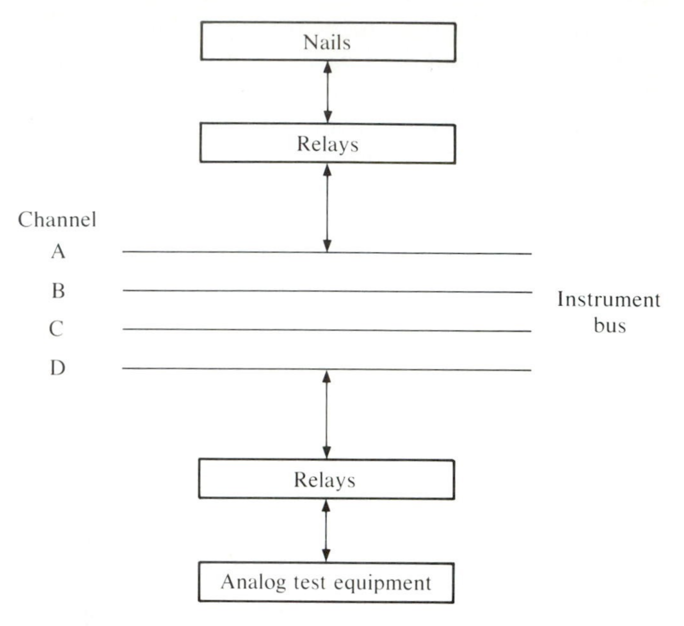

FIGURE 15–4
Communications between nails and test equipment

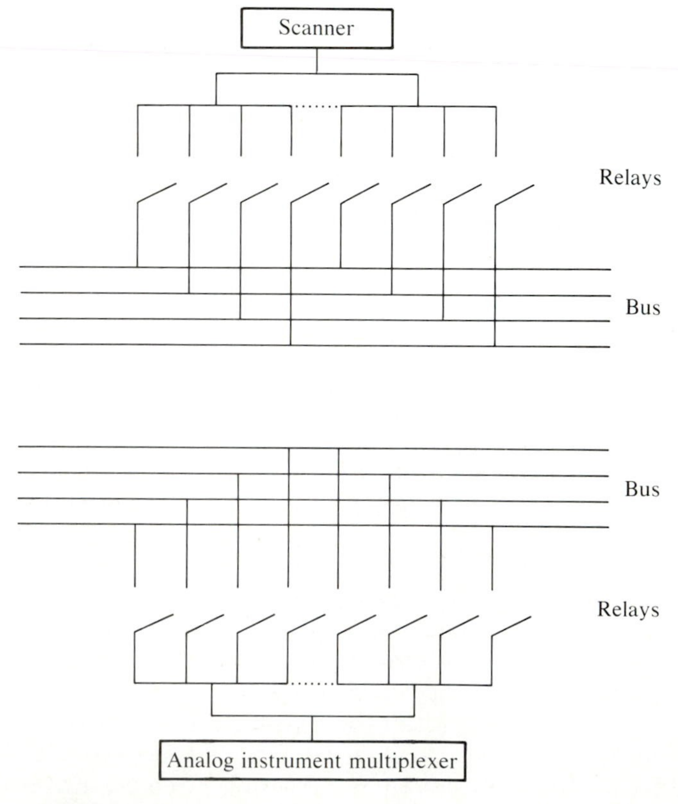

FIGURE 15–5
How ATE works

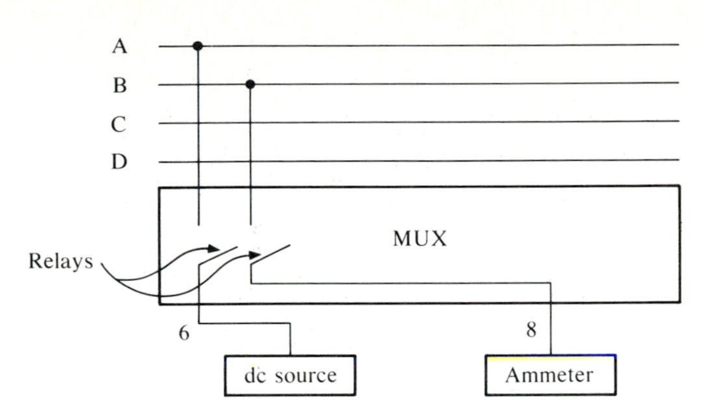

FIGURE 15–6
Relays activated by command 1

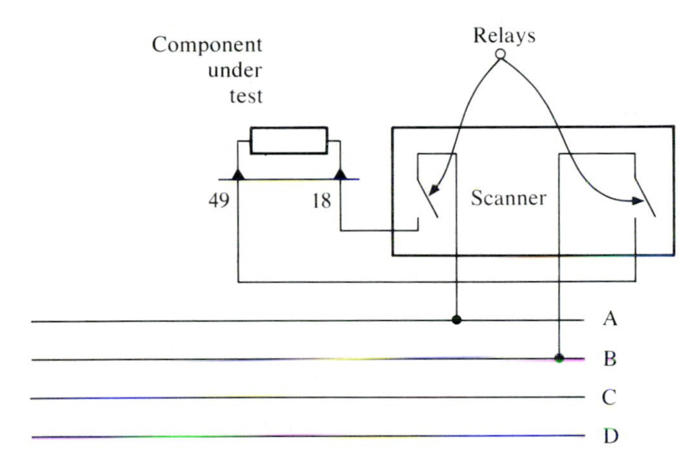

FIGURE 15–7
Relays activated by command 2

DIGITAL TESTING

The tester has digital drivers that drive the integrated circuit (IC) inputs to the desired logic levels. Digital sensors check the logic levels at the IC outputs. Each nail must have both a driver (D) and a sensor (S). Each driver and sensor is controllable by software. The following programming commands illustrate this control:

IC—Input connect
ID—Input disconnect
OS—Output sense
OI—Output ignore
IH—Input high
IL—Input low
OH—Output high
OL—Output low

The command IC(13,53) enables the driver associated with nails 13 and 53. IH(13,53) forces a high logic level on these nails. OS(86) enables the sensor from nail 86. OL(86) causes the sensor to check nail 86 for a logic low output. This sequence of instructions is illustrated in Figure 15–8.

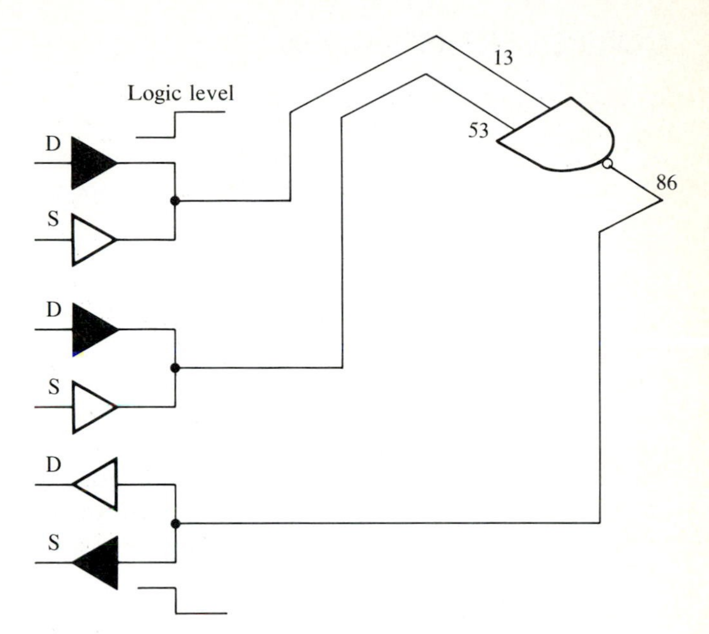

FIGURE 15–8
Inputs and outputs are set.

Figure 15–9 illustrates the process involved in performing the preceding test. The program is loaded into the driver/sensor (D/S) memory. The test is started. The test patterns are transferred to the drivers, and the sensor results are stored in the driver/sensor memory. Special techniques are required to perform in-circuit tests. Testing analog and digital components requires unique tests. Programming commands are usually specific to a particular manufacturer's equipment. The processor control of the testing is achieved through the same means in all ATE.

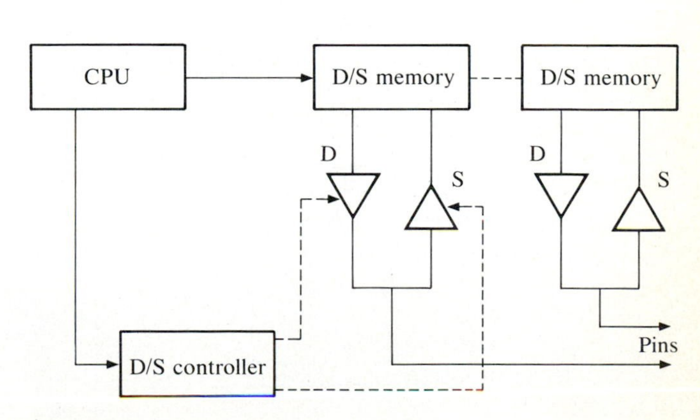

FIGURE 15–9
Flow chart illustrating test flow in ATE

COMPONENT ISOLATION

Analog Testing

In Figure 15–10, resistor R_x is part of a complex circuit. A voltage is to be applied to the source node, and the current through resistor R_x is to be measured by connecting an ammeter to the measure node. A problem arises because components are usually not mounted in isolation. Parallel paths occur through R_1, R_2, R_3, and R_4. If all paths around R_x are grounded (by test nails), producing a guard node, the circuit is reduced to that in Figure 15–11A. (The guard node "guards" against shunt currents.) This circuit can be considered equal to the circuit in Figure 15–11B. The input to an op-amp is at virtual ground. This characteristic is used in the isolation of analog components (Figure 15–12). If the voltage at the measure node is zero volts and the voltage at the guard node is also zero volts, none of the measurement current will flow between the two points, effectively isolating R_x for measurement purposes.

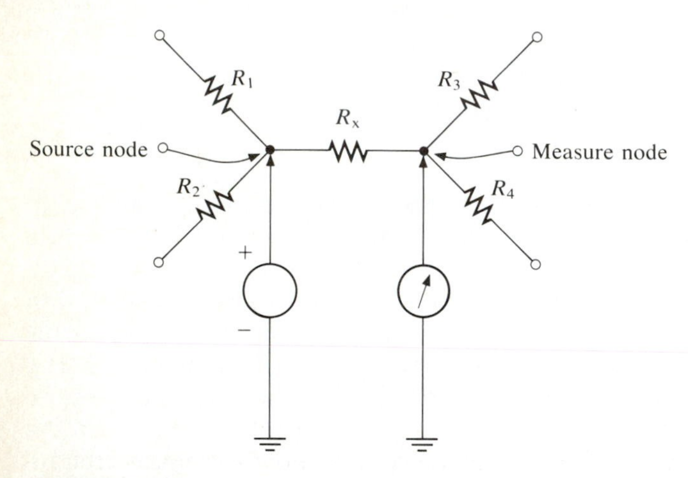

FIGURE 15–10
R is part of a complex network.

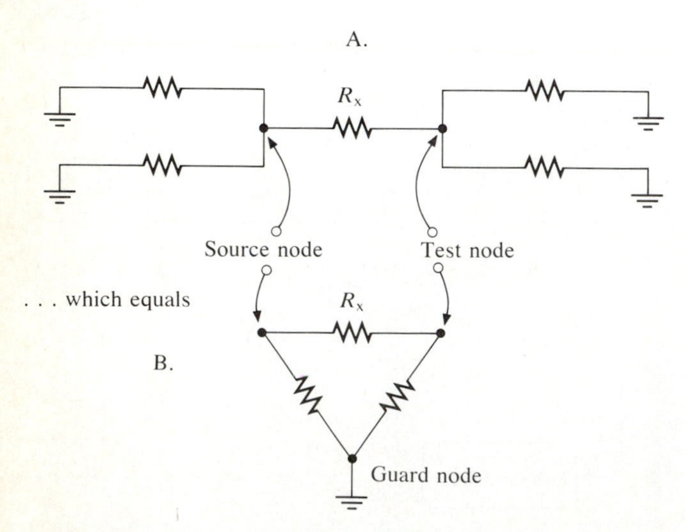

FIGURE 15–11
Software causes certain points to be grounded.

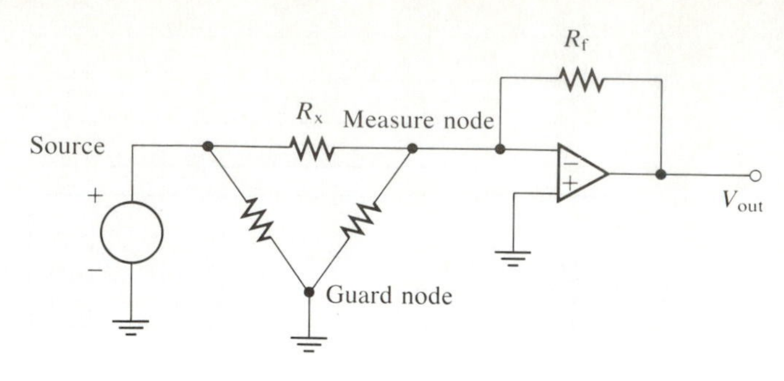

FIGURE 15–12
Op-amp "pulls" the measure node to virtual ground

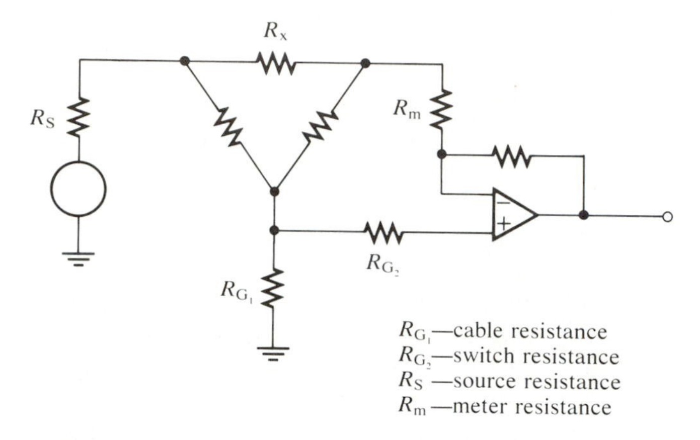

FIGURE 15–13
Eliminating stray resistances

Some problems occur in low-resistance measurements. Cable resistance and switch contact resistance could result in inaccurate data. Connecting the op-amp across the measure node and guard node eliminates this problem (Figure 15–13). The same voltage appears at the measure node as at the guard node, and no current flows into the op-amp. Therefore, no voltage will occur across R_{G_1} and R_{G_2}, which represent the cable and switch contact resistances, respectively. Except in low-resistance measurements, R_x (the resistance of the voltage source) and R_m (the resistance of the meter) can be ignored.

Digital Testing

To illustrate the testing of digital components, consider the NAND gate shown in Figure 15–14. The IC must be "powered up" before it can be tested. If

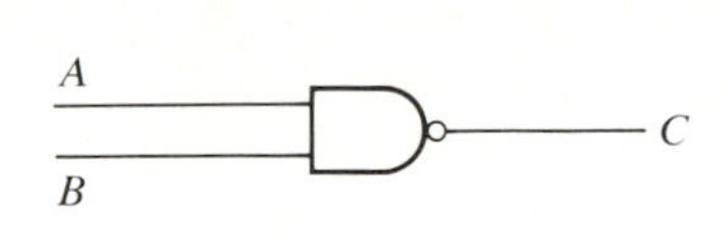

A	B	C
0	0	1
0	1	1
1	0	1
1	1	0

FIGURE 15–14
NAND gate to be tested

an IC is being driven high or low by another IC, the tester must handle this problem. The technique of controlling the logic level of ICs in-circuit is called *backdriving* (Figure 15–15).

In Figure 15–15, transistor Q_1 is in an "on" or *saturated* state, resulting in a low output. To force the output of Q_1 to a high state, a 200-mA current is forced back through Q_2. The current is forced through the emitter-collector resistance, resulting in a low at that point. The low is held only momentarily so that the gate will not burn out.

Figure 15–16 shows Q_1 in an "off" or *cutoff* state, producing a high logic output. A momentary 200-mA current forced back through Q_1 reduces the voltage across Q_1 and forces a temporary low logic output. If a backdriving current is applied for a very short time, the IC will not be damaged. Sense circuits in the drivers automatically reduce the current if the driver remains on too long. These techniques permit isolation and testing of digital components wired into circuitry.

Consider the program

```
IC(13,28)
IH(13,28)
OS(66)
OL(66)
IL(28)
OH(66)
IL(13)
IH(28)
IL(28)
```

The conditions imposed on the NAND gate to be tested, and the expected results, are as follows:

PIN	13	28	66
	1	1	0
	1	0	1
	0	1	1
	0	0	1

IC(13,28) connects the drivers to nails 13 and 28. IH(13,28) forces a logic high on these nails. OS(66) enables the sensor on nail 66. OL(66) "looks for" a low on nail 66. IH and IL commands drive specified nails high or low, and OH and OL commands indicate the expected high or low logic levels at the output nails.

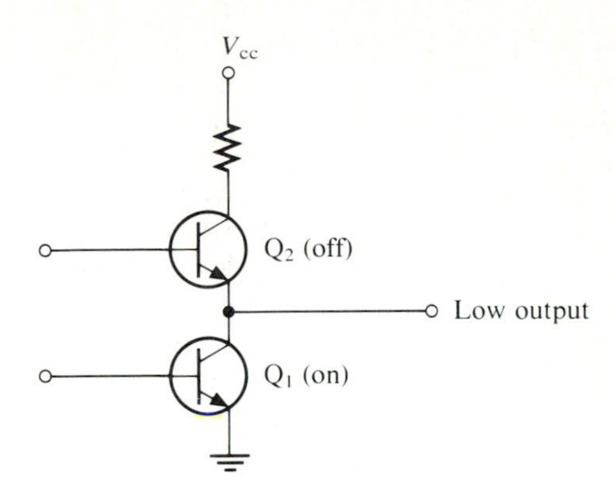

FIGURE 15–15
Backdriving forces the output low.

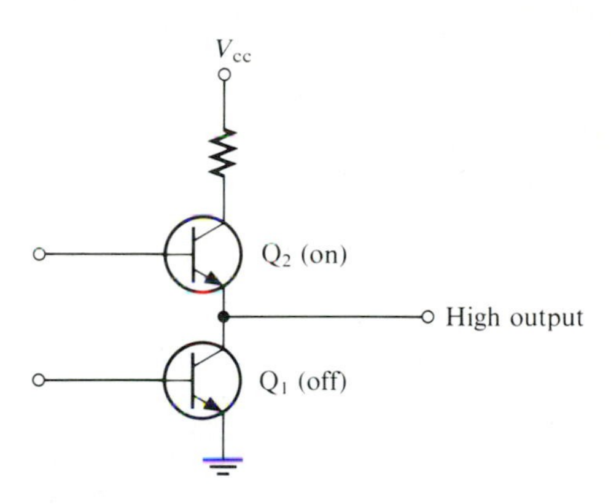

FIGURE 15–16
Backdriving forces the output high.

FLIP-FLOPS

If a flip-flop is initially cleared and set up to toggle to the opposite state whenever a high is applied to the clock input (Figure 15–17), problems can result.

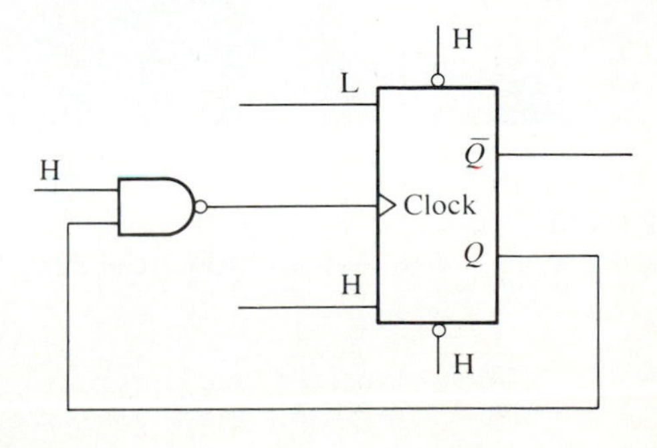

FIGURE 15–17
Flip-flop with a feedback loop.

When the flip-flop is pulsed high–low, the feedback loop to the clock line drives it low. The driver will then (after a delay) drive the clock line high again, producing a glitch that can be misleading. To protect against this, testers are designed to disable all three-state devices by putting them in their high-impedance state and thus inhibit all other devices except the one under test.

BUSSED DEVICES

Bussed devices require a pull-up, pull-down test to isolate each device to see if it can control the state of the bus. To accomplish this, every other device is temporarily disabled. Logic is applied to the enabled device. This procedure isolates faulty devices attached to the bus.

Analog and digital devices can be isolated by software-controlled hardware. Conditions can be set and outputs evaluated through these procedures. "Complete" PC boards can be tested and faults detected before systems are assembled (Figure 15–18).

SUMMARY

Automatic test equipment (ATE) is becoming increasingly popular in the electronics industry. These computer-controlled hardware test systems can be programmed to perform in-circuit component testing. Passive elements such as resistors and active elements such as logic circuits can be tested while in operation on printed circuit boards. The characteristic bed-of-nails fixture allows all or part of a board to be tested as dictated by programmed software without rearranging test leads or fixtures. Hardcopy test data can be obtained by adding a printer. In high-volume industries, using ATE for repetitive testing can save time and wear and tear on test equipment while ensuring quality control.

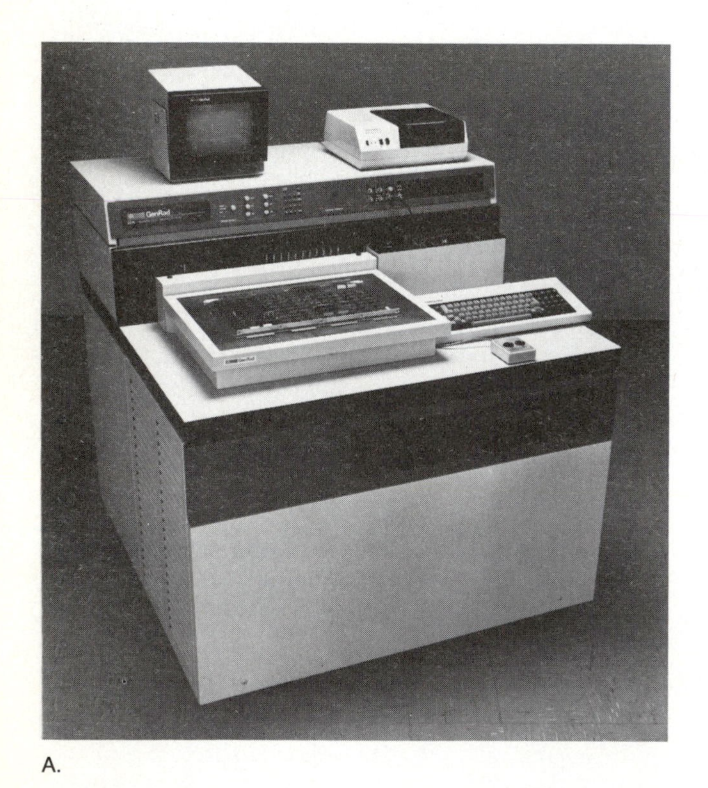

A.

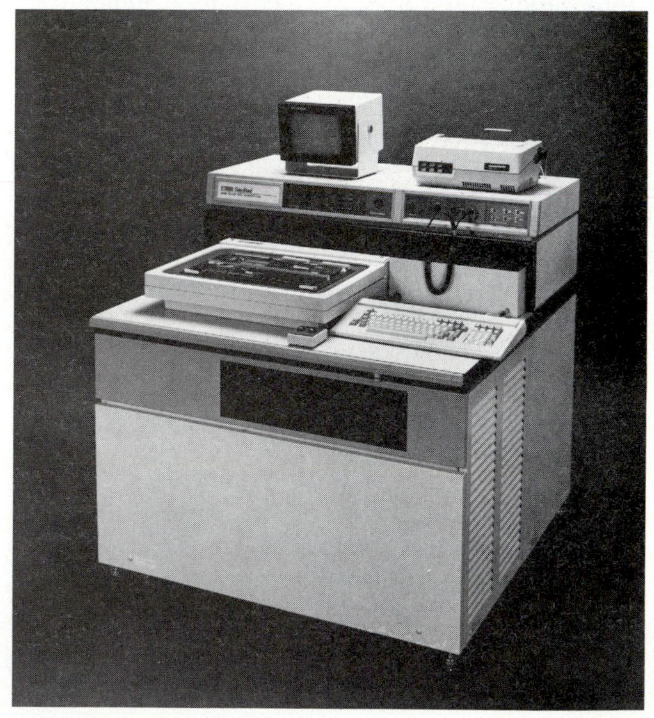

B.

FIGURE 15–18
Automatic board testers. A. GenRad Model 2275. B. GenRad Model 2276.

EXERCISES

1. What is the most significant reason for using ATE in industry?
2. Play the role of a supervisor in industry. Write a memorandum convincing your manager to invest part of the budget in ATE.
3. Play the role of a supervisor in industry. Write a memorandum convincing your manager not to invest part of the budget in ATE.
4. What faults can occur during the assembly of a PC board?
5. How does an automatic tester isolate a resistor for testing?
6. What is backdriving? How is it used for in-circuit testing?
7. Briefly describe the programming used to test a digital gate circuit.
8. Using the programming statements given in the text, write a program to test an exclusive-OR gate.

LAB PROBLEM 15–1
ATE Isolation

Equipment

One 10-V power supply; one 1-kΩ, one 2.2-kΩ, one 3.3-kΩ, one 5.6-kΩ, and one 10-kΩ resistor; one μA741 op-amp; one milliammeter; other resistors as required.

Object

To demonstrate how automatic test equipment isolates a component for measurement. The device being measured is a resistor. In this simulation, resistance will be calculated as *V*/*I* using Ohm's law. (In an ATE system, the calculation would be done by the controlling computer.)

One of the functions of the microcomputer-controlled ATE through the bed of nails is to pull test points to ground, resulting in the isolation of the component to be tested. Manually grounding points *A*, *B*, *C*, and *D* in the circuit in Figure 15–19 and adding the appropriate test equipment results in the circuit in Figure 15–20. This is a simulation of ATE.

FIGURE 15–19

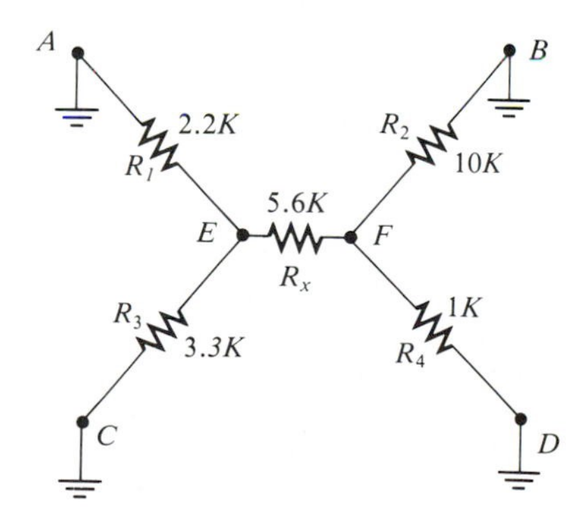

Procedure

Wire the circuit shown in Figure 15–20.

FIGURE 15–20

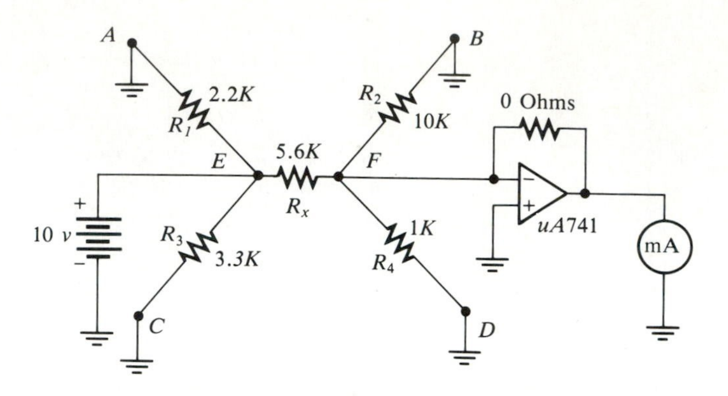

Ten volts are applied across the test node, E. Resistors R_1, R_2, R_3, R_4, and R_x are essentially in parallel and have the same voltage across them. The op-amp places the measure node, F, at virtual ground and thus causes the current through R_x to be measured. The value of R_x can then be computed using Ohm's law:

$$R_x = \frac{V}{I}$$

Apply 10 V to point E and measure the current flowing from point F using the milliammeter. Replace R_x with several values. Enter the results in Table 15–2. Comment on this method of component isolation and evaluation.

TABLE 15–2
Complete the table.

Trial	$R_{x\ (measured)}$	$R_{x\ (actual)}$

APPENDIX A
Thevenin's Theorem—Three Applications

Any two-terminal network can be thought of as a series combination of resistance (impedance) and a source of emf. This open-circuit analysis will be illustrated in three common examples: a ladder network, a Wheatstone bridge, and a circuit with two sources of emf.

LADDER NETWORK

Problem: Develop a Thevenin equivalent of the circuit enclosed by the dashed lines in Figure A–1.

FIGURE A–1
Ladder network

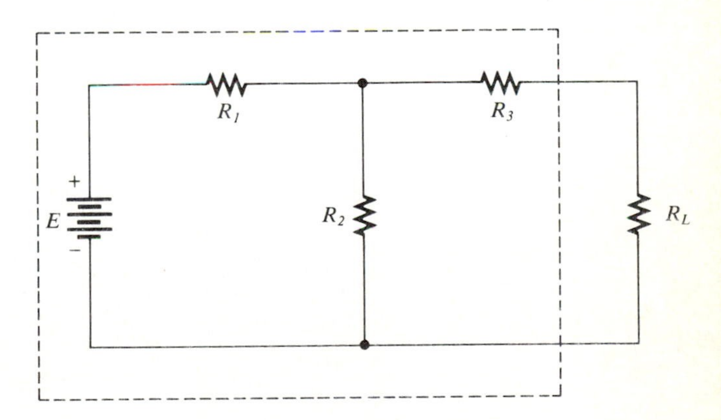

Solution: First, remove the load, R_L, which gives the circuit in Figure A–2. Then,

FIGURE A–2

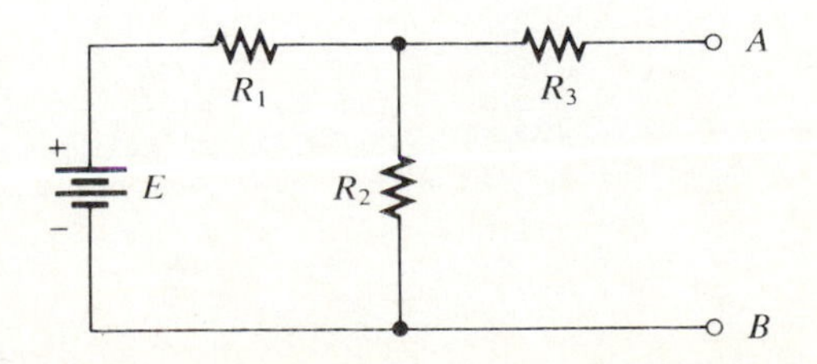

from points A and B—as it would be "seen" by a voltmeter—the potential across the terminals found by voltage division, is

$$V_{AB} = E\left(\frac{R_2}{R_1 + R_2}\right) \quad \textbf{(A–1)}$$

By Thevenin's theorem,

$$V_{\text{Th}} = V_{AB} \quad \textbf{(A–2)}$$

Since the circuit is open, the current through R_3 is zero amperes and no voltage drop occurs between points A and B. Therefore, the voltage across R_3 is zero volts and the potential at point A equals the voltage at point B.

Next, replace the source of emf with its internal resistance (assumed here to be zero ohms), which gives the circuit in Figure A–3. The resistance between points A and B (as seen by an ohmmeter) is

$$R_{\text{Th}} = R_3 + R_1 \parallel R_2 \quad \textbf{(A–3)}$$

FIGURE A–3

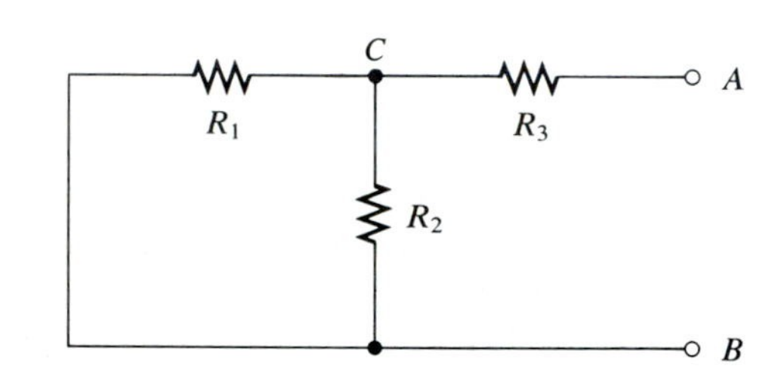

The Thevenin equivalent appears in Figure A–4. If load R_{L} is connected to terminals A and B, it can be shown that

$$I_{\text{L}} = \frac{V_{\text{Th}}}{R_{\text{L}} + R_{\text{Th}}} \quad \textbf{(A–4)}$$

and that

$$V_{\text{L}} = V_{\text{Th}}\left(\frac{R_{\text{L}}}{R_{\text{L}} + R_{\text{Th}}}\right) \quad \textbf{(A–5)}$$

FIGURE A–4
Thevenin equivalent of the ladder network in Figure A–1

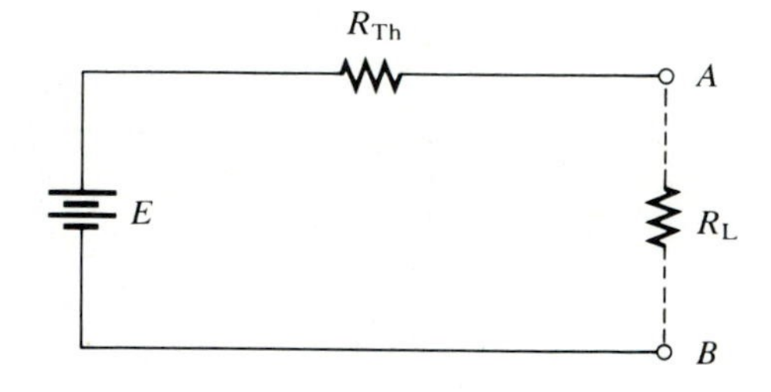

WHEATSTONE BRIDGE

Problem: Find the Thevenin equivalent of the Wheatstone bridge circuit in Figure A–5.

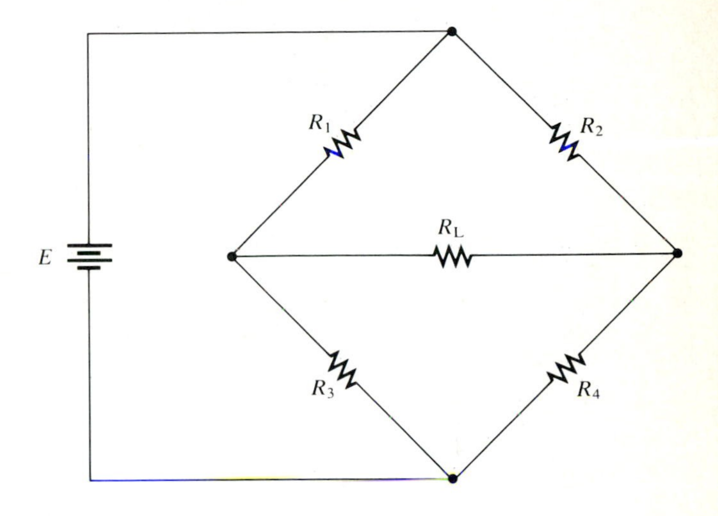

FIGURE A–5
Wheatstone bridge

Solution: Removing the load gives the circuit in Figure A–6. The potential difference between points A and B as seen by a voltmeter can be shown to be

$$\begin{aligned} V_{AB} &= V_A - V_B \\ &= E\left(\frac{R_3}{R_1 + R_3}\right) - E\left(\frac{R_4}{R_2 + R_4}\right) \\ &= E\left(\frac{R_3}{R_1 + R_3} - \frac{R_4}{R_2 + R_4}\right) \end{aligned} \quad \textbf{(A–6)}$$

This is the Thevenin equivalent voltage (V_{Th}).

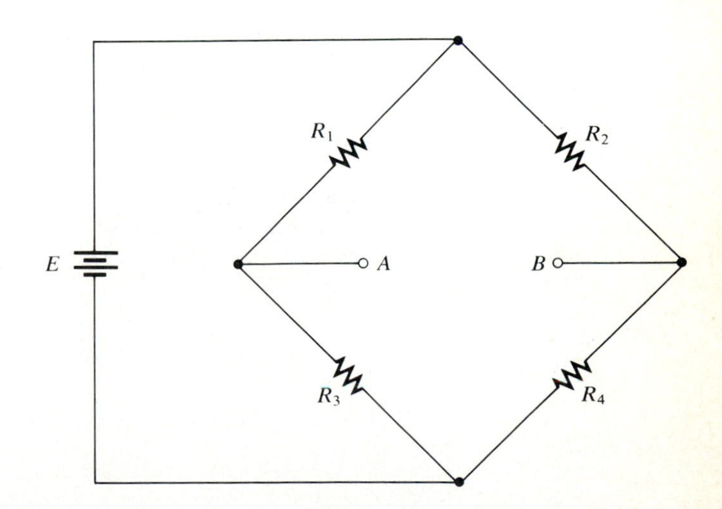

FIGURE A–6

Replacing the source of emf with its internal resistance (assumed here to be zero ohms) leads to Figure A–7. The resistance from A to B as seen by an ohmmeter is

$$R_{Th} = R_1 \parallel R_3 + R_2 \parallel R_4 \tag{A-7}$$

FIGURE A–7

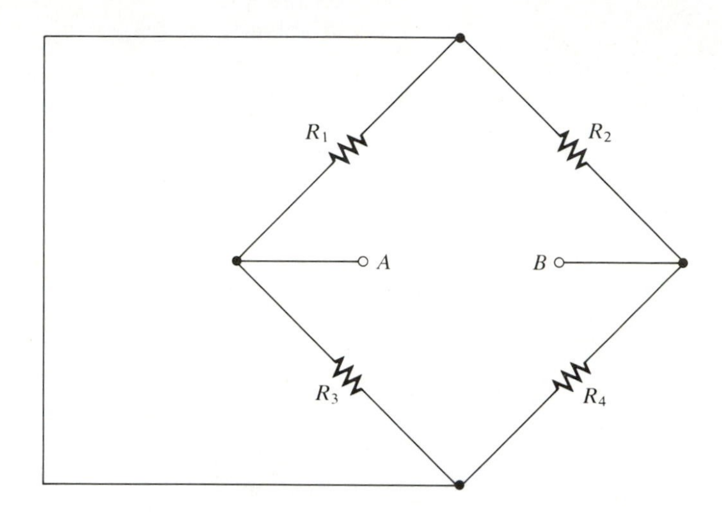

Figure A–8 is the Thevenin equivalent resistance (R_{Th}) of the circuit.

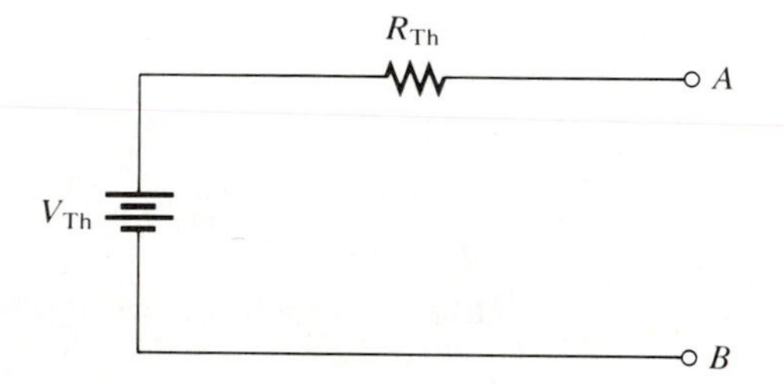

FIGURE A–8
Thevenin equivalent of the Wheatstone bridge in Figure A–5

CIRCUIT WITH TWO SOURCES OF emf

Problem: Find the Thevenin equivalent of the circuit in Figure A–9.

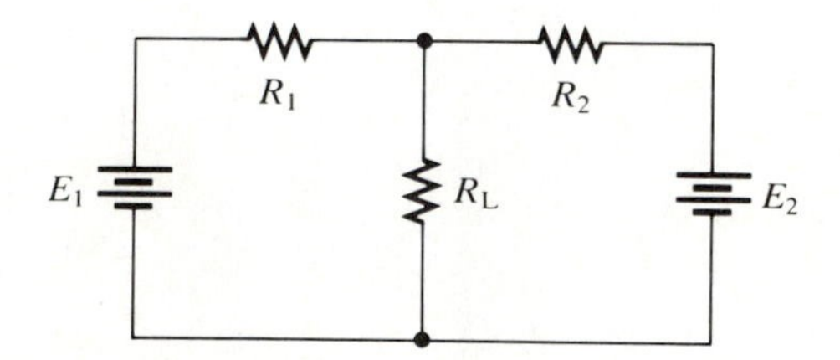

FIGURE A–9
Circuit with two sources of emf

Solution: Removing load R_L yields the circuit in Figure A–10. The potential difference between points A and B can be calculated using either $E_1 - V_{R_1}$ or $E_2 - V_{R_2}$. The polarity of the potential differences across R_1 and R_2 will be determined by the polarity of $E_1 - E_2$:

$$V_{R_1} = (E_1 - E_2)\frac{R_1}{R_1 + R_2} \quad \text{(A–8)}$$

$$V_{R_2} = (E_1 - E_2)\frac{R_2}{R_1 + R_2} \quad \text{(A–9)}$$

FIGURE A–10

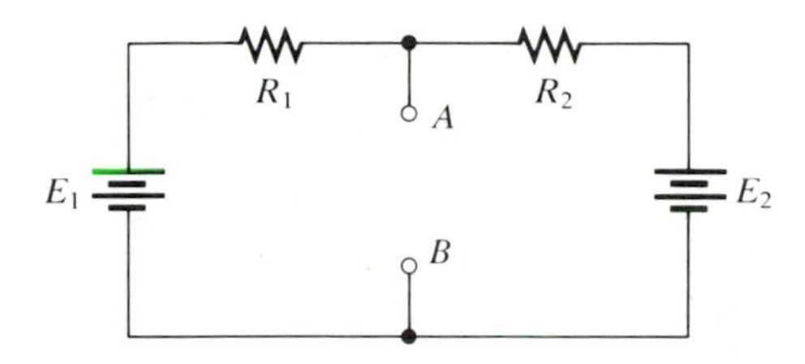

Assume that $E_1 > E_2$. The voltage from point B to point A through E_1 and R_1 is the same as the voltage from point B to point A through E_2 and R_2 (remember that parallel branches have the same potential across them). Quantity V_{R_1} can be calculated using

$$V_{R_1} = (E_1 - E_2)\frac{R_1}{R_1 + R_2} \quad \text{(A–10)}$$

Since

$$\begin{aligned} V_{AB} &= E_1 - V_{R_1} \\ &= E_1 - \left((E_1 - E_2)\frac{R_1}{R_1 + R_2}\right) \\ &= \frac{E_1R_1 + E_1R_2 - E_1R_1 + E_2R_1}{R_1 + R_2} \\ &= \frac{E_1R_2 + E_2R_1}{R_1 + R_2} \end{aligned} \quad \text{(A–11)}$$

then

$$R_{Th} = R_1 \parallel R_2 \quad \text{(A–12)}$$

This will result in the Thevenin equivalent circuit shown in Figure A–11.

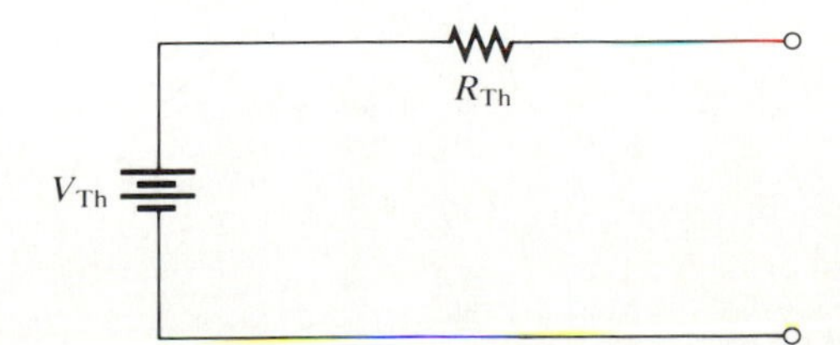

FIGURE A–11
Thevenin equivalent of the circuit in Figure A–9

APPENDIX B
SI Units and Dimensional Analysis

Most human activities involve some means of measurement. The more complex these activities, the greater the degree of measurement precision required.

A quantity can be measured only if an appropriate unit has been defined. To specify the distance between two points, it is necessary to make a measurement of length, that is, to compare the distance to a unit of length such as the meter (or yard or mile). Length is the dimension to be measured, and the meter (or yard or mile) is the unit in which the dimension is measured.

The International Electrotechnical Commission (IEC) represents the chief electrical engineering institutions throughout the world. It defines the electrical engineering units to be used internationally. The General Conference of Weights and Measures (CGPM) establishes international standards consistent with IEC decisions, and members of the IEC report back to their governments, which put into effect the decisions of the CGPM. The International Bureau of Weights and Measures (BIPM) maintains international standards. See Figure B–1.

FIGURE B–1

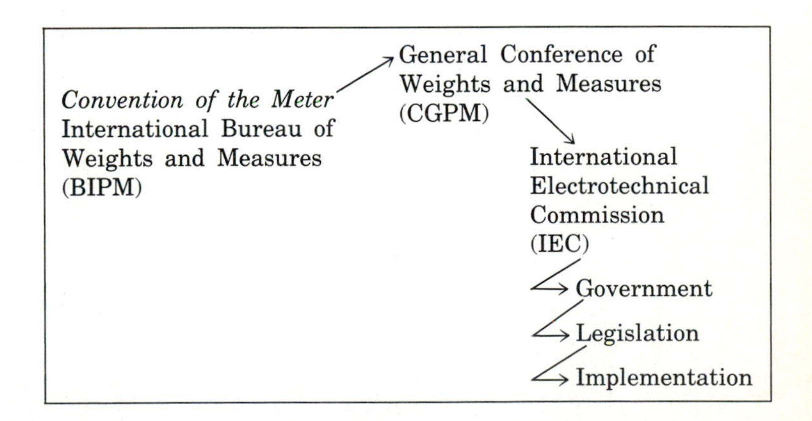

In 1954, the CGPM adopted a rationalized and coherent system of units based on the meter, the kilogram, the second, the ampere, the kelvin, the candela, and the mole. In 1960, the CGPM formally gave it the title Système International d'Unités, for which the international abbreviation is SI.

BASIC SI UNITS

Unit of length—meter (m)

The unit of length called the *meter* is 1,650,763.73 wavelengths in vacuo of the light emitted by the atmospheric gas krypton (11th CGPM—1960).

Unit of mass—kilogram (kg)

The unit of mass called the *kilogram* is the mass of the international prototype, which is in the custody of the BIPM at Sèvres, near Paris, France (3rd CGPM—1901).

Unit of time—second (s)

The unit of time called the *second* is the duration of 9,192,631,770 periods of the radiation corresponding to the transition between the energy levels F = 4, Mf = 0 and F = 3, Mf = 0 of the cesium atom of atomic weight 133 (1983 Conference of CGPM).

Unit of electric current—ampere (A)

The unit of electric current called the *ampere* is the constant current that if maintained in two parallel rectilinear conductors of infinite length, of negligible cross section, and placed one meter apart in a vacuum, would produce a force equal to 2×10^{-7} newtons per meter length (9th CGPM—1948).

Definitions of the other three basic units, that is,

Unit of thermodynamic temperature—kelvin (K)
Unit of luminous intensity—candela (cd)
Unit of amount of substance—mole (mol)

can be found in U.S. Bureau of Standards publications.

In addition to these basic units, the SI system contains the nondimensional units for plane angle and solid angle—the *radian* (rad) and *steradian* (sr), respectively—which are called supplementary units.

The SI is a coherent system of units. For example, unit area results from multiplying unit length times unit length, unit velocity results from dividing unit length by unit time, and unit force results from multiplying unit mass times unit acceleration. In a coherent system with the foot as the unit of length, the square foot is the unit of area (but the acre is not).

As stated, the meter (m) is the unit in which the dimension length (denoted L) is measured, the kilogram (kg) is the unit in which the dimension *mass* (M) is measured, the second (s) is the unit in which the dimension *time* (T) is measured, and the ampere (A) is the unit in which the dimension *current* (I) is measured. To find the dimensions of other quantities in the SI system (i.e., derived units), a defining relationship is needed.

Force—Newton's Law

$$\text{Force} = \text{mass} \times \text{acceleration} \qquad \textbf{(B–1)}$$

$$F = M \times A$$

$$\text{Dimensions of force} = [M] \times \left[\frac{L}{T^2}\right] = [MLT^{-2}]$$

The SI unit of force, called the *newton* (N), is the force that when applied to a mass of one kilogram gives the mass an acceleration of one meter per second squared.

Energy (Work Done)

$$\text{Energy} = \text{force} \times \text{distance} \tag{B–2}$$
$$E = F \times d$$
$$\text{Dimensions of energy} = [MLT^{-2}] \times [L] = [ML^2T^{-2}]$$

The SI unit of energy, called the *joule* (J), is the work done when the point of application of a force of one newton is displaced through a distance of one meter in the direction of the force.

Power (Rate of Doing Work)

$$\text{Power} = \frac{\text{energy}}{\text{time}} \tag{B–3}$$
$$P = \frac{E}{t}$$
$$\text{Dimensions of power} = \frac{[ML^2T^{-2}]}{[T]} = [ML^2T^{-3}]$$

The SI unit of power, called the *watt* (W), is the rate of working of one joule per second.

Electrical Potential

$$\text{Electrical potential} = \frac{\text{power}}{\text{current}} \tag{B–4}$$
$$E = \frac{P}{I}$$
$$\text{Dimensions of electrical potential} = \frac{[ML^2T^{-3}]}{[I]} = [ML^2T^{-3}I^{-1}]$$

The SI unit of electrical potential, called the *volt* (V), is the difference in potential between two points of a conducting wire carrying a constant current of one ampere, when the power dissipated between the points is one watt.

All electrical engineering quantities can be expressed in terms of the four basic quantities—length, mass, time, and current. The dimensions and units of other electrical quantities are given in Table B–1.

TABLE B–1
Dimensional analysis of electrical quantities

Quantity	Defining Relationship	Dimensions	Unit
Electric resistance	$\frac{\text{potential difference}}{\text{current}}$	$[ML^2T^{-3}I^{-2}]$	ohm (Ω)
Charge	current × time	$[TI]$	coulomb (C)
Flux density	$\frac{\text{charge}}{\text{area}}$	$[L^{-2}TI]$	coulombs/meter^2 (C/m^2)
Capacitance	$\frac{\text{charge}}{\text{voltage}}$	$[M^{-1}L^{-2}T^4I^2]$	farad (F)
Force	$\frac{\text{voltage}}{\text{distance}}$	$[MLT^{-3}I^{-1}]$	volts/meter (V/m)
Permittivity	$\frac{\text{electric flux density}}{\text{electric force}}$	$[M^{-1}L^{-3}T^4I^2]$	farads/meter (F/m)
Admittance	$\frac{\text{current}}{\text{voltage}}$	$[M^{-1}L^{-2}T^3I^2]$	siemen (S)
Inductance	$\frac{\text{voltage}}{\text{rate of current change}}$	$[ML^2T^{-2}I^{-2}]$	henry (H)
Magnetomotive force	current change × turns	$[I]$	ampere-turn (A-t)
Magnetizing force	$\frac{\text{mmf}}{\text{length}}$	$[L^{-1}I]$	ampere-turns/meter (A-t/m)
Magnetic flux	voltage × time	$[ML^2T^{-2}I^{-1}]$	weber (Wb)
Magnetic flux density	$\frac{\text{magnetic flux}}{\text{area}}$	$[MT^{-2}I^{-1}]$	tesla (T)
Permeability	$\frac{\text{magnetic flux density}}{\text{magnetizing force}}$	$[MLT^{-2}I^{-2}]$	henrys/meter (H/m)
Frequency	$\frac{1}{\text{period}}$	$[T^{-1}]$	hertz (Hz)

The following are additional examples of derived units:

The ampere = 1 coulomb/second = s^{-1} **(B–5)**

The volt:

$$E = \frac{W}{q} = \frac{FS}{q} = \frac{\text{kgm}}{\text{s}^2} \times \frac{\text{m}}{\text{C}} = \frac{\text{kgm}^2}{\text{s}^2} \tag{B–6}$$

The ohm:

$$R = \frac{E}{I} = \frac{\text{kgm}^2/\text{s}^2}{1/\text{s}} = \frac{\text{kgm}^2}{\text{s}^2} \times \text{s} = \frac{\text{kgm}^2}{\text{s}} \tag{B–7}$$

The farad:

$$C = \frac{q}{V} = \frac{\text{C}}{\text{kgm}^2/\text{s}^2} = \frac{\text{s}^2}{\text{kgm}^2} \tag{B–8}$$

RC time constant:

$$RC = \frac{\text{kgm}^2}{\text{s}} \times \frac{\text{s}^2}{\text{kgm}^2} = \text{s} \tag{B–9}$$

L/R time constant:

$$L/R = \frac{\text{kgm}^2}{\text{kgm}^2/\text{s}} = \text{s} \tag{B–10}$$

$$\text{The watt} = 1 \text{ joule/second} = \frac{\text{kgm}^2/\text{s}^2}{\text{s}} = \frac{\text{kgm}^2}{\text{s}^3} \tag{B–11}$$

Verification

$$IE = \frac{\text{kgm}^2}{\text{s}^2} \times \frac{1}{\text{s}} = \frac{\text{kgm}^2}{\text{s}^3} \tag{B–12}$$

APPENDIX C Derivation of Amplitude Modulation and Frequency Modulation Formulas

AMPLITUDE MODULATION

Given: $v = E_c\cos \omega_c t + E_s\cos \omega_s t$. Assume that the characteristic curve of a diode is a parabola with the formula $i = A + Bv + Cv^2$. By substitution,

$$i = A + B(E_c\cos \omega_c t + E_s\cos \omega_s t) + C(E_c\cos \omega_c t + E_s\cos \omega_s t)^2 \quad \textbf{(C–1)}$$

Expansion gives

$$i = A + BE_c\cos \omega_c t + BE_s\cos \omega_s t + CE_c^2\cos^2\omega_c t + 2CE_c\cos \omega_c tE_s\cos \omega_s t + CE_s^2\cos^2\omega_s t \quad \textbf{(C–2)}$$

Using the trigonometric identity

$$\cos^2 x = \frac{1 + \cos 2x}{2} \quad \textbf{(C–3)}$$

The term $CE_c^2\cos^2\omega_c t$ becomes

$$\frac{CE_c^2}{2} + \frac{CE_c^2\cos 2\omega_c t}{2} \quad \textbf{(C–4)}$$

and $CE_s^2\cos^2 \omega_s t$ becomes

$$\frac{CE_s^2}{2} + \frac{CE_s^2\cos 2\omega_s t}{2} \quad \textbf{(C–5)}$$

Rearranging the terms yields

$$i = A + \frac{CE_c^2}{2} + \frac{CE_s^2}{2} + BE_s\cos \omega_s t + \frac{CE_c^2\cos 2\omega_c t}{2} + \frac{CE_s^2\cos 2\omega_s t}{2} + \ldots + BE_c\cos \omega_c t + 2CE_c\cos \omega_c tE_s\cos \omega_s t \quad \textbf{(C–6)}$$

The first three terms are dc since they have no t terms. The fourth term is the fundamental modulating signal. The fifth term is the second harmonic of the carrier (double angle). The sixth term is the second harmonic of the modulating signal (double angle). Factoring $BE_c \cos \omega t$ out of the next two terms yields

$$\frac{BE_c(1 + 2E_s \cos \omega_s t)}{2} \cos \omega_c t \quad \text{(C–7)}$$

The amplitude of the $\omega_c t$ carrier wave is time varied according to the frequency of the modulating signal. But

$$BE_c\left(\frac{1}{2} + \frac{2E_s C \cos \omega_s t}{2}\right)\cos \omega_c t \quad \text{(C–8)}$$

becomes

$$\frac{BE_c \cos \omega_c t}{2} + \frac{BE_c E_s \cos \omega_s t \cos \omega_c t}{2} \quad \text{(C–9)}$$

Using the basic trigonometric identities

$$\cos(\omega_c t + \omega_s t) = \cos \omega_c t \cos \omega_s t - \sin \omega_c t \sin \omega_s t \quad \text{(C–10)}$$

and

$$\cos(\omega_c t - \omega_s t) = \cos \omega_c t \cos \omega_s t + \sin \omega_c t \sin \omega_s t \quad \text{(C–11)}$$

and adding gives

$$\cos \omega_c t \cos \omega_s t = {}^1\!/\!_2\cos(\omega_c t + \omega_s t) + {}^1\!/\!_2\cos(\omega_c t - \omega_s t) \quad \text{(C–12)}$$

The final term is then written

$$\frac{BE_c E_s}{2}[\cos(\omega_c t + \omega_s t) + \cos(\omega_c t - \omega_s t)] \quad \text{(C–13)}$$

This term shows the upper and lower side frequencies of the amplitude-modulated waveform.

FREQUENCY MODULATION

Given: $v = A \sin(\omega_c t + m \sin \omega_m t)$, where $\omega_c t$ represents the carrier wave, $\omega_m t$ represents the modulating signal or the information to be carried by the carrier wave, and m is the deviation index. Using the trigonometric identity

$$\sin(x + y) = \sin x \cos y + \cos x \sin y \quad \text{(C–14)}$$

and MacLaurin's theorem

$$\sin x = x - \frac{x^3}{3!} + \frac{x^5}{5!} - \ldots \quad \text{(C–15)}$$

and

$$\cos x = 1 - \frac{x^2}{2!} + \frac{x^4}{4!} - \ldots \quad \text{(C–16)}$$

yields

$$v = A\left[\sin \omega_c t\left(1 - \frac{m^2\sin^2\omega_m t}{2!} + \frac{m^4\sin^4\omega_m t}{4!} - \ldots\right) + \cos \omega_c t\left(m \sin \omega_m t - \frac{m^3\sin^3\omega_m t}{3!} + \ldots\right)\right] \quad \textbf{(C–17)}$$

If all terms above the second power are neglected,

$$v = A\left[\sin \omega_c t\left(1 - \frac{m^2\sin^2\omega_m t}{2} + \cos \omega_c t(m \sin \omega_m t)\right)\right] \quad \textbf{(C–18)}$$

But since by a trigonometric identity

$$\sin^2 x = 1/2(1 - \cos 2x) \quad \textbf{(C–19)}$$

then

$$\begin{aligned} v &= A\left\{\sin \omega_c t\left[1 - \frac{m^2}{2}\left(\frac{1}{2}(1 - \cos 2\omega_m t)\right)\right] + \cos \omega_c t(m \sin \omega_m t)\right\} \\ &= A\left\{\sin \omega_c t - \frac{m^2}{4}\sin \omega_c t + \frac{m^2}{4}\sin \omega_c t \cos 2\omega_m t + m \sin \omega_m t \cos \omega_c t\right\} \quad \textbf{(C–20)} \\ &= A\left\{\sin \omega_c t\left(1 - \frac{m^2}{4}\right) + \left(\frac{m^2}{4}\sin \omega_c t \cos 2\omega_m t\right) + m(\sin \omega_m t \cos \omega_c t)\right\} \end{aligned}$$

By the trigonometric identities

$$\sin(x + y) = \sin x \cos y + \cos x \sin y \quad \textbf{(C–21)}$$

and

$$\sin(x - y) = \sin x \cos y - \cos x \sin y \quad \textbf{(C–22)}$$

The sum . . .

$$\sin x \cos y = 1/2[\sin(x + y) + \sin(x - y)] \quad \textbf{(C–23)}$$

and the difference . . .

$$\cos x \sin y = 1/2[\sin(x + y) - \sin(x - y)] \quad \textbf{(C–24)}$$

Therefore,

$$v = A\left(1 - \frac{m^2}{4}\right)\sin w_c t + \frac{Am^2}{4}\left(\frac{1}{2}\sin(\omega_c t + 2\omega_m t)\right) - \frac{Am^2}{(4)(2)}\sin(\omega_c t - 2\omega_m t) + \frac{Am}{2}\sin(\omega_c t + \omega_m t) - \frac{Am}{2}\sin(\omega_c t - \omega_m t) \quad \textbf{(C–25)}$$

Since there are an infinite number of terms, there are an infinite number of sidebands.

APPENDIX D Impedance Bridge Derivation

This appendix solves a balanced impedance bridge using two methods, the derived formula and parallel-to-series formula from Chapter 10 (see Figure D–1).

FIGURE D–1

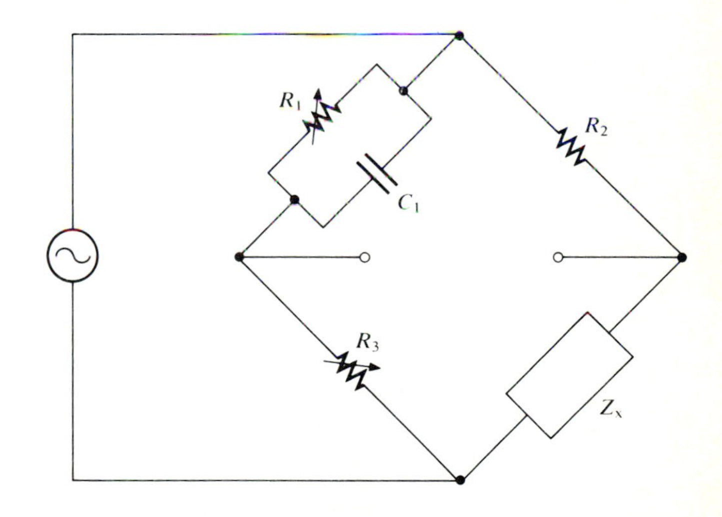

DERIVED FORMULA

Impedance $\mathbf{Z_x}$ must be inductive to provide a positive phase angle that will combine with the negative phase angle of $\mathbf{Z}_1$ ($R_1 \parallel C_1$) to yield a phase angle of zero to match the phase angle of R_2 and R_3. At balance,

$$\mathbf{Z}_1\mathbf{Z}_4 = \mathbf{Z}_2\mathbf{Z}_3 \qquad \textbf{(D–1)}$$

which can be rewritten

$$\mathbf{Z}_4 = \frac{\mathbf{Z}_2\mathbf{Z}_3}{\mathbf{Z}_1} = \mathbf{Z}_2\mathbf{Z}_3\mathbf{Y}_1 \qquad \textbf{(D–2)}$$

Substitution yields

$$R_s + jX_s = R_2R_3\left(\frac{1}{R_1} - j\frac{1}{X_1}\right) \tag{D–3}$$

Since $X_s = 2\pi fL$ and $X_1 = ½\pi f\ C_1$,

$$R_s + j2\pi fL_s = R_2R_3\left(\frac{1}{R_1} + j2\pi fC_1\right) \tag{D–4}$$

Expansion gives

$$R_s + j2\pi fL_s = \frac{R_2R_3}{R_1} + j2\pi fC_1R_2R_3 \tag{D–5}$$

Since the real terms on the left side of the equation must equal the real terms on the right side of the equation, and the imaginary terms must also be equal,

$$R_s = \frac{R_2R_3}{R_1} \tag{D–6}$$

and

$$L_s = C_1R_2R_3 \tag{D–7}$$

As an example, assume that $R_1 = 10\ \text{k}\Omega$, $R_2 = 2.2\ \text{k}\Omega$, $R_3 = 1\ \text{k}\Omega$, $C_1 = 0.1\ \mu F$, and $f = 1000$ Hz. Then,

$$R_s = \frac{(2.2\ \text{k}\Omega)(1\ \text{k}\Omega)}{10\ \text{k}\Omega} = 220\ \Omega \tag{D–8}$$

and

$$L_s = (0.1 \times 10^{-6}\ \text{F})(10^4\ \Omega)(10^3\ \Omega) = 0.22\ \text{H} \tag{D–9}$$

PARALLEL-TO-SERIES FORMULA

Converting the R_1C_1 combination to a series equivalent would result in

$$X_p = \frac{1}{2\pi(10^3\ \text{Hz})(10^{-7}\ \text{F})} = -j1590\ \Omega \tag{D–10}$$

Q for parallel components is defined as

$$Q_p = \frac{R_p}{X_p} = 6.28 \tag{D–11}$$

From

$$R_s = \frac{R_p}{1 + Q^2} = \frac{10{,}000\ \Omega}{1 + 6.28^2} = 247\ \Omega \tag{D–12}$$

and

$$X_s = \frac{X_p}{1 + 1/Q^2} = \frac{1590\ \Omega}{1 + 1/6.28^2} = -j1550\ \Omega \tag{D–13}$$

the real terms are equal:

$$(247\ \Omega)R_s + (1550\ \Omega)X_s = 2.2 \times 10^6\ \Omega \tag{D–14}$$

and the imaginary terms are equal:

$$(-j1550\ \Omega)R_s = (-j247\ \Omega)X_s \tag{D–15}$$

Solving for R_s gives

$$R_s = \frac{(247\ \Omega)X_s}{1550\ \Omega} = 0.159X_s \tag{D–16}$$

Substitution into equation (D–14) yields

$$247\ \Omega\left(\frac{(247\ \Omega)X_s}{1550\ \Omega}\right) + (1550\ \Omega)X_s = 2.2 \times 10^6\ \Omega \tag{D–17}$$

Simplification gives

$$\begin{aligned} 39X_s + 1550X_s &= 2.2 \times 10^6\ \Omega \\ 1589X_s &= 2.2 \times 10^6\ \Omega \end{aligned} \tag{D–18}$$

The reactance is

$$X_s = 1384\ \Omega = 2\pi fL \tag{D–19}$$

Therefore,

$$L = \frac{1384\ \Omega}{2\pi(10^3\ \text{Hz})} = 0.22\ \text{H} \tag{D–20}$$

From equation (D–15),

$$X = \left(\frac{1550\ \Omega}{247\ \Omega}\right)R_s \tag{D–21}$$

Substituting into equation (D–14),

$$(247\ \Omega)R_s + \left(\frac{(1550\ \Omega)\ (1550\ \Omega)}{247\ \Omega}\right)R_s = 2.2 \times 10^6\ \Omega \tag{D–22}$$

and so

$$\begin{aligned} 9974R_s &= 2.2 \times 10^6\ \Omega \\ R_s &= 220\ \Omega \end{aligned} \tag{D–23}$$

The results of this example show that the parallel-to-series conversion formulas derived in Chapter 10 are valid.

APPENDIX E
BASIC Syntax

BASIC TERMS

ADDITION +
ARRAY
ASTERISK *
(multiplication)
CHANGE
DATA
DEF
DIM
DIVISION /
E-- (3E3 = 3 × 10^3)
END
EQUALS =
EXPONENTIATION
** or (2**3 =
2^3 = 2 × 2 × 2)
FILE
FILES
FNEND
FOR-TO
GOSUB
GOTO
GREATER THAN >
(A>B means A is
greater than B)
GREATER THAN OR
EQUAL TO ≥ (A≥B
means A is greater
than or equal to B)
IF END-THEN
IF-GOTO
IF-THEN
INPUT
LESS THAN < (A<B
means A is less than
B)
LESS THAN OR
EQUAL TO ≤ (A≤B
means A is less than
or equal to B)
LET
MAT-CON
MAT EQUALS =
MAT-IND
MAT-INPUT
MAT-INV
MAT-MINUS –
MAT-
MULTIPLICATION *
MAT-PLUS +
MAT PRINT
MAT READ
MAT SCALER
MULTIPLICATION
MAT SIZ
MAT-TRN
MAT-ZER
MINUS –
MULTIPLICATION *
NEXT
NOT EQUAL TO <>
(A<>B means A is
not equal to B)
ON-GOTO
ON THEN
PARENTHESES ()
PLUS +
PRINT
PRINTUSING
QUOTE
RANDOMIZE
READ
REM
RESTORE
RESTORE$
RESTORE*
RETURN
SCRATCH
SET
SETDIGITS
STOP
SUBTRACTION –
WRITE

MICROSOFT BASIC RESERVED WORDS

$
%
!
#
&
'
AND
AUTO
CALL
CDBL
CHAIN
CINT
CLEAR (can control
stack size)
CLOAD
CLOSE
COMMON
CONSOLE
CONT
CSAVE
CSGN
CVD
CVI
CVS
DEF DBL
DEF INT
DEF SNG
DEF STR
DEF VSR
EDIT
EOF
EQV
ERASE
ERL
ERR
FIELD
FIX
FRE
GET
HEX$
IF THEN ELSE
IMP
INKEY$
INP
INPUT#
INPUT$
INSTR
KILL

LEFT$
LEN
LINE INPUT
LINE INPUT#
LLIST
LOAD
LPOS
LPRINT
LPRINT USING
LSET
MERGE
MID$
MKD$
MKI$
MKS$
MOD
NAME
NOT
NOTRACE
NULL
OCT$
ON ERROR GOTO
ON-GOSUB
OPEN
OR
OUT
PEEK
POKE
POS
PRINT#
PRINT# USING
PUT
RENUM
RESUME
RESUME NEXT
RIGHT$
RND
RSET
SPACE$
SPC
STR$
STRING$
SWAP
TRACE
TROFF
TRON
VAL
VALPTR
WAIT
WEND
WHILE
WIDTH
WRITE#
XOR

BASIC LIBRARY FUNCTIONS

ABS
ASC
ATN
CHR$
CLG
CON
COS
COT
DEL
DET
EXP
FRE (string variable)
IDN
INT
INV
LGT
LOC
LOF
LOG
NUM
PRINT FRE(0)
RND
SGN
SIN
SPC()
SQR
TAB
TAN
TRN
ZER

BASIC SYSTEM COMMANDS

BYE
CATALOG
GOODBYE
LIST
MON
NEW
OLD
RENAME
RUN
SAVE
SCRATCH
SYSTEM
UNSAVE

APPENDIX F
The Laboratory Notebook

The laboratory notebook is a good alternative to writing standard laboratory reports. This notebook, based on the logbooks kept by technicians in industry, allows students to keep chronological records of all experiments performed in the lab. In addition, it helps to prevent the loss of data, calculations, and other materials pertinent to lab work.

The notebook should be bound, not looseleaf, to prevent the loss of pages, and pages should be numbered consecutively. Entries should be made as follows:

1. Include your name and the date. If two or more students are allowed to work as a group, they should keep individual notebooks, but all names should be listed.
2. Include the title of the experiment and list any references to textbooks or other sources with page numbers.
3. Outline the objectives of the experiment.
4. List materials and equipment to be used. Include serial or other identification numbers to increase the reproducibility of the results.
5. Outline the procedure to be followed step-by-step.
6. Draw wiring diagrams and schematics neatly.
7. Present data tables, waveforms, graphs, and calculations neatly and in logical order.
8. Conclude with a summary. Discuss any problems that may have occurred as well as their solutions.

All corrections should be made neatly with a single line through the error and the correction written above. A sample laboratory notebook entry appears in Figure F–1.

FIGURE F–1
Laboratory notebook example

J. Smith
9/10
11:00 a.m.

Experiment #1
Electronics Lab Manual p.5
Electronics Text ch. 1, sec. 2, p. 30

Objective:
Prove experimentally that, in a simple resistive circuit, the current is inversely proportionate to the resistance as stated by Ohm's law.

Equipment / Materials:
dc Power Supply #31250
Breadboard #7412
Resistance Decade Box #56321
Multimeter #73185

Procedure:

1. Wire the following circuit:

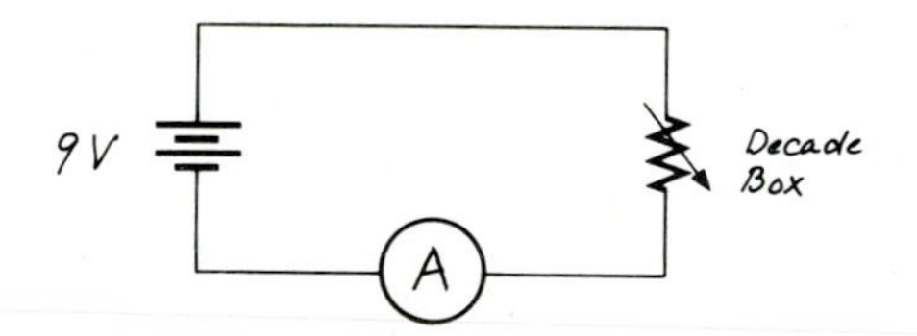

2. Set the resistance to 1 kΩ, then measure the current. Record the current in tabe I. Calculate the current and record in table II.

3. Increase the resistance by 1 kΩ, then measure and calculate the current as for step 1. Repeat until resistance equals 10 kΩ

4. Graph experimental values.

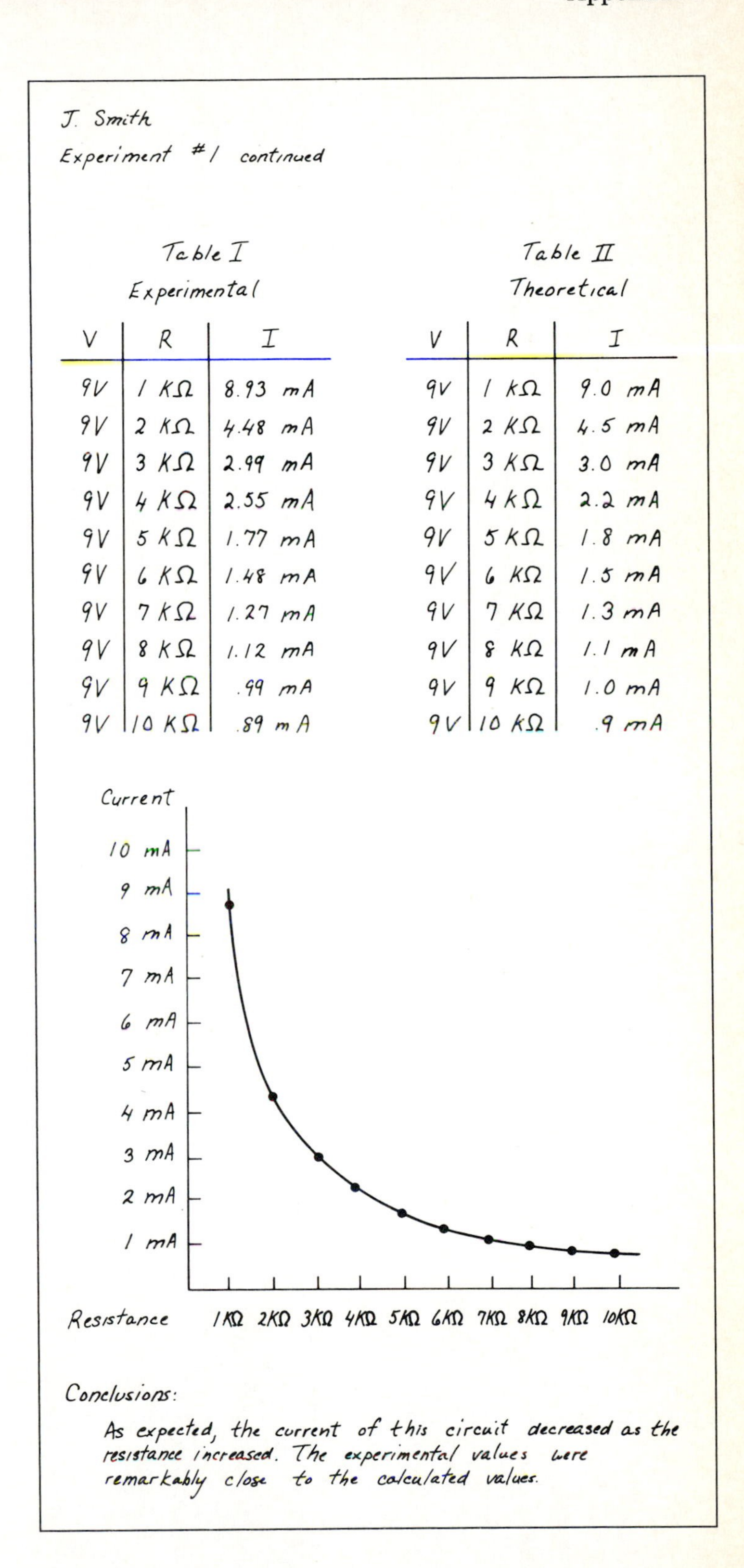

J. Smith
Experiment #1 continued

Table I
Experimental

V	R	I
9V	1 KΩ	8.93 mA
9V	2 KΩ	4.48 mA
9V	3 KΩ	2.99 mA
9V	4 KΩ	2.55 mA
9V	5 KΩ	1.77 mA
9V	6 KΩ	1.48 mA
9V	7 KΩ	1.27 mA
9V	8 KΩ	1.12 mA
9V	9 KΩ	.99 mA
9V	10 KΩ	.89 mA

Table II
Theoretical

V	R	I
9V	1 KΩ	9.0 mA
9V	2 KΩ	4.5 mA
9V	3 KΩ	3.0 mA
9V	4 KΩ	2.2 mA
9V	5 KΩ	1.8 mA
9V	6 KΩ	1.5 mA
9V	7 KΩ	1.3 mA
9V	8 KΩ	1.1 mA
9V	9 KΩ	1.0 mA
9V	10 KΩ	.9 mA

Current
10 mA
9 mA
8 mA
7 mA
6 mA
5 mA
4 mA
3 mA
2 mA
1 mA
Resistance
1KΩ 2KΩ 3KΩ 4KΩ 5KΩ 6KΩ 7KΩ 8KΩ 9KΩ 10KΩ

Conclusions:

As expected, the current of this circuit decreased as the resistance increased. The experimental values were remarkably close to the calculated values.

FIGURE F–1
Laboratory notebook example, *continued*

APPENDIX G
Important Formulas

$$\sigma = \sqrt{\frac{\sum d^2}{n - 1}} \quad \textbf{(2–7)}$$

$$A = 1 + \frac{R_f}{R_1} \quad \textbf{(3–29)}$$

$$A = -\frac{R_f}{R_1} \quad \textbf{(3–30)}$$

V_{pk} = amplitude **(5–35)**

$V_{pk\text{-}pk} = 2V_{pk}$ **(5–36)**

$V_{avg} = 0.318V_{pk}$ for a half-wave rectified sinewave **(5–37)**

$V_{avg} = 0.636V_{pk}$ for a full-wave rectified sinewave **(5–38)**

$V_{rms} = 0.707V_{pk}$ for a pure sinewave **(5–39)**

$$S = \frac{0.45}{I_{fs}} \quad \textbf{(5–40)}$$

$$S = \frac{0.9}{I_{fs}} \quad \textbf{(5–41)}$$

$$S = \frac{1}{I_{fs}} \quad \textbf{(5–42)}$$

$$\sin\theta = \frac{y_0}{y_{max}} \quad \textbf{(8–14)}$$

$$\theta = \arcsin\frac{y_0}{y_{max}}$$

$$R_4 = \frac{R_2R_3}{R_1} \quad \textbf{(10–9)}$$

$$V_{BA} = \frac{E\Delta R}{4R} \quad \textbf{(10–19)}$$

$$\mathbf{Z}_2\,\mathbf{Z}_3 = \mathbf{Z}_1\,\mathbf{Z}_4 \quad \textbf{(10–30)}$$

$$R_p = \frac{R_s^2 + X_s^2}{R_s} \quad \textbf{(10–61)}$$

$$X_p = \pm\frac{R_s^2 + X_s^2}{X_s} \quad \textbf{(10–62)}$$

$$R_s = \frac{R_p}{1 + Q^2} \quad \textbf{(10–69)}$$

$$X_s = \frac{X_p}{1 + 1/Q^2} \quad \textbf{(10–72)}$$

Glossary

accelerometer A transducer used to convert changes in velocity to measurable electrical quantities.

accuracy The maximum amount of error in the measurement by a particular instrument in relation to full-scale measurement, expressed as a percentage.

active device An electronic component whose output depends on a power source other than the main input signal.

admittance The ease with which ac flows in a circuit. The reciprocal of reactance.

alternating mode A mode of oscilloscope operation that allows two signals to be viewed through the rapid switching of the processing circuits.

analog signal The representation of a numerical quantity as a continuous wave that changes in proportion to the input.

analog-to-digital (A/D) converter A device that produces a digital output from a continuous-wave input.

angstrom unit A unit of measurement of the wavelength of light. Equal to 10^{-8} cm.

aquadag A conductive graphite coating on the side walls of a CRT that serves as a path to ground for spent electrons when applied to the inner surface and as a capacitive filter for the applied high voltage when on the outer surface.

ATE (*A*utomatic *T*est *E*quipment) A programmable measuring system capable of testing an entire electronic circuit or network to determine faults and provide parameter data needed for quality control.

attenuator An adjustable resistive network for reducing the amplitude of an electronic signal without causing phase or frequency distortion.

autoranging The automatic switching of a multirange instrument from one scale to another without exceeding the maximum scale limit.

average The value under the curve of an ac wave divided by the period. Equal to 0.636 × peak value.

ayrton shunt A low-resistance parallel connection used to increase the measuring capacity of a meter.

beam modulation *See* *z*-axis modulation.

bleeder resistor A resistor used to draw a fixed amount of current to improve voltage regulation or to discharge filter capacitors after the power supply is turned off.

carbon button microphone A microphone that depends on the variations of the resistance of its carbon contacts for operation.

chop mode A mode of oscilloscope operation that allows two signals to be viewed by rapid alternate sampling.

clamper A diode circuit that adds a fixed bias to a signal to hold the voltage or current to a specified level.

class-A amplifier A linear amplifier with an operating point that allows output current to flow at all times.

class-AB amplifier A linear amplifier with an operating point that allows output current to flow for more than one-half but not the entire cycle.

class-B amplifier A nonlinear amplifier with an operating point that allows current to flow one-half cycle (180°).

class-C amplifier A nonlinear amplifier with an operating point that allows output current to flow less than one-half the entire cycle (usually 120°).

clipper A diode circuit that limits the peak value of a signal.

closed-loop amplifier An operational amplifier

application that uses a feedback loop to continuously feed output back to the input for comparison.

coherent light A monochromatic light beam with parallel rays in phase with one another.

comparator An active circuit that compares two input signals and indicates agreement or disagreement.

condenser (capacitor) microphone A device that produces an audio-frequency current dependent on variations of its electrostatic capacitance. Also called *electrostatic microphone.*

coupling The relationship between two circuits or networks that allows energy to be transferred efficiently from one to the other.

CRO (*C*athode *R*ay *O*scilloscope) An instrument that uses a CRT to allow visualization of an ac sinewave as well as other instantaneous values and waveforms of an electrical nature.

CRT (*C*athode *R*ay *T*ube) A tube into which an electronic beam may be focused to produce a visible pattern on a luminescent field.

cursor An electronically generated line that can be moved over an oscilloscope trace to highlight all or part of a signal.

Darlington driver A transistor circuit in which two transistors of the same type are connected together to act as a single transistor (collector to collector, emitter 1 to base 2) producing an overall current gain that is the product of both current gains.

delayed sweep In a CRO, a sweep that is not allowed to begin for a predetermined amount of time after the triggering pulse occurs.

delay line A circuit in the oscilloscope that delays the beginning of the vertical signal long enough to synchronize it with the horizontal sweep.

deviation The difference between the actual and measured values of a quantity.

differentiation A calculus process by which the slope of the graph of a mathematical function can be found.

digital Related to digits or discrete quantities.

digital-to-analog (D/A) converter A device that produces a continuous-wave output from a digital pulse input.

digitizing The process of converting an analog measurement into digits.

dual trace A mode of oscilloscope operation in which two input signal channels use a single electron beam.

dynamic microphone A device in which a diaphragm attached to a moving coil in a fixed magnetic field produces an audio-frequency ac current. Also called *moving-coil microphone.*

dynode A secondary electrode capable of giving off secondary emission electrons when impacted by an electron beam.

effective value *See* rms.

electron beam A stream of electrons moving in the same direction under the influence of an electric or magnetic field to produce an image on the screen of a CRT.

electron gun The portion of a CRT that produces, focuses, and controls the electron beam to produce an image.

electrostatic microphone *See* condenser microphone.

ESD (electrostatic discharge) An undesirable discharge of static electricity.

feedback loop The transmission path from the loop output signal to the input.

flow meter A transducer for measuring the movement of fluids (liquids, gases, etc.).

focus The sharp definition of an electron beam to produce a clear image on a CRT screen.

frequency compensation A method of expanding the higher or lower limits of the passband of an amplifier.

frequency synthesizer A highly accurate frequency source that produces a range of frequencies by combining a standard frequency produced by a crystal with the frequency of a free-running oscillator.

full-duplex A communication system that allows simultaneous two-way communication.

full-scale deflection The largest possible reading on a meter scale or digital display.

gallium arsenide light-emitting diode *See* injection laser diode.

graticule A calibrated screen etched on or placed in front of the face of an oscilloscope CRT for measurement purposes.

gross errors Measurement errors caused by improper techniques or faulty or inappropriate equipment.

half-duplex A communication system that allows two-way transmission, but not both ways simultaneously.

Hall-effect transducer A semiconductor device used for measuring ac power and magnetic field strength.

heterodyning The mixing of two signals to produce a signal that is the sum or difference of the inputs.

hold off Term used for the blanking of the electron beam during retrace.

impedance The total opposition a circuit offers to the flow of alternating current, measured in ohms.

indication In electronic instruments, a value representing the amount of energy present rather than a comparison to a standard value.

infrared detector A transducer that uses a semiconductor, a thermocouple, or a photo-sensing device to detect infrared (invisible) radiation.

injection laser diode A semiconductor device in which radiation is produced when electrons recombine with holes.

integration A calculus process by which the area under the graph of a mathematical function can be found.

intensity modulation *See* *z*-axis modulation.

internal resistance The inherent resistance of a circuit component, battery, or generator.

interpolation The process of determining a value that lies between two known values on a fixed scale.

laser (*l*ight *a*mplification by *s*timulated *e*mission of *r*adiation) A device for transforming incoherent light to a very intense beam of coherent light.

LCD (*L*iquid *C*rystal *D*isplay) An alphanumeric display created by applying current to two glass plates separated by a nematic liquid crystal solution.

leakage current A small, unwanted current flow.

leakage resistance The resistance path through which leakage current flows.

LED (*L*ight-*E*mitting *D*iode) A semiconductor device that emits light when a forward bias current is applied.

linear amplifier An amplifier that produces an output directly proportional to the input.

linearity The ability of a meter to provide equal angular deflections proportional to the amount of applied current. Expressed as a percentage of full-scale deflections.

Lissajous patterns Patterns produced in a CRO when two sinewaves of different amplitude and frequency are applied separately to the vertical and horizontal inputs.

local oscillator An oscillator in a mixing circuit whose output is combined with its input to produce an intermediate signal, which is a sum or difference frequency.

LRC bridge A test device that can be used to determine inductance *(L)*, resistance *(R)*, capacitance *(C)*, or the dissipation factor *(D)*, of the component or circuit being tested.

make-before-break contact switch Multiple contacts arranged so that a new circuit is established before the old one is broken.

Maxwell bridge A four-arm bridge normally used for measuring inductance in terms of resistance and capacitance, or capacitance in terms of resistance and inductance. The null balance is independent of frequency.

measurement The comparison of a value to a standard value.

measurement acquisition chip (MAC) A microchip found in some digital multimeters that measures a conditioned input with an analog-to-digital converter and controls the power supply and continuity tone generator.

measuring system Any device or set of devices that combine to input, process, and measure some kind of signal.

modulation The process of transferring information by using one information signal to vary another.

moving-coil microphone *See* dynamic microphone.

network A system of electronic components and transmission lines that provides connections between signal sources and loads.

nixie tube A glow tube with stacked metal elements that convert a combination of electric pulses to a visual display.

normalization A calibration technique that involves restoring a slightly unbalanced potentiometric bridge to its balanced condition.

null method A method of measurement whereby the circuit is balanced, in order to bring the pointer of the indicating instrument to zero, before a measurement is taken.

open-loop amplifier An operational amplifier application without a feedback loop.

operational amplifier (op-amp) A high-gain dc amplifier with high input and low output impedance that can be used to perform a variety of mathematical operations.

optical transducer A measuring device used to convert light energy to an electrical signal.

parallax An error caused by an apparent change in value due to a change in the viewer's position.

paraphase amplifier A two-stage amplifier that produces two antiphase signals of equal amplitude.

passive device An electronic device that has no gain or control and requires no other input than a signal to function.

peak The maximum instantaneous value of voltage or current.

peak detector A device that measures and indicates the peak voltage value of a sinewave by measuring the maximum charge of a capacitor.

peak-peak The algebraic difference between the positive and negative maximum values of a sinewave.

peak value The maximum instantaneous value of a time-varying quantity such as a voltage, current, or power, measured from a zero reference point.

persistence The time required for the phosphor of a CRT to stop glowing after it has been energized by an electron beam.

phase splitter A device that uses a single input to produce two or more output waveforms that are out of phase with one another.

phosphor A layer of luminescent material applied to the inner surface of a CRT screen that glows when energized by the electron beam.

photodiode A semiconductor device that displays an increased reverse current when exposed to light.

phototransistor A germanium semiconductor device in which holes are generated by light absorption, causing an increased photocurrent at the collector.

photovoltaic cell A self-generating semiconductor device that converts light energy to electricity when exposed to light.

potentiometer A three-terminal resistor with an adjustable contact or wiper as its center terminal that allows the resistance to be varied.

potentiometric bridge An instrument in which a potentiometer is used as a voltage divider to produce a known voltage that can be balanced against an unknown voltage.

precision A measure of the consistent accuracy of an instrument.

primary standard A unit of measurement specifically defined and established by the U.S. National Bureau of Standards against which secondary standards are calibrated.

pull-up resistor A resistor that provides a small voltage differential above ground.

push-pull amplifier A balanced amplifier with two identical signal branches that operate in phase opposition.

quantizing Expressing an analog value as the nearest one of a discrete set of prechosen values.

random errors Errors that occur for no apparent reason.

reactance Opposition to the flow of ac. Capacitive reactance (X_C) is the opposition by a capacitance; inductive reactance (X_L) is the opposition by an inductance.

reference angle An angle measuring between 0° and 90°.

reflection The phenomenon that causes a waveform striking a different medium to be returned to the original medium with angles of incidence and reflection equal and in the same plane.

refraction A change in the propagation of a waveform resulting from the waveform's passing from one medium to another.

resistance thermometer A temperature-measuring device that indicates a change in temperature as a function of the change of resistance it causes.

response time The time required for an instrument to provide an indication from a zero position when a quantity to be measured is applied.

retrace The line traced by the electron beam of a CRO as it travels from the end of one sweep to the beginning of the next.

ribbon microphone A microphone that uses a ribbon driven by sound waves as a moving conductor.

rms (*r*oot *m*ean *s*quare value) The value of alternating current or voltage that will produce the same heating effect as an equal amount of direct current or voltage. Equal to 0.707 × peak value. Also called *effective value.*

sampling oscilloscope An oscilloscope that uses progressive samples of adjacent repetitive waves to produce a replica of the original signal.

secondary standard A commercially available unit of measurement that has been calibrated to a primary standard for guaranteed accuracy.

sensitivity The ratio of response of a measuring device to the magnitude of the input.

shunt A precision resistor placed in parallel with an ammeter to increase the range of the ammeter.

signal injection A troubleshooting technique that involves the application of a test signal monitored to isolate faulty stages of an electronic circuit.

signal tracing A troubleshooting technique that involves following a signal through various stages of an electronic circuit to determine which stage is faulty.

simplex A communication system that allows only one-way transmission.

solder bridge An inadvertent flow of solder between connections which may cause short circuits or other electronic faults.

standard cell A primary cell that serves as a standard of voltage.

standard deviation A measure of the variation of data from the average of the data.

storage oscilloscope An oscilloscope that can store an instantaneous CRT image, allowing it to be observed for any required amount of time.

strain gage A measuring device for converting force, pressure, and tension to an electrical signal.

stray capacitance An unwanted capacitance caused by the leads and wires connecting circuit components.

successive approximations A method of measurement by which the parameter to be measured is compared to a preset group of standards, each one closer to the real value.

sweep The horizontal movement of the electron beam in a CRO.

systematic errors Errors caused by limitations of the equipment being used.

thermistor A semiconductor device used to measure temperature changes as its resistance varies with changes in temperature.

thermocouple A temperature-measuring transducer composed of two dissimilar metals. When a difference in the temperature of a test junction and a reference junction occurs, electric current flows in the conductors proportional to this difference in temperatures.

transducer A device that transfers energy from one system to another, especially from one form to another to enable measurement.

trigger A circuit in a CRO that initiates the horizontal sweep.

vector The term or symbol for a directed quantity consisting of both magnitude and direction.

wave soldering A method of soldering the preinserted components of a printed circuit board uniformly and instantaneously by passing the board over a flow of liquid solder.

Wheatstone bridge A null-type resistance-measuring device consisting of four resistance arms and powered by ac or dc voltage.

Wien bridge An ac bridge used to measure inductance or capacitance in terms of resistance and frequency.

***z*-axis modulation** The variation of the electron beam of a CRT through the application of a pulse or waveform to the control grid or cathode. Also called *beam modulation* or *intensity modulation.*

Solutions to Odd-Numbered Problems

CHAPTER 2

1. 85%
3. mean = 5795.8, standard deviation = 18.7249, probable error = 12.63
5. mean = 29.4, standard deviation = 5.26, probable error = 3.55
7. a. 3 b. 2 c. 4 d. 1 e. 2
9. a. 650 b. 590,000 c. 2.0 d. 2100 e. 1
11. 0.88 W ± 5%, or 0.88 W ± 0.044 W
13. 180 Ω ± 11%, or 180 Ω ± 19.8 Ω

CHAPTER 3

3. The negative sign indicates a 180° phase shift, and the 10 means that the output has 10 times the amplitude of the input.
5. 5 V − 3 V = +2 V
7. $V_{out} = -\frac{60\ \text{k}\Omega}{30\ \text{k}\Omega}(-5\text{V} + 3\text{V} - 2\text{V}) = -2(-4\text{V}) = +8\ \text{V}$
9. $x = 2 - \frac{2}{3}y$, $y = 5 + 3x$. Figure S–1 is one possible solution.

FIGURE S–1

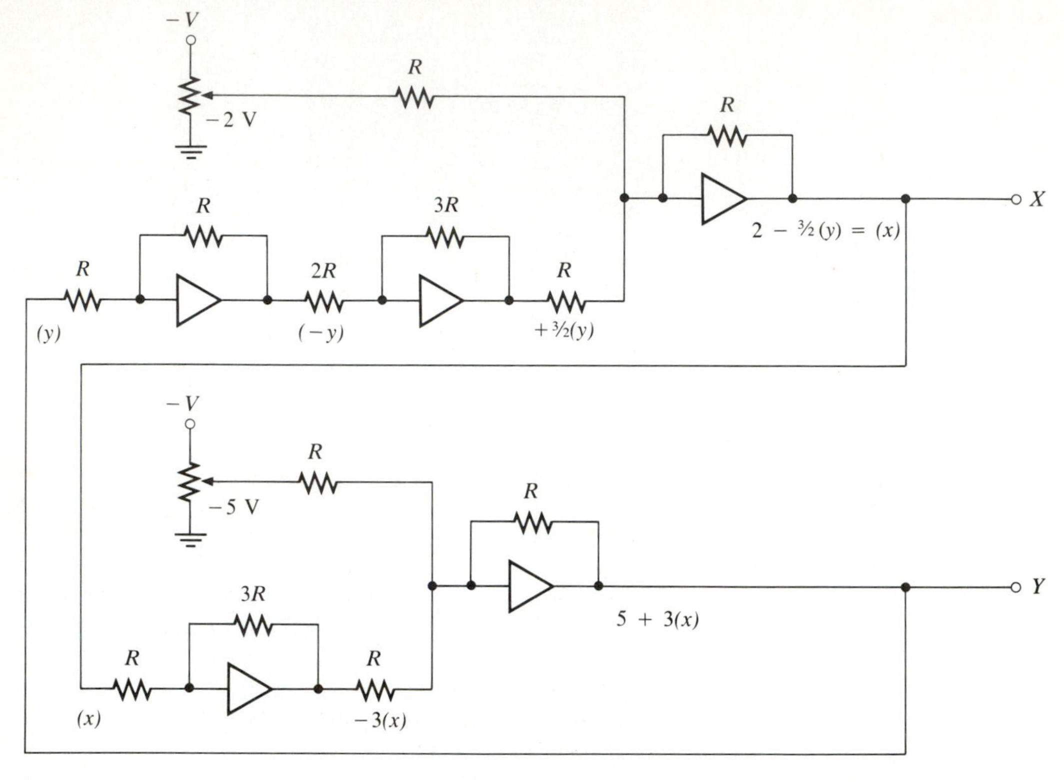

CHAPTER 4

1. 75 Ω
3. R_{sh} = 1.32 Ω, R_m = 1.3 Ω
13. R_s = 49,950 Ω
15. $R_{s(1\text{ V})}$ = 18 kΩ. The other resistors will be added in series. $R_{5\text{ V}}$ = 80 kΩ, $R_{10\text{ V}}$ = 100 kΩ, $R_{50\text{ V}}$ = 800 kΩ
17. 50 kΩ
19. >99%
21. >99%
23. quarter-scale = 36 kΩ − 9 kΩ = 27 kΩ, three-quarter-scale = 12 kΩ − 9 kΩ = 3 kΩ
25. $R_{multiplier}$ = 1450 Ω, R_{sh} = 500 Ω

CHAPTER 5

1. 87,900 Ω
3. for 95% accuracy, 0.1 μF; for 99% accuracy, 0.5 μF
7. S_{ac} = 18 kΩ/V, S_{dc} = 20 kΩ/V
9. 15 V rms = 21.2 V peak = 42.4 V peak-peak = 6.75 V average. The average voltage for a half-wave rectifier is 6.75 V, and the average voltage for a full-wave rectifier is 13.5 V.

CHAPTER 8

11. 150° or 210°

13. a. amplitude = 15 V peak-peak, frequency = 5 kHz
b. amplitude = 30 V peak-peak, frequency = 250 kHz
c. amplitude = 30 V peak-peak, frequency = 1 MHz

CHAPTER 9

1. 0.58 mA

3. $R_{\mathrm{Th}} = 5.45\ \mathrm{k\Omega}$, $V_{\mathrm{Th}} = 2.72\ \mathrm{V}$

5. 0.488 V

7. a. 10 mA
b. E_s
c. 1.6 V
d. $\dfrac{2.6195\ \mathrm{V}}{49.98\ \Omega + R_m}$
e. If the full-scale deflection current of the galvanometer is 50 μA, the sensitivity resistance is 52,350 Ω − R_m.

CHAPTER 10

1. 7.5 kΩ

3. 500–5000 Ω

5. $R_{\mathrm{Th}} = 5813\ \Omega$, $V_{\mathrm{Th}} = -1.69\ \mathrm{V}$

7. $R_p = 250\ \mathrm{k\Omega}$, $X_p = 6.3\ \mathrm{M\Omega}$

9. $R_x = 102\ \Omega$, $X_L = 427.2\ \Omega$, $L = 68\ \mathrm{mH}$

CHAPTER 11

1. $R_1 = 5\ \mathrm{k\Omega}$, $R_2 = 95\ \mathrm{k\Omega}$

3. 4 doublers, 2 triplers

CHAPTER 14

3. 7.5×10^{14} Hz

5. 3.5 lb

Index

A alternating B, 98
A − B, 114–115
A + B, 114–115
ac analog meter
  balanced bridge electronic voltmeter, 70–71
  frequency considerations, 69–70
  frequency response, 69–70
ac bridge, 146–155
ac detectors, 72
ac measurement with an oscilloscope, 107–108
Accelerometers, 196
Accuracy, 14
Acoustical transducers, 201–203
Active, 93
Ammeter, 47–50
Amplification
  buffer amplifier, 29
  closed-loop, 27–29
  open-loop, 25–26
Amplifiers. *See* Operational amplifiers, 163
  IF, 170
  vertical, 92–94
Amplitude, 63–64
Amplitude modulation, 215, 241–242
Analog meter movement, 45
Analog-to-digital (A/D) conversion
  converters, 166
  counter, 75
  cycle, 79
  dual-slope integrator type, 76–77
  integrator, 166
  multirange, 48–49
  ramp-type, 75
  successive approximation type, 75–80
Analyzers, 185–187
  logic, 185
  network, 185–186
  network/system, 186
  system, 186–187
  waveform, 186
Aquadag, 91
Attenuator, 93–94, 163–164
Automatic test equipment (ATE), 219–227
Avalanche effect, 7
Average or dc equivalent, 64–65
Ayrton shunt, 49–51

Backdriving, 225
Bandwidth measurement, 72
BASIC syntax, 249–250
Beam finder, 90
Bed of nails, 221–222
Bench testing, 181
Biological transducers, 205–206
Blanking, 94–95, 97
Bridge
  ac, 144–155
  automatic, 154
  balanced, 141–142
  capacitance comparison, 147–148, 152
  electronic voltmeter, 70
  Hay, 152
  impedance, 153–155
  manual, 153–154
  Maxwell, 151
  normalizing, 132
  potentiometric, 131–134
  slide-wire, 131–132

Bridge, *continued*
slightly unbalanced, 130, 144–146
Thevenin equivalent, 142–146, 231–232
Wheatstone, 141–146
slightly unbalanced, 144
Thevenin equivalent, 231–232
Wien, 152–153
Brightness control, 89
Busses
GPIB, 180
IEEE, 180

Calibration
curve, 15
of an oscilloscope, 106
Carbon button microphones, 202–203
Carrier, 166, 211
Cathode ray oscilloscope, 88
Chemical transducers, 205
Chopped mode, 97–98
CHU, 188
Clampers, 161–162
Clippers, 161–162
Coherent, 210
Condenser microphone, 202
Constructive interference, 210
Controller, IEEE-488, 222
Conventional oscilloscope, 91
CRT, 88–89
Current sensing, 93

dc ammeter, 45–51
Ayrton shunt, 49–51
multirange ammeter, 45–51
sensitivity, 52
shunting, 48
dc analog meter, 45
full-scale deflection, 47–48
internal resistance, 46–47
meter movement, 46
dc bridges, 141–146
dc equivalent, 64–65
dc measurement with an oscilloscope, 106–107
dc voltmeter, 51–53
loading, 53, 54
multirange voltmeter, 51–52
sensitivity, 52–53
series dropping resistor, 51
Delayed sweep, 98–99, 115–116
Delay line, 94
Demodulation, 168
Destructive interference, 210
Detector, 170
infrared, 199
Deviation, 17
standard, 17–19
Differential-pressure transducer, 203

Digital multimeter (DMM), 73–85
A/D conversion cycle, 79
conductance measurement, 81
continuity measurement, 81
criteria for selection, 74–75
frequency considerations, 82
frequency measurement, 82
history, 81–82
measurement acquisition chip, 79
microcomputer, 78–79
resistance measurement, 81
safety, 82
typical DMM, 77–82
voltage measurement, 79–80
Digital-to-analog converters (d/a), 166–168
Digital-versus-analog, 75
Display, 91–92
Dissipation factor, 153
Domed emitter, 214
Dual trace, 97–98
alternating mode, 97–98
chopped mode, 97
Lissajous, 97
Dynamic microphones, 201–202
Dynodes, 207

Electric current, 236
effect on human body, 3
Electric potential, 237
Electric shock, 2
Electron beam, 88–89
Electron gun, 88
Electrostatic discharge, static charge, 6–7
Energy, 235
Errors
gross, 15
probable, 19–20
random, 15
systematic, 15, 47–48

Fiber optics
advantages, 212–213
disadvantages, 213
history, 209–210
uses, 214–215
Field service, 116
Floating grounds, 3–4
Flow meters, 196
Frequency, 63
considerations, 69–70
counters, 183–184
dividers, 164
intermediate (IF), 170
measurements, 109–114
multipliers, 164–165
Frequency modulation, 169–170, 242–243
Frequency synthesizers, 182–183

Full-duplex transmission, 211
Full-scale deflection, 47

Gages, 194–195
  bonded resistance strain, 194
  dummy, 195
  strain, 194–195
Geiger tube, 206
Generator
  microwave, 185
  Selsyn, 205
  sweep, 184–185
Germanium gamma ray detector, 207
GPIB bus, 180
Graticules, 91

Half-duplex transmission, 212
Hall-effect transducers, 204
Heterodyning, 165
Holdoff, 97
Horizontal
  plates, 87
  section, 94–99
  sweep, 97–99

IEEE bus, 180
Impedance bridges, 153–155
  automatic, 154
  derivation, 245–247
  manual, 153–154
  theory of operation, 154–155
Impedance conversion formulas, 148–151
Infared detector, 199
Injection laser diode, 214
Input attenuator, 93–94
Interference
  constructive, 210
  destructive, 210
Intermediate frequency (IF), 170
Interpolation, 16
Isolation transformer, 4, 8

Lab procedures, 5–7
Laboratory notebook, 251–253
Ladder network, 229–230
Laser, 210–211
LCD (liquid crystal display), 7
Leakage, 67, 69–70
LED (light-emitting diode), 214
Light wave parameters, 211
Linearity, 66, 67
Linear-variable differential transformer, 204
Lissajous figures, 97, 111–114
Loading, 8, 53–54
  circuit, 93
  problems, 53–54
Logbook, 5, 11, 251–253

Logic analyzer, 185
Logic probes, 183
Logic pulsers, 183

Magnetic transducers, 203–205
Make-before-break switches, 49
Maxwell bridge, 151
Measurement, 13
  conductance, 81
  continuity, 81
  current, 80
  frequency, 81
  resistance, 81
  voltage, 79
Measurement acquisition chip (MAC), 79–82
Measuring system, 179
Microphone,
  ratings, 203
  ribbon, 202
Microprocessor control, 184
Microwave, 185
Modulating wave, 168
Modulation, 168–169
  amplitude, 241–242
  frequency, 242–243
  phase, 169–171
  pulse, 170
  pulse amplitude (PAM), 170
  pulse frequency (PFM), 170
  pulse position (PPM), 170
  pulse width (PWM), 170
Multimeters, digital, 73–82

National Bureau of Standards, 14
  WWV, 14, 114
Network analyzer, 185–186
Network/spectrum analyzer, 186
Noncoherent light, 210
Notebook, laboratory, 251–253
Nuclear transducers, 206–207
Null, 132

Ohmmeters, 54–56
  series-type, 54–56
  shunt-type, 54–56
Operational amplifiers
  buffer amplifier, 29
  characteristics of, 24
  feedback-loop amplifier, 27–29
  history, 24
  inverting with a feedback loop, 27–29
  mathematical operations, 30–31
    addition, 30
    differentiation, 33
    division, 30
    integration, 34
    multiplication, 30
    subtraction, 31

Operational amplifiers, *continued*
  non-inverting with feedback, 27–28
  open-loop amplifier, 25–27
  uses, 24
Optical fiber links, 211
Optical transducers, 199–201
Oscilloscopes, 85, 87–116
  conventional, 91
  sampling, 91, 99
  storage, 92
    analog, 99–100
    digital, 99

Parallax, 16
Passive, 93
Peak, 63
Peak detector, 67–68
Peak-peak, 64
  detector, 68
  displacement, 108–109
Persistence, 90
Personal safety, 1–2
Phase measurement
  dual trace method, 110–111
  using Lissajous figures, 111–112
Phase modulation, 170
Phase splitter, 94
Photoconductive cell, 199–201
Photodiode, 199
Photomultiplier tube, 207
Phototransistor, 199–200
Photovoltaic cell, 200–201
Potentiometer, 126–131
  definition, 126
  linear precision, 134
  slide wire, 131
  slightly unbalanced, 130
  Thevenin equivalent, 127, 128
Potentiometric bridge, 131–134
  definition, 132
  measuring potentials, 132
  normalizing, 132
  sensitivity control, 132
  slide wire, 131
  Thevenin equivalent, 132
  uses, 134
Power, 237
Precision, 14, 74
Pressure transducers, 195
Primary standard, 15
Probes, 7, 92–93
  × 10, 92–93
Pulse amplitude modulation, 170, 215
Pulse frequency modulation, 170
Pulse position modulation, 170
Pulse width modulation, 170

Q factor, 150–154

Radiation detector, 206–207
Range selector, 93–94
Rectifiers, rectification, 64–68
Resistance measurement, internal of meter movement,
    46, 59
Resistance thermometers, 197–198
Rms, 64
  detector, 68
Rounding off, 17

Safety, 1–5
  high-voltage probe, 85
  multimeter, 82
Sampling oscilloscope, 91, 99
Secondary standard, 14
Selsyn generator, 205
Sensitivity, 52–53, 79,
Series dropping resistor, 51
Shunt, 48
  Ayrton, 49–51, 59–61
SI units, 235–237
Signal
  generators, 180
  injection, 9–10
  sources, 180–181
  tracing, 8–9
Significant figures, 16
Simplex transmission, 211
Space optical systems, 213–214
Standard deviation, 17–19
Standards, 12–13
Static electricity, 15
Statistical analysis, 6–7, 17–19
Storage oscilloscope, 92, 99–100
Strain gages, 194–195
Stray capacitance, 69
Sweep, horizontal, 97
Sweep generator, 95, 184–185
Synthesizers, frequency, 182–183
System analyzer, 186–187
Systematic error, 15

Technique, 15
Thermal transducers, 196–199
Thermistors, 198–199
Thermocouples, 196–197
Thermometers, resistance, 197–198
Thevenin's Theorem, 229–233
Time
  base, 94
  measurement, 109
Times, 92
Tolerance, 19–20
Transducers, 193–208
  acoustical, 201–203
  active, 193
  biological, 205–206
  chemical, 205

Transducers, *continued*
magnetic, 203–205
mechanical, 194–196
optical, 199–201
passive, 193–194
pressure, 195
thermal, 196–199
Transformer, linear-variable differential, 204
Transformerless chassis, 4
Transmission
full-duplex, 211
half-duplex, 211
simplex, 211
Transmission line, 212
Triggering, 94–96
external, 96
internal, 96
line, 96
Trimmer capacitor, 92
Troubleshooting, 7–8
True RMS detector, 68
Tube-driven equipment, 6

Ultrasonic ranging, 203

Vertical
amplifier, 94
attenuator, 93–94
plates, 88
section, 92–94
Voltage doubler, 68
Voltage sensing, 93
Voltmeters, multirange, 51–52

Wave soldering, 220
Waveform analyzer, 187
Wheatstone bridge, 195
null method, 141–142
ratio arms, 144
Thevenin equivalent, 231–233
unbalanced bridge, 142
Workbench safety, 3–7
WWV, 14, 188

Z-axis modulation, 116
Zero adjust, 7